Green Innovations in Supply Chain Management

IEEE Press
445 Hoes Lane
Piscataway, NJ 08854

Green Innovations in Supply Chain Management

Case Studies and Applications

Edited by

Marcia Mkansi
Department of Operations, Management College of Economic and Management Sciences, University of South Africa, Pretoria, South Africa
and
Office of Graduate Studies and Research, University of South Africa, Pretoria, South Africa

Godfrey Mugurusi
Department of Industrial Economics, and Technology Management, Gjøvik, Norwegian University of Science and Technology, Trondheim, Norway
and
College of Economic and Management Sciences, University of South Africa, Pretoria, South Africa

Aaron L. Nsakanda
Sprott School of Business, Carleton University, Ottawa, Ontario, Canada
and
College of Economic and Management Sciences, University of South Africa, Pretoria, South Africa

Temidayo Akenroye
Department of Supply Chain and Analytics, College of Business Administration, University of Missouri St. Louis, St. Louis, MO, USA

Frank Tietze
Institute for Manufacturing, University of Cambridge, Cambridge, UK

IEEE Press
WILEY

Published by John Wiley & Sons, Inc., Hoboken, New Jersey.

Library of Congress Cataloging-in-Publication Data applied for:

Hardback ISBN: 9781394289004

Cover Design: Wiley
Cover Image: © Kmatta/Getty Images

Set in 9.5/12.5pt STIXTwoText by Straive, Pondicherry, India

Contents

List of Contributors

Temidayo Akenroye
Department of Supply Chain and Analytics
College of Business Administration
University of Missouri St. Louis
St. Louis, MO, USA

Ernest Marfo Asiedu
Department of Operations Management
University of South Africa
Pretoria, South Africa

Elizma S. Bok
Department of Operations Management
University of South Africa
Pretoria, South Africa

Eko Ruddy Cahyadi
Department of Management
Faculty of Economics and Management
IPB University
Bogor, Indonesia

Wellington Chakuzira
Department of Operations Management
College of Economics and Management Sciences
University of South Africa
Unisa, South Africa

Tom Courtright
Centre for Transport Studies
University of Cape Town
Cape Town, South Africa

Anthony Halog
School of Environment
Faculty of Science
The University of Queensland
Brisbane, Australia

Leonie B. Louw
Department of Operations Management
University of South Africa
Pretoria, South Africa

Frank Ranganai Matenda
Department of Operations Management
College of Economic and Management Sciences
University of South Africa
Pretoria, South Africa

Ntwanano Mboweni
Department of Operations Management
College of Economic and Management Sciences
University of South Africa
Pretoria, South Africa

Marcia Mkansi
Department of Operations Management
College of Economic and Management Sciences
University of South Africa
Pretoria, South Africa
and
Office of Graduate Studies and Research
University of South Africa
Pretoria, South Africa

Godfrey Mugurusi
Department of Industrial Economics and Technology Management
Gjøvik, Norwegian University of Science and Technology
Trondheim, Norway
and
College of Economic and Management Sciences
University of South Africa
Pretoria, South Africa

Julius Mugwagwa
Department of Science
Technology, Engineering and Public Policy (STEaPP)
University College London
London, UK

Pravashen Naidoo
Electro-Mechanical Engineering
University of Cape Town
Cape Town, South Africa

Cynthia Siona Ndeh
London Graduate School
London, UK

Derrick Joseph Njuba
Operations Department
Green Heat (Uganda) Limited
Wakiso, Uganda

Aaron L. Nsakanda
Sprott School of Business
Carleton University
Ottawa, Ontario, Canada
and
College of Economic and Management Sciences
University of South Africa
Pretoria, South Africa

Gabriel Okello
Institute for Sustainability Leadership
University of Cambridge
Cambridge, UK
and
Department of Research and Innovation
African Centre for Clean Air
Kampala, Uganda

John Michael Maxel Okoche
Department of Operations Management
College of Economics and Management Sciences
University of South Africa
Unisa, South Africa

Masilela Phumlani
College of Economic and Management Sciences
University of South Africa
Pretoria, South Africa

Arushi Sharma
Sprott School of Business
Carleton University
Ottawa, Ontario, Canada

Richard Nelson Sserunjogi
AirQo
Department of Computer Science
Makerere University
Kampala, Uganda

Ita Sualia
School of Environment
Faculty of Science
The University of Queensland
Brisbane, Australia

Frank Tietze
Institute for Manufacturing
University of Cambridge
Cambridge, UK

Vianney Tumwesige
Operations Department
Green Heat (Uganda) Limited
Wakiso, Uganda

Mika Westerlund
Sprott School of Business
Carleton University
Ottawa, Ontario, Canada

Ehab A. Yaseen
Faculty of Management Technology
German University in Cairo
Cairo, Egypt

Esraa Osama Zayed
Faculty of Economics and Business Administration
German International University
Cairo, Egypt
and
Collage of Business Administration
American University in the Emirates
Dubai, UAE

About the Editors

Marcia Mkansi, Ph.D., is the Head of Research, Post-Graduate Studies, Innovation, and Commercialization, and Professor of Operations and Supply Chain Management, College of Economic & Management Sciences, University of South Africa.

Godfrey Mugurusi is an Associate Professor in Logistics and Supply Chain Management, Department of Industrial Economics and Technology Management, Norwegian University of Science and Technology, Norway.

Aaron L. Nsakanda is an Associate Professor in Supply Chain Management and Business Analytics, Sprott School of Business at Carleton University, Canada.

Temidayo Akenroye, Ph.D., is an Associate Professor, Supply Chain Management at the University of Missouri-St Louis, USA, Academic Advisor for the Kuehne Foundation, Germany, a Senior Visiting Fellow at Lagos Business School, Nigeria, and a member of the Peer Review College of the British Academy of Management.

Frank Tietze leads the Innovation and Intellectual Property Management Laboratory. He is the President of the European Policy for Intellectual Property Association.

Preface

Innovation plays a critical role in today's evolving global supply chains, especially amid growing environmental concerns and the pressing need to advance the global sustainability agenda. This book, Green Innovations in Global Supply Chains: Cases and Applications, emerges from a growing recognition that technology-enabled green innovations are not only desirable but essential for achieving most of the United Nations Sustainable Development Goals (SDGs), particularly SDG13 on climate actions.

The motivation behind this book arises from the urgent need to address the growing environmental concerns caused by unsustainable industrial practices, particularly in the manufacturing, transportation, and logistics sectors. As global greenhouse gas (GHG) emissions continue to rise, the imperative to reimagine supply chains through the lens of sustainability becomes increasingly urgent. In response, this book presents a diverse collection of evidence-based case studies that showcase how digital technologies are being harnessed to drive green innovation across the globe.

What sets this book apart is its global perspective and its emphasis on practical, real-world applications. From autonomous delivery robots in North America to crowd logistics platforms in East Africa, from carbon tracking apps in South Africa to sustainable aquaculture in Indonesia, the chapters in this volume illustrate the breadth and depth of sustainability-motivated innovations in supply chain management. Each case study is grounded in rigorous research and offers insights into the challenges, successes, and lessons learned from implementing "green" technologies in diverse industrial and social contexts.

The book is organized thematically, with each chapter exploring a specific innovation or set of innovations within a particular geographical or sectoral context. The cases span a wide range of industries—including food delivery, public transportation, heavy-duty logistics, aquaculture, e-waste management, retail, and healthcare—demonstrating the versatility and adaptability of green technologies. Crucially, the book explores the policy, organizational, and societal dimensions of green innovation, recognizing that technological solutions require enabling environments and inclusive governance structures.

A recurring theme throughout the book is the importance of context-specific solutions. As highlighted in several chapters, there is no one-size-fits-all approach to green innovation. Instead, successful initiatives are those tailored to local needs, resources, and capabilities. These insights are particularly relevant for developing regions of the world, where the challenges of climate change are often most acute, but where opportunities for leapfrogging to sustainable technologies are also significant.

This book is intended for a broad audience. It will be of particular interest to scholars and students in the fields of supply chain management, sustainability studies, environmental policy, and innovation management. Practitioners, including supply chain professionals, technology developers, and sustainability officers, will find valuable insights into how green innovations can be implemented and scaled in practice. Policymakers and development agencies will also benefit from the book's evidence-based approach to understanding the enablers and barriers to green innovation.

Ultimately, this book serves as a compelling call to action, urging readers to rethink the role of supply chains in shaping a sustainable future and to consider how digital technologies can be leveraged to create more resilient, inclusive, and environmentally responsible supply chains. By showcasing a wide array of green innovations from around the world, this book aims to inspire further research, investment, and collaboration in the pursuit of sustainable supply chain transformation.

We hope that the insights and experiences shared in these pages will contribute to a deeper understanding of the pathways toward sustainable industrialization and encourage the adoption of green innovations that benefit not only businesses and economies but also people and the planet.

A special thank you to all the authors and contributors of the book, including: Anthony Halog (*University of Queensland, Australia*), Arushi Sharma (*Carleton University, Canada*), Cynthia Siona Ndeh (*Cardiff Metropolitan University, UK*), Derrick Joseph Njuba (*Green Heat Limited, Uganda*), Ehab A. Yaseen (*German International University, Egypt*), Eko Ruddy Cahyadi (*IPB University, Bogor, Indonesia*), Elizma S. Bok (*University of South Africa, South Africa*), Ernest Marfo Asiedu (*University of South Africa, South Africa*), Esraa Osama Hassan Gohary (*German International University, Egypt*), Frank Ranganai Matenda (*University of South Africa, South Africa*), Gabriel Okello (*University of Cambridge, UK* and *African Centre for Clean Air, Uganda*), Ita Sualia (*University of Queensland, Australia*), John Michael Maxel Okoche (*University of South Africa, South Africa*), Julius Mugwagwa (*University College London, UK*), Leonie B. Louw (*University of South Africa, South Africa*), Masilela Phumlani (*University of South Africa, South*

Africa), Mika Westerlund (*Carleton University, Canada*), Ntwanano Mboweni (*University of South Africa, South Africa*), Pravashen Naidoo (*University of Cape Town, South Africa*), Richard Nelson Sserunjogi (*Makerere University, Uganda*), Tom Courtright (*University of Cape Town, South Africa*), Vianney Tumwesige (*African Centre for Clean Air, Uganda*), and Wellington Chakuzira (*University of South Africa, South Africa*).

Marcia Mkansi (*University of South Africa, South Africa*)
Godfrey Mugurusi (*Norwegian University of Science and Technology, Norway*)
Aaron L. Nsakanda (*Carleton University, Canada*)
Temidayo Akenroye (*University of Missouri St Louis, USA*)
Frank Tietze (*University of Cambridge, UK*)

1

Introduction

Green Innovations in the Supply Chain Management: Cases and Applications

Marcia Mkansi[1], *Godfrey Mugurusi*[2], *Aaron L. Nsakanda*[3], *Temidayo Akenroye*[4], *and Frank Tietze*[5]

[1] *Department of Operations Management, University of South Africa, Pretoria, South Africa*
[2] *Department of Industrial Economics and Technology Management at Gjøvik, Norwegian University of Science and Technology, Trondheim, Norway*
[3] *Sprott School of Business, Carleton University, Ottawa, ON, Canada*
[4] *Department of Supply Chain and Analytics, College of Business Administration, University of Missouri St. Louis, St. Louis, MO, USA*
[5] *Institute for Manufacturing, University of Cambridge, Cambridge, UK*

1.1 Motivation

Global environmental challenges, particularly those driven by the rise in greenhouse gas (GHG) emissions such as CO_2, CH_4, N_2O, and F-gases, have been widely recognized by the United Nations Framework Convention on Climate Change (UNFCCC) and the 23rd Conference of the Parties (CoP23) as the primary drivers of global warming. There is a global need for innovative systems, as the overexploitation of nonrenewable resources, excessive consumption, and unsustainable modes of production threaten the triple bottom line thinking in most organizations. Global supply chain demands across various sectors are contributing to unsustainable trends in industrialization, storage, passenger transport, inventory and production (Castillo et al., 2018; Kafle et al., 2017). These practices accelerate environmental degradation and exacerbate poverty and global socioeconomic inequalities. However, Green technologies appear to offer ways to reconfigure the trajectory of socioeconomic inequalities and environmental degradation toward more sustainable industrialization.

Green Innovations in Supply Chain Management: Case Studies and Applications,
First Edition. Edited by Marcia Mkansi, Godfrey Mugurusi, Aaron L. Nsakanda,
Temidayo Akenroye, and Frank Tietze

Published 2026 by John Wiley & Sons, Inc.

In Africa for example, the effects of GHG emissions are becoming increasingly ubiquitous (DeWitt et al., 2019). The consequences include the decline in air quality that, together with other types of pollution, is linked to thousands of cases of child mortality (The Big Debate, 2014). Technology-enabled green innovation promises to safeguard the environment and, in some cases, reverse past environmentally harmful trends (Sahoo et al., 2022; Shafiei & Abadi, 2017; Walker et al., 2022). Various approaches have been identified and in some cases implemented to address GHG emissions, as reviewed in reports such as the Emissions Gap Report 2019 (UNEP, 2019) and the Intergovernmental Panel on Climate Change (IPCC) Special Report on the impact of global warming (Masson-Delmotte et al., 2019).

Most of these approaches appear to cut across three key areas: policies, industrial action, and societal issues. Policies encompass emission targets, carbon emission reporting standards, reductions in fossil fuel subsidies, and energy-specific target taxes (Christensen & Olhoff, 2019). The UN has already instituted various policies, platforms, protocols, and agreements that have been adopted by its member nations over the years. Most of these were implemented under the UNFCCC, and include but are not limited to the following: the Kyoto Protocol introduced in 1997 and formally adopted in 2005, the Copenhagen Accord in 2009, the Cancun Pledge in 2010, and the Paris Agreement in 2015 (French, 1998; United Nations, 2019). The goal of these frameworks is to bind industrialized nations to commit to certain agreed-upon targets, which are individualized for each member nation in order to keep global temperatures within 2 °C, and that ultimately aim to keep global temperatures within 1.5 °C of preindustrial levels (Mulvaney et al., 2012).

In addition to policies, scientists have estimated that focusing on a few key industrial areas could reduce CO_2 emissions enough to reach the 2030 Paris Agreement targets of 56.24 $GtCO_2eq$ (Christensen & Olhoff, 2019). These areas include renewable energy sources and transport. However, significant hurdles remain. Many countries are not equipped or prepared for a shift to low-carbon societies, and the costs can be prohibitive. Additionally, the technical readiness of cleaner energy technologies is often insufficient for large-scale implementation or currently too expensive to make a significant impact (Christensen & Olhoff, 2019).

Other carbon emission reduction strategies include electrification and the use of biofuels instead of oil-based products or even hydrogen (Bednar-Friedl et al., 2015). The work by Mkansi et al. (2019) refers to an approach called Design for Environment (DfE) as a carbon reduction approach, whereby the transport industry makes use of hybrid drives and alternative energy sources to mitigate carbon emissions. The abovementioned authors again highlight the importance of electrification of the transport industry as an important DfE approach in sustainable development. Yet today, we find very few studies that have explored green innovations or clearly documented, through case studies, practical approaches to reducing GHG emissions, hence the goal of this book.

Lifestyle-based strategies including carpooling, environmental education, and support for green transport behaviors are often mentioned; however, their effectiveness needs to be well documented (Froehlich et al., 2009). This highlights the need for more evidence-based research, including well-documented case studies and practical applications of green innovations, especially in the domain of supply chain management.

To address research gaps, this book focuses on trends in green innovation, and presents evidence-based case studies that tackle practical problems associated with implementing green technologies in supply chain and operations management. The book provides clarity on the questions such as: "What are green innovations across the logistics elements of the supply chain?," "What problems are they addressing?," "How is green technology substituting or complementing conventional operations management methods?," and "How do green technologies contribute to reduction in global GHG targets?" The book contends that the content provided by current case studies provides clear insights into some or all the above questions. Each chapter approaches these questions from a different thematic lens. The result is a diversified interpretation of the content, contexts, and adaptations of contemporary green technologies and how they enable and augment current perspectives on green supply chain management practices.

The book also considers the arguments presented by Walker et al. (2022), who highlighted that technological adaptation will increasingly become a necessity for the world's rapidly growing population, as well as for natural ecosystems in every region. The book acknowledges that there is no "one-size-fits-all" approach to technological adaptation; solutions must be tailored to local contexts and needs (Wang, 2022).

1.2 Overview of Chapters and Cases

1.2.1 Chapter 2: Robotics and Applications in Canada and the United States

The study of Westerlund and Sharma (2026) focuses on the Uber Eats case and their adoption of autonomous delivery robots (ADRs). Uber Eats is a popular global online food delivery platform that connects customers with local restaurants. In Canada, for example, Uber Eats partnered with Coco to introduce four-wheeled ADRs, also known as "Coco." Coco robots are compact, four-wheeled autonomous vehicles that navigate sidewalks to deliver food orders directly to customers' doors. They are equipped with advanced sensors and navigation systems to ensure safe and efficient delivery. These robots are designed to enhance the efficiency and sustainability of food delivery services. The "Coco" robots serve as an environmentally friendly alternative to traditional vehicle-based food delivery, helping to reduce carbon emissions and traffic congestion in urban areas.

Specifically, the chapter written by Mika Westerlund and Arushi Sharma (2026) examines the use of ADRs for sustainable last-mile delivery. Their study aims to explore two interrelated things: one, the main themes in consumers' sustainability perceptions of ADRs, and two, how companies can address these themes in their robotic delivery service models. Their findings based on social media posts about Uber Eats' robotic delivery from Reddit show both positive benefits of ADR, such as reduction in emissions, energy efficiency, and reduced traffic congestion, and negative themes including ADR's inferiority compared to drones and notable societal harms of ADRs. Their adoption of the SD-X-BM framework by Musulin and Strahonja (2023) allows them to identify different solutions across the "service design, business model, and user experience."

1.2.2 Chapter 3: Crowd Logistics Applications in Uganda

In the study of Okello et al. (2026), the case of Easy Matatu in Uganda is explored. Easy Matatu is a mobile application that allows commuters to book and pay for seats in advance on public transport buses locally called "*Matatus*." *Matatus* are 14-seater public minibuses commonly used in Uganda. This system helps reduce wait times and minimizes delays, making the daily commute more predictable and efficient. This system optimizes routes and vehicle utilization, potentially reducing the number of vehicles needed and lowering overall emissions.

The study by Gabriel Okello et al. (2026) examines the impact of using Easy Matatu app in the Kampala, Entebbe, Wakiso, and Mukono districts of Uganda. Specifically, the study investigates if app-based scheduling contributed to reducing air pollution, travel time, and fuel usage by reducing stoppages. The study finds that matatus operating on the app produced 13.07 kg of CO_2 per trip compared to 13.61 kg per trip from matatus operating without the app. Yet an independent *t*-test ($p > 0.05$) showed no statistically significant difference between average CO_2 emissions per trip between matatus on Easy Matatu app and those not using the app. In general, their study concludes that online platforms/apps for transport planning and route scheduling have the potential to reduce transport-related emissions from idling, reduce travel time, and increase profit for matatu drives. This was because on-app matatus made savings on fuel costs on a per-kilometer basis because they made fewer stops per trip, had less drop-offs, and registered an overall higher average speed compared to offline matatus.

1.2.3 Chapter 4: Dilivari Crowd Logistics in South Africa

Mkansi et al. (2026c) present the Dilivari app in South Africa. The Dilivari app is an unstructured crowd sourcing logistics mobile app based in South Africa. The app aggregates live traffic data and geographical coordinates on routes using artificial intelligence and an Elasticsearch engine and assigns them to people traveling along the route to the order's destination. The app connects available delivery capacity with delivery needs. This approach optimizes resource utilization, potentially reducing empty trips and unnecessary fuel consumption.

The study by Mkansi et al. (2026c) examines the impact of digitally enabled crowdsourcing logistics—based on a mobile application—on the environmental challenges as measured by CO_2 emissions, noise, and pollution as well as economic goals such as the creation of shared value. This study builds on literature on how firms can leverage unused capacity on already planned journeys in a model described as "piggyback delivery" to improve their environmental performance. The study finds that the piggyback delivery model eliminates the aggregation step, which improves the delivery speed, particularly for long distance deliveries. Additionally, as the cost of delivery through a piggyback driver is lower than using a courier, the point-to-point piggyback model offers notable benefits. It reduces delivery costs, allowing people (buyers) to save money. It enables ordinary people (piggyback drivers) to earn additional income through the gig economy. In general, the findings highlight the role of green crowdsourcing applications in reducing carbon emissions and quantify this impact through reductions of both distance traveled and CO_2 emissions.

1.2.4 Chapter 5: Amitruck Logistics Platform in East Africa

The study by Okoche and Chakuzira (2026) presents the case of Amitruck, a logistics platform based in Kenya that connects cargo owners with trucking logistics professionals through a digital platform in order to eliminate middlemen, reducing costs and increasing transparency in the logistics sector across the country and the broader East African region. Amitruck uses a digital competitive bidding process to match cargo owners with vetted delivery vehicle drivers, ensuring cost-effective and reliable service. By cutting out intermediaries, Amitruck reduces shipping costs by up to 40%, making logistics more affordable for businesses. The platform supports economic growth by improving the efficiency of the supply chain and enabling businesses to expand their reach.

The study by Okoche and Chakuzira (2026) examined the role of the crowdsourcing platform for heavy-duty transportation such as Amitruck in Kenya on sustainability performance in different industries. The study used an exploratory case study methodology that included document analysis and key informant interviews in Kenya. Their study finds that a heavy goods transportation platform is not different from other crowdsourcing platforms; however, their impact on environmental efforts is more broader due to collaboration between the other trucking sectors such as agriculture, waste management, manufacturing, construction, and energy.

1.2.5 Chapter 6: Carbon Tracker Innovation

The study by Okoche and Mkansi (2026) presents a case of carbon dioxide (CO_2) Emission Tracker in South Africa. The mobile-based application measures, monitors, and records individual GHG by transport mode, enabling individuals and organizations to understand their carbon footprint and identify areas for improvement. The CO_2 Tracker collects and records individual carbon emissions, which are then summarized and analyzed for personalized use. The app targets environmentally sensitive corporate organizations that have good social contracts with employees in order to measure and reduce carbon emissions of their work and lifestyles. The personalized individual carbon emissions information can form the basis for aggregate employee carbon emissions, which can facilitate the development of carbon emissions reduction strategy for an organization.

The study by Okoche and Mkansi (2026) implements a multilevel assessment of carbon emission tracking technology in South Africa, focusing on individual and organizational factors. The study focuses on the mobile carbon dioxide (CO_2) Emission Tracker for measuring the carbon footprint in South Africa, which measures, monitors, and records individual GHG by transport mode. The study builds on the Technology Acceptance Model (TAM) framework by developing a holistic model for the adoption of carbon emission tracking technology for understanding the absorption of green technologies, particularly in the context of carbon emission reduction. They conclude that the app itself offers a deep understanding of carbon credit implementation by highlighting issues of data reliability, organizational integration of carbon tracking systems, and strategies for effective individual and corporate engagement.

1.2.6 Chapter 7: The Case of Eco.mio Innovation in Germany

The study by Mkansi, Matenda, and Mboweni (2026a) presents the Eco.mio case. Eco.mio is a German software solution that aims to promote sustainable business travel by integrating seamlessly with existing online booking tools, guiding employees and organizations toward greener travel options. The platform displays CO_2 savings and highlights sustainable booking options, making it easier for employees to choose eco-friendly travel alternatives. Eco.mio incorporates gamification elements such as scores, rewards, and peer comparisons to motivate employees to make sustainable travel choices. The platform also uses behavioral nudges at the point of decision-making to encourage employees to opt for greener travel options. At the organizational level, the platform provides detailed reports on the company' environmentally friendly travel behavior, helping businesses track and report their performance toward sustainability goals.

The study by Mkansi, Matenda, and Mboweni (2026a) examines how eco.mio solution encourages sustainable business travel practices and what impact it has on corporate emissions reduction. The findings show that the eco.mio solution drives behavioral change through its simplicity and by giving travelers an opportunity (to reduce their environmental impact) and personal motivation. By adopting tools for sustainable business travel, corporates can guide their workers toward more sustainable travel choices, thereby contributing to environmental conservation and aligning with corporate sustainability goals. The study offers empirical evidence on how technology-driven solutions enable reductions in carbon emissions connected to business travel, thus advancing the comprehension of sustainable technology applications in real-world settings.

1.2.7 Chapter 8: UNIDO Indonesian Global Quality Standards Program

The study by Sualia et al. (2026) presents a case of greening the Indonesian seafood sector through the United Nations Industrial Development Organization (UNIDO) Indonesian Global Quality and Standards Program (GQSP). UNIDO's GQSP in Indonesia is a collaboration between the Ministry of Marine Affairs and Fisheries (MMAF) and the National Standardization Agency (BSN) to strengthen quality infrastructure and services for the seaweed, shrimp, and milkfish markets, ensuring that these products meet international market requirements. The program has several actors, but the most notable ones include *Jala Tech*, which provides Internet-based water quality monitoring solutions for shrimp farming. Their Baruno device measures various water quality parameters such as pH, salinity, dissolved oxygen, temperature, and oxidation-reduction potential (ORP). The data is stored in the cloud and can be accessed anytime via the JALA app, helping farmers make informed decisions to optimize shrimp cultivation. Another actor is called *e-Fishery*, which offers an automatic feeding device called eFeeder that can be controlled via a mobile phone. Farmers can set feeding schedules, receive recommendations, and record feed data in real time. The eFeeder helps accelerate growth, improve feed conversion ratios (FCR), and maintain water quality by reducing feed waste. GQSP has promoted sustainable practices such as integrated shrimp-mangrove farming and the cultivation of new seaweed species like *Ulva* and *Caulerpa*, which have contributed to climate change adaptation and mitigation, and helped optimize water resources, improve nutrient cycles, and support the coexistence of aquaculture with local ecosystems.

The study by Sualia et al. (2026) examined the role and challenges of greening of the Indonesian seafood supply chain and how implementing circular economy principles will enhance the environmental and economic resilience of Indonesia' seafood supply chain. The study is based solely on the lessons from the GQSP initiatives such as Good Aquaculture Practices (GAP) for shrimp production, real-time water quality monitoring, and automatic feeders and Least Cost Feed (LCF) formulation in pangasius farming implemented in Indonesia between 2014 and 2023 that aimed to increase productivity and expand Indonesia's seafood global market access. A regression model assessed the relationship between the GAP adoption and FCR, survival rates, and cost savings. Their findings showed positive impacts from implementing a combination of shrimp GAP and digital innovations, with every 10-point (out of 100 points) increase in GAP adoption correlating to an additional 54 kg in shrimp production and a 9% improvement in survival rates. In pangasius farming, adopting LCF formulation resulted in feed cost savings of approximately IDR 6000 per m^2 and production increase by 14.1 kg/m^2 compared to 9.7 kg/m^2 non-LCF ponds, demonstrating the effectiveness of LCF in enhancing cost-efficiency and productivity.

1.2.8 Chapter 9: E-waste Innovation in Africa

The study by Asiedu et al. (2026) presents the case of E-waste Africa. E-waste Africa is a South African company that has developed several eco-friendly waste management technologies to address the growing problem of electronic waste. The company has an Luminophosphor Powder Extraction (LPX) plant, which uses hydrometallurgical processes to extract rare earth metals from luminous phosphor powder found in lighting waste. This plant not only recovers valuable rare earth elements but also captures and safely processes mercury, achieving a 99.9% diversion of waste from landfills. Also, E-waste Africa has established a solar panel recycling plant to handle the end-of-life disposal of photovoltaic panels. This facility recycles the various components of solar panels, including glass, aluminum, and silicon, ensuring that these materials are reused rather than discarded. From recycled glass, the company has developed CONVAL pavers used in construction, providing a sustainable alternative to traditional building materials.

The study by Asiedu et al. (2026) analyzes whether the business model of E-Waste Africa has specific sustainability benefits and challenges. Their findings show that the company uses a dynamic and evolving business model that embeds fee-for-service, extended producer responsibility (EPR) compliance, material recovery, and strategic partnerships. They conclude that the company's operations clearly contribute to positive environmental outcomes (reduced CO_2 emissions, resource conservation), social outcomes (job creation, skills development), and economic outcomes (new revenue streams, green economy growth). The company faces challenges like bureaucratic hurdles, client onboarding, cash flow management, and the need to build trust.

1.2.9 Chapter 10: Waste-to-Energy and Environmental Protection (WEEP) Initiative

The study by Mkansi, Asiedu, and Matenda (2026b) presents the case of Basadi E-Waste. Basadi E-Waste is a South African company that transforms e-waste plastic into valuable by-products through its Waste-to-Energy and Environmental Protection (WEEP) initiative. The company repurposes e-waste plastic into several products, including high-quality 3D printing filament, which involves shredding the plastic, melting it at controlled temperatures, and extruding it into filaments, providing an affordable and sustainable alternative for 3D printing enthusiasts; digital crutches for people with impaired mobility, designed to be lightweight, durable, and cost-effective, providing essential mobility aids for individuals with disabilities; and roof tiles for low-cost housing, made from recycled e-waste plastic, designed to be durable, weather-resistant, and affordable, addressing housing needs while promoting environmental sustainability.

The study by Mkansi, Asiedu, and Matenda (2026b) analyzes the operations of Basadi E-waste, their motivations, the core elements of the innovations, and challenges involved in e-waste recycling. Their findings show that innovations in e-waste management are not only driven by environmental concerns but also powered by social and economic motivations. Social drivers of green motivation include inclusion of women in a male-dominated e-waste industry (i.e., fostering economic opportunities within the township economy), bridging the gap between IT and e-waste sectors, equitable access to green innovation opportunities, and addressing industry disparities. The environmental drivers of green innovation comprise environmental consciousness, identifying gaps in e-waste management, and commitment to true recycling, while economic drivers include problem-solving mindset, driving innovation through solutions, and passion for environmental impact. Their findings also show challenges associated with financial barriers to green technology adoption, securing consistent e-waste feedstock, and low public awareness of e-waste issues.

1.2.10 Chapter 11: Planet Points (Plan3T) for Sustainable Lifestyle in Denmark and Germany

Mkansi and Matenda (2026) present the case of Planet Points (Plan3T). Plan3T is a rewards system for sustainable lifestyle that operates in Germany and Denmark. The platform incentivizes sustainable lifestyle choices through a cashback program where users earn rewards on purchases deemed environmentally friendly, such as products with low carbon footprints, sustainable packaging, and ethical sourcing. The platform incorporates gamification elements like points, badges, and leaderboards to motivate users to make greener choices. Additionally, Plan3T provides detailed information on the environmental impact of various products and services, helping users make informed decisions. The goal of Plan3T is to encourage individuals to make sustainable purchasing decisions by offering rewards for eco-friendly actions. The platform integrates with various retailers and service providers to track and reward sustainable behaviors.

The study by Mkansi and Matenda (2026) presents the innovation, challenges, and impact of Plan3T's free application. The authors find that technologies from start-ups play a crucial role in bringing consumption to more sustainable levels. Also, strategic partnerships among diverse stakeholders across multiple sectors in different countries allow Plan3T's application to link carbon emissions from different consumption points in real time.

1.2.11 Chapter 12: Retail Sustainability Initiatives in South Africa

The study by Bok and Louw (2026) presents sustainability initiatives of South Africa's top five retailers. These include Shoprite, Woolworths, Pick n Pay, SPAR, and Mass Mart. **Case 12a** is Shoprite, a South African retailer that has implemented several sustainability initiatives to promote environmental responsibility and reduce its ecological footprint. These initiatives include offering R1.00 back for customers who reuse planet bags, encouraging the reduction of single-use plastic bags; a hanger recycling program to minimize waste and promote circular economy practices; using 100% recycled materials for plastic bags and hangers to contribute to waste reduction and resource conservation; employing bio-digester bins at distribution centers to process wet and organic waste, converting it into useful by-products like compost; and providing e-waste recycling services and in-store recycling bins to make it easier for customers to dispose of electronic waste responsibly.

Case 12b is Woolworths, another South African retailer that has implemented several innovative sustainable packaging solutions to reduce its environmental footprint. These initiatives include pioneering the use of recyclable materials for coffee bags and sauce pouches, ensuring that these packaging items can be recycled and reused; introducing collection points for used reusable shopping bags at its various stores, encouraging customers to return their worn bags for recycling into new products; collecting used hangers from all its stores for recycling or refurbishment and reuse, supporting circular economy practices; and utilizing sustainable materials like paper, cardboard, and 100% recycled plastics for its packaging, minimizing the need for nonrenewable resources and promoting recycling. **Case 12c** is Pick n Pay, another South African retailer that has implemented several sustainable materials and waste management initiatives to reduce its environmental impact. The company has a comprehensive hanger recycling program, where used hangers are collected, rescued, and recycled. For footwear hangers, Pick n Pay uses paper and cardboard, while plastic hangers are made from 100% recycled materials. Additionally, the retailer's plastic bags are also produced from 100% recycled materials. To manage organic waste efficiently, Pick n Pay employs bio-digester bins at its distribution centers. These bins process wet and organic waste, converting it into useful by-products like compost. The company also focuses on e-waste recycling, providing in-store recycling bins to make it easier for customers to dispose of electronic waste responsibly. **Case 12d** is SPAR, which operates in the grocery market in South Africa. It has also implemented several recycling and sustainable materials programs. One of their key initiatives is the collection program for old reusable shopping bags. Customers are encouraged to return their used bags to SPAR stores, where they are collected and recycled into new products. SPAR also implements a hanger recycling program that focuses on reusing and recycling hangers made from sustainable materials. This program ensures that hangers do not end up as waste and supports circular economy practices. Furthermore, SPAR uses 100% recycled materials for its plastic products, including plastic bags and other packaging items. These recycled materials help divert significant amounts of plastic waste from landfills and reduce the carbon footprint associated with producing new plastic. **Case 12e** is Mass Mart, which has also implemented several waste reduction and renewable energy programs to enhance sustainability profile in South Africa. These initiatives include a comprehensive hanger recycling program that focuses on reusing and recycling hangers made from sustainable materials, helping to reduce waste and support circular economy practices. At its distribution centers, Mass Mart employs bio-digester bins to process wet and organic waste efficiently, converting it into useful by-products like compost and biogas, thereby reducing landfill waste and GHG emissions. The retailer also provides e-waste recycling services and in-store recycling bins, making it easier for customers to dispose of electronic waste responsibly and promoting recycling. Mass Mart has been described as "the biggest producer" in its sector for solar power, with solar power plants installed at several locations, including a 480-kilowatt peak plant at Makro Strubens Valley in Gauteng, which helps reduce the reliance on nonrenewable energy sources and lower carbon emissions.

The study by Bok and Louw (2026) aimed to analyze and compare the sustainability strategies and approaches adopted by the top five retailers in South Africa and evaluate the effectiveness of their sustainability strategies in addressing environmental, social, and governance (ESG) challenges using the triple bottom line framework. In general, all the five retailers identified place greater emphasis on sustainable initiatives, focusing on environmentally responsible practices to drive long-term impact. Most appear to favor partnerships and collaborations to achieve their environmental goals. Also, the authors' findings show that South African retailers are increasingly adopting green initiatives to (i) improve operational efficiency, (ii) become more innovative in waste management,

(iii) benefit from collaborating on sustainability initiatives and transparent reporting, and (iv) ensure their supply chains adhere to environmental standards due to reporting requirements.

1.2.12 Chapter 13: Blockchain Technology Diffusion

This study by Zayed and Yaseen (2026) seeks to identify the factors influencing blockchain technology (BCT) diffusion in supply chain management (SCM) by deploying a quantitative survey in Egypt. Their study adopts the technology-organization-environment (TOE) framework and diffusion of innovation (DoI) theory to identify the factors that influence BCT implementation and whether such innovation supports better sustainability performance at the SCM level. This is achieved through integrating TOE framework and also through an exploratory research approach focusing on Egypt as a sample for an emerging economy from the Middle East and North Africa (MENA) region. Their findings show different factor impacts. At the top level are governmental policies, awareness about BCT benefits, and complexity of setup and use, which are considered factors of high influence and are not or are weakly influenced by other factors. Intermediate-level factors include data security, maturity of BCT, collaboration and coordination among supply chain partners, and cultural differences. These factors are highly interconnected and represent the core technical and operational challenges of implementing BCT in SCM. These were followed by factors such as customer demand and influence, which can accelerate the adoption of BCT by pushing companies to improve security, maturity, and collaboration. However, there were factors that were considered critical for decision-makers for the adoption of BCT. They included the cost of technology adoption, top management support, and organizational culture.

1.2.13 Chapter 14: Green Investment

The study by Nsakanda, Mkansi, Asiedu, and Matenda (2026d) resulted in five cases on how to secure funding for scaling green innovations in Africa. The authors argue that the African funding landscape is diverse and fragmented, which makes it extremely challenging for organizations and individuals to raise green funding and even increase awareness about green innovation funding.

The outcome of their study is a six-stage guide to securing funding for scaling green innovations in Africa with the following stages: (i) understanding funding methodologies, (ii) defining the sustainable nature of the business, (iii) decoding funding schemes, (iv) meeting funding access eligibility criteria, (v) showcasing impacts by businesses accessing the fund, and (vi) understanding typical funding amounts to set realistic expectations.

1.2.14 Chapter 15: Green Tech in Healthcare Supply Chain

The study by Mugwagwa and Ndeh (2026) explores the adoption and integration of green innovations in African health supply chains, focusing on sustainability, efficiency, and resilience. Their study explores examples of different green innovations such as solar-powered vaccine refrigeration, green logistics, and eco-friendly waste management and identifies the barriers and policy implications for scaling these green innovations.

In their findings, they show that adoption of green innovations in healthcare in Africa is impeded by several barriers, including financial constraints, technical limitations, institutional fragmentation, social resistance, and infrastructure deficiencies.

1.3 Conclusion and Value Proposition

In sum, each of these cases and applications demonstrates at least one green innovation impact on the (i) environment, (ii) local businesses, (iii) governments, (iv) universities, and (v) communities. The impacts are realized through pro-environment policy, increased B2B and B2C networking, creation of new knowledge and innovations, improved public services, new green jobs, and enhanced quality of life and health. In particular, the book has a strong focus on the impact of green innovations in the transport sector. It is a well-known fact that the share of transportation in global CO_2 emissions is increasing daily, and in the global south, the debate on this issue is almost unheard of. Through these innovations and practice-based focus, the book raises awareness of the cumulative impact of green innovation that will certainly further advance the triple bottom line concept, especially across different industries and communities. This contribution supports global GHG mitigation efforts and aligns with key sustainability frameworks, including UN Sustainable Development Goal 13 (climate action), the African Union Agenda 2063, and the UN 2030 Agenda. The book is primarily relevant to academic audiences in the social and human sciences, particularly those interested in sustainability theory, applications, innovations, and emerging developments within the field of SCM.

The book also appeals to digital technological innovators who seek to develop and enhance environmentally sound practices within SCM. Furthermore, the book is a resource for educators and curriculum developers in higher education institutions who wish to improve the students' understanding of how economic activities, such as transport, contribute to socioeconomic and environmental problems (including noise pollution, smog etc), and their broader impact on global climate change.

References

Asiedu, E.M., Matenda, F.R., & Naidoo, P. (2026). Business model innovation and technological advancement for sustainable e-waste management: A case study of ewaste Africa in South Africa. In Mkansi, M., Mugurusi, G., Nsakanda, A.L., Akenroye, T., & Tietze, F. (Eds.), *Green innovations in supply chain management: Case studies and applications* (pp. 179–207). Wiley.

Bednar-Friedl, B., Wolkinger, B., König, M., Bachner, G., Formayer, H., Offenthaler, I., & Leitner, M. (2015). Transport. In Steininger, K.W., König, M., Bednar-Friedl, B., Kranzl, L., Loibl, W., & Prettenthaler, F. (Eds.), *Economic evaluation of climate change impacts* (pp. 279–300). Springer Climate. https://doi.org/10.1007/978-3-319-12457-5_15

Bok, E.S., & Louw, L.B. (2026). Green initiatives of South Africa's leading retailers: A comparative analysis of sustainability strategies. In Mkansi, M., Mugurusi, G., Nsakanda, A.L., Akenroye, T., & Tietze, F. (Eds.), *Green innovations in supply chain management: Case studies and applications* (pp. 243–267). Wiley.

Castillo, V.E., Bell, J.E., Rose, W.J., & Rodrigues, A.M. (2018). Crowdsourcing last mile delivery: Strategic implications and future research directions. *Journal of Business Logistics*, *39*(1), 7–25. https://doi.org/10.1111/jbl.12173

Christensen, J. & Olhoff, A. (2019). *Lessons from a decade of emissions gap assessments* (pp. 1–14). http://www.un.org/Depts/.

DeWitt, H.L., Gasore, J., Rupakheti, M., Potter, K.E., Prinn, R.G., Ndikubwimana, J.D.D., Nkusi, J., & Safari, B. (2019). Seasonal and diurnal variability in O_3, black carbon, and CO measured at the Rwanda Climate Observatory. *Atmospheric Chemistry and Physics*, *19*, 2063–2078, https://doi.org/10.5194/acp-19-2063-2019.

French, D. (1998). Kyoto protocol to the United Nations Framework Convention on Climate Change. *Journal of Environmental Law*, *10*(1), 215–224. https://doi.org/10.1093/jel/10.1.215

Froehlich, J., Dillahunt, T., Klasnja, P., Mankoff, J., Consolvo, S., Harrison, B. & Landay, J.A. (2009). *UbiGreen: Investigating a mobile tool for tracking and supporting green transportation habits*. http://www.ecorio.org/.

IPCC, Masson-Delmotte, V., Zhai, P., Pörtner, H.-O., Roberts, D. C., Skea, J., Shukla, P. R., Pirani, A., Moufouma-Okia, W., Péan, C., Pidcock, R., Connors, S., Matthews, R., Chen, Y., Zhou, X., Gomis, M. I., Lonnoy, E., Maycock, T., Tignor, M., Tabatabaei, M. (2019). *Global warming of 1.5°C. An IPCC Special Report on the impacts of global warming of 1.5°C above pre-industrial levels and related global greenhouse gas emission pathways, in the context of strengthening the global response to the threat of climate change*. www.environmentalgraphiti.org.

Kafle, N., Zou, B., & Lin, J. (2017). Design and modeling of a crowdsource-enabled system for urban parcel relay and delivery. *Transportation Research Part B: Methodological*, *99*, 62–82. https://doi.org/10.1016/j.trb.2016.12.022

Mkansi, M., & Matenda, F.R. (2026). Toward sustainable consumption: A case of PLAN3T application. In Mkansi, M., Mugurusi, G., Nsakanda, A.L., Akenroye, T., & Tietze, F. (Eds.), *Green innovations in supply chain management: Case studies and applications* (pp. 229–241). Wiley.

Mkansi, M., McLennan, N.K., & de Villiers, G. (2019). *Contemporary issues in operations and supply chain management*. Van Schaik. https://books.google.co.za/books?id=jxfTxgEACAAJ

Mkansi, M., Matenda, F.R., & Mboweni, N. (2026a). Toward sustainable business travel: A case of eco.mio software solution. In Mkansi, M., Mugurusi, G., Nsakanda, A.L., Akenroye, T., & Tietze, F. (Eds.), *Green innovations in supply chain management: Case studies and applications* (pp. 137–156). Wiley.

Mkansi, M., Asiedu, E.M., & Matenda, F.R. (2026b). A sustainable electronic waste recycling: A case study of basadi e-waste. In Mkansi, M., Mugurusi, G., Nsakanda, A.L., Akenroye, T., & Tietze, F. (Eds.), *Green innovations in supply chain management: Case studies and applications* (pp. 207–228). Wiley.

Mkansi, B., Nsakanda, A.L., Mugurusi, G. *et al.* (2026c). Mobile application-enabled logistics crowdsourcing and the triple bottom line in South Africa's e-grocery industry. In Mkansi, M., Mugurusi, G., Nsakanda, A.L., Akenroye, T., & Tietze, F. (Eds.), *Green innovations in supply chain management: Case studies and applications* (pp. 51–71). Wiley.

Mugwagwa, J., & Ndeh, C.S. (2026). Green innovations in African health supply chains: Case studies and policy implications. In Mkansi, M., Mugurusi, G., Nsakanda, A.L., Akenroye, T., & Tietze, F. (Eds.), *Green innovations in supply chain management: Case studies and applications* (pp. 341–367). Wiley.

Mulvaney, D., Robbins, P., & Kte'pi, B. (2012). UN framework convention on climate change. In Mulvaney, D.R. (Ed.), *Green politics: An A-to-Z guide*. https://doi.org/10.4135/9781412971867.n129

Musulin, J., & Strahonja, V. (2023). User experience, business models, and service design in concert: Towards a general methodological framework for value proposition enhancement. *Sustainability*, *15*, 12509. https:// doi. org/ 10. 3390/ su151612509

Nsakanda, A.L., Mkansi, M., Asiedu, E.M., & Matenda, F.R. (2026d). A guide to securing funding for scaling green innovations in Africa. In Mkansi, M., Mugurusi, G., Nsakanda, A.L., Akenroye, T., & Tietze, F. (Eds.), *Green innovations in supply chain management: Case studies and applications* (pp. 299–339). Wiley.

Okello, J., Tumwesige, V., Njuba, D.J. *et al.* (2026). Assessing the impact of an app-based scheduling platform on the overall CO_2 emissions and travel time of minibus taxis in four cities (municipalities) in Uganda. In Mkansi, M., Mugurusi, G., Nsakanda, A.L., Akenroye, T., & Tietze, F. (Eds.), *Green innovations in supply chain management: Case studies and applications* (pp. 33–49). Wiley.

Okoche, J.M.M., & Chakuzira, W. (2026). Exploring the use of crowdsourcing platform truck transport in the specialized sectors in Kenya: The case of Amitruck application. In Mkansi, M., Mugurusi, G., Nsakanda, A.L., Akenroye, T., & Tietze, F. (Eds.), *Green innovations in supply chain management: Case studies and applications* (pp. 73–102). Wiley.

Okoche, J.M.M., & Mkansi, M. (2026). A multilevel analysis of carbon emission tracking technology adoption in South Africa. In Mkansi, M., Mugurusi, G., Nsakanda, A.L., Akenroye, T., & Tietze, F. (Eds.), *Green innovations in supply chain management: Case studies and applications* (pp. 103–135). Wiley.

Sahoo, S., Kumar, A., & Upadhyay, A. (2022). How do green knowledge management and green technology innovation impact corporate environmental performance? Understanding the role of green knowledge acquisition. *Business Strategy and the Environment, 32*, 551–569.

Schiederig, T., Tietze, F., & Herstatt, C. (2012). Green innovation in technology and innovation management – An exploratory literature review. *R&D Management, 42*(2), 180–192. https://doi.org/10.1111/j.1467-9310.2011.00672.x

Shafiei, M.W.M., & Abadi, H. (2017). The importance of green technologies and energy efficiency for environmental protection. *International Journal of Applied Environmental Sciences, 12*(5), 937–951.

Sualia, I., Cahyadi, E.R., & Halog, A. (2026). Greening the seafood supply chain through circular economy and digital transformation: A case study of the UNIDO Indonesian Global Quality and Standards Program. In Mkansi, M., Mugurusi, G., Nsakanda, A.L., Akenroye, T., & Tietze, F. (Eds.), *Green innovations in supply chain management: Case studies and applications* (pp. 157–175). Wiley.

The Big Debate. (2014). The big debate on energy. *eNCA*.

UNEP. (2019). *Emissions gap report.* https://wedocs.unep.org/bitstream/handle/20.500.11822/30797/EGR2019.pdf?sequence=1&isAllowed=y.

United Nations. (1992). *United Nations Framework Convention on Climate Change.* Treaty Series, 1771 UNTS 107; 31 ILM 849. New York. US.

Walker, T., Wendt, S., Goubran, S., & Schwartz, T. (2022). Climate change adaptation: An overview. In Walker, T., Wendt, S., Goubran, S., & Schwartz, T. (Eds.), *Business and policy solutions to climate change: From mitigation to adaptation* (pp. 3–12).

Wang, C. (2022). Green technology innovation, energy consumption structure and sustainable improvement of enterprise performance. *Sustainability, 14*(16), 10168.

Westerlund, M., & Sharma, A. (2026). Sustainable last-mile delivery with autonomous robots. In Mkansi, M., Mugurusi, G., Nsakanda, A.L., Akenroye, T., & Tietze, F. (Eds.), *Green innovations in supply chain management: Case studies and applications* (pp. 17–32). Wiley.

Zayed, E.O., & Yaseen, E.A. (2026). Exploring blockchain technology diffusion in supply chain management to enhance sustainability performance: Insights from Egypt an emerging Economy. In Mkansi, M., Mugurusi, G., Nsakanda, A.L., Akenroye, T., & Tietze, F. (Eds.), *Green innovations in supply chain management: Case studies and applications* (pp. 269–298). Wiley.

2

Sustainable Last-Mile Delivery with Autonomous Robots

Mika Westerlund and Arushi Sharma

Sprott School of Business, Carleton University, Ottawa, Ontario, Canada

2.1 Introduction

The importance of effective and sustainable last-mile delivery in logistics has grown rapidly due to the rise of e-commerce and increasing customer demand for home delivery (Ayyildiz & Erdogan, 2024; De Maio et al., 2024). According to a 2018 study by USPS, consumers place significant value on receiving parcel deliveries when and where they choose (Figliozzi, 2020). Last-mile delivery, defined as the final stage of the supply chain where goods reach the end consumer or destination (Jaller et al., 2021), has become critical for customer satisfaction and retention. Today, the post-purchase experience is as influential as the pre-purchase experience in shaping customer loyalty (Ha et al., 2023).

Changing online consumer habits further underscore the importance of last-mile delivery. For example, more than 20% of U.S. shoppers now purchase groceries online at least once a month (Figliozzi, 2020). However, the rising volume of goods and food deliveries has created new societal challenges, including road congestion, traffic safety concerns, and increasing carbon dioxide (CO_2) emissions. In fact, last-mile delivery is widely regarded as the most polluting, inefficient, and costly part of the supply chain (Ha et al., 2023). To make matters worse, the number of delivery vehicles in urban areas is projected to increase by 40% by 2030, potentially causing heightened congestion and over six million additional tons of CO_2 emissions (Han et al., 2024). Fortunately, some customers are willing to wait

Green Innovations in Supply Chain Management: Case Studies and Applications,
First Edition. Edited by Marcia Mkansi, Godfrey Mugurusi, Aaron L. Nsakanda,
Tomidayo Akenroye, and Frank Tiotso.

Published 2026 by John Wiley & Sons, Inc.

longer, pay more, or select a less convenient location in exchange for a more environmentally or socially sustainable delivery (Ha et al., 2023).

The introduction of autonomous delivery robots (ADRs) is a recent attempt to reduce carbon emissions and lower parcel delivery costs (Figliozzi, 2020). ADRs are small, self-driving vehicles that deliver packages, groceries, or service goods in urban environments characterized by short distances and pedestrian-focused, traffic-limited areas (Garus et al., 2024; Han et al., 2024). Equipped with advanced technological features such as cameras, 3D and ultrasonic sensors, Light Detection and Ranging (LiDAR), and Global Positioning System (GPS), ADRs navigate complex urban settings autonomously (Garus et al., 2024). Typically battery-operated, ADRs can respond to consumer expectations regarding convenience, speed, and sustainability from deliveries at the right price (Ha et al., 2023).

Previous research has examined ADRs from the perspective of consumer acceptance, measuring user acceptance and/or consumers' intention to adopt ADRs (e.g., Koh et al., 2024; Wu et al., 2024), as well as from the perspective of social acceptance, focusing on nonuser acceptance and tolerance to the use of robotic delivery technologies in people's local communities (Westerlund, 2024). However, consumer acceptance of ADRs for last-mile delivery needs to be studied to promote delivery automation (Ha et al., 2023; Koh et al., 2024), especially from the perspective of environmental concerns and motivations (Han et al., 2024). Hence, this chapter aims to research (i) the main themes in consumers' sustainability perceptions of ADRs and (ii) how companies can address these themes in their robotic delivery service models. For this purpose, the chapter will analyze a corpus of social media data drawn from Reddit.

2.2 The State of the Art: Introduction of ADRs for Last-Mile Delivery

According to Ayyildiz and Erdogan (2024), today's retailers and food businesses are continually seeking innovative and efficient ways to deliver products to customers. Companies like Walmart and Amazon, which offer diverse delivery options, including two-hour deliveries for those who value quick commerce, have reshaped customer expectations. Consumers now demand not only rapid delivery but also a seamless user experience from all online businesses (Ha et al., 2023). Consequently, challenges such as traffic congestion, parking limitations, driver shortages, rising fuel and fleet management costs, and narrow drop-off time windows have pushed retailers to experiment with automated goods delivery in urban areas (Hossain, 2023; Westerlund, 2024).

There are three types of autonomous last-mile delivery vehicles: (i) drones or unmanned aerial vehicles (UAVs), and two types of ADRs, namely (ii) sidewalk autonomous delivery robots (SADRs) and (iii) road autonomous delivery robots

(RADRs) (Figliozzi, 2020). ADRs, particularly SADRs, are primarily used for grocery and food deliveries (Jennings & Figliozzi, 2020). Their adoption accelerated during the COVID-19 pandemic, as social distancing and self-isolation measures spurred e-commerce growth and drove demand for contactless delivery methods (Ayyildiz & Erdogan, 2024; Koh et al., 2024). Over time, automated goods delivery using ADRs is expected to account for 80% of all consumer market deliveries (Chen et al., 2021), highlighting their potential sustainability impact.

Ha et al. (2023) point out that sustainability in the context of last-mile delivery should be considered from the triple bottom line perspective, addressing economic, social, and environmental sustainability at the same time. Economic sustainability focuses on improved economic performance through additional delivery options, cost reductions, and increased speed. Social sustainability, on the other hand, addresses the impacts of automated delivery technologies on employment, working conditions, and societal benefits, as well as its effects on traffic congestion, accidents, and extended waiting times in public transportation (Ha et al., 2023). Environmental sustainability deals with aspects like reductions in air pollution, emission of greenhouse gases (GHGs), noise pollution, and use of plastics for product packaging (Ha et al., 2023).

While autonomous delivery technologies like ADRs—small electric vehicles—can significantly reduce the carbon footprint of deliveries, no single autonomous delivery vehicle type excels universally in terms of energy and emissions efficiency (Figliozzi, 2020). For example, SADRs are highly efficient in reducing carbon emissions for short-distance deliveries near depots, whereas drones are better suited for time-sensitive, low-density deliveries. RADRs perform well for midrange distances with fewer delivery stops (Figliozzi, 2020). One frequently discussed option is the use of a hybrid model, such as the mothership model, where a human-driven truck transports goods and releases a fleet of robots to complete deliveries to customers' doors (Ostermeier et al., 2022). As for drones, robotic deliveries face additional challenges, such as restrictive airspace traffic regulations and disruptive drone noise in urban environments (Westerlund, 2024).

2.3 Case Examples

The following four cases exemplify how some large companies have experimented with ADRs.

2.3.1 Case 1: Domino's Pizza Delivery

According to Jennings and Figliozzi (2020), Domino's Pizza Inc. introduced an autonomous pizza delivery robot, Domino's Robotic Unit (DRU), in Australia and New Zealand in 2016. This four-wheeled robot features separate compartments with a sliding roof to keep pizzas hot and drinks cold while traveling at safe speeds on sidewalks from the store to customers' door (Domino's, 2024). While DRU has since been introduced to many other countries, Domino's has also been experimenting with other types of unmanned delivery vehicles. In the United Kingdom, Domino's and Starship Technologies launched an ADR pilot project to test pizza order delivery (Jennings & Figliozzi, 2020). In Texas, United States, a partnership with the robotics company Nuro led to trials with the "R2," a mini-sized autonomous delivery van capable of carrying multiple pizzas. Similarly, in Michigan, United States, Domino's tested pizza delivery using Ford's full-sized autonomous cars (Adapted from CBC.ca, 2017).

2.3.2 Case 2: Amazon's Parcel Delivery

The retail giant Amazon has been experimenting with robotic deliveries for a decade. Following several autonomous drone delivery projects, Amazon began testing Amazon Scout, a six-wheeled ground-based ADR in Washington and California, United States, in 2019. The rollout expanded to other U.S. states in 2020 (Amazon Scout, 2025). Unlike other retailers pursuing robotic delivery, Amazon acquired the robotics company Dispatch in 2017 to develop Scout internally (Amazon Scout, 2025). Amazon's robotic delivery strategy gained heightened relevance during the COVID-19 pandemic, as isolation and social distancing restrictions increased demand for contactless delivery. However, Scout's technology was still in its pilot phase, relying on "Scout Ambassadors"—human operators tasked with ensuring the smooth functioning of the robotic delivery system (Heater, 2022). Unfortunately, in 2023, Amazon discontinued Scout's real-world pilots as part of a larger belt-tightening at the company and the increased attention of regulatory bodies to several of Amazon's major corporate acquisitions (Amazon Scout, 2025; Heater, 2022).

2.3.3 Case 3: Co-op's Grocery Delivery

In 2018, the British supermarket chain Co-op, in partnership with Starship Technologies, launched a commercial service using Starship's six-wheeled ADRs for grocery delivery in the United Kingdom. In 2023, Co-op announced a major expansion of its robotic delivery service to cover Leeds and other areas, many of which were already added during the pandemic (Co-op, 2023). This service supports Co-op's quick commerce model, offering consumers deliveries within an hour for a small fee starting at just 99p (Co-op, 2023). Customers can also track the ADR's journey in real time through an interactive map. Starship's ADRs are powered by renewable electricity, positioning them as an environmentally friendly option. Remarkably, an average robotic delivery only consumes as little energy as boiling a kettle to make one cup of tea (Co-op, 2023). An executive member for sustainable development and infrastructure at Leeds City Council commented that Co-op's ADR service helps reduce short car trips to local shops, supporting the city's goal of becoming carbon neutral (Adapted from Co-op, 2023).

2.3.4 Case 4: Uber Eats' Food Delivery

Uber Eats has been one of the most active companies experimenting with ADRs in recent years, piloting numerous projects worldwide. In 2023, Serve Robotics announced plans to deploy potentially thousands of ADRs in partnership with Uber Eats across multiple cities in the United States and Canada to deliver food orders (Serve Robotics, 2023). In 2024, Uber Eats partnered with Coco Robotics, a robotic delivery company that operates in multiple countries and has completed over 300,000 zero-emissions deliveries with their four-wheeled "Coco" robots after 2020 through its partnerships; Uber Eats planned to test Coco in California (The Robot Report Staff, 2024), with major restaurant chains adopting Uber Eats' robotic deliveries with Coco include, for example, Shake Shack in the United States. At the same time, Uber Eats is collaborating with Mitsubishi Electric and Cartken to experiment with ADR-based food delivery in Japan; their four-wheeled ADR is fully autonomous but operates under remote monitoring by a human operator (Uber, 2024). In late 2024, Uber Eats announced a partnership with Avride, an autonomous vehicle company, to pilot food delivery using ADRs in Texas, United States (Salgado, 2024).

Table 2.1 provides a comparison of the cases based on key characteristics.

Based on the comparison table, companies collaborate with robotics firms to test ADR technology. Features like real-time tracking (Co-op) and remote monitoring (Uber Eats) enhance reliability and customer service, while specialized designs, such as Domino's hot-and-cold compartments, improve delivery quality. Projects are often piloted regionally before expansion, adapting to local conditions. Although some initiatives ended due to corporate financial constraints, ADR projects emphasize environmental benefits, such as zero-emissions vehicles (Uber Eats, Co-op) and renewable energy-powered robots (Co-op). However, public perception of these technologies, particularly regarding their sustainability impacts, remains unclear.

2.4 The Impact and Challenges of ADRs as a Delivery Innovation

Although ADRs remain unfamiliar to many, with encounters often spontaneous, unplanned, and unsolicited (Hossain, 2023), emerging research indicates a generally positive user reception and sentiment toward these robots, suggesting ADRs a promising alternative for last-mile deliveries in the future (Ostermeier et al., 2022).

Table 2.1 Case summary with key characteristics

	Domino's	**Amazon**	**Co-op**	**Uber Eats**
Launch year	2016 (Australia and New Zealand)	2019 (U.S.)	2018 (U.K.)	2023 (with Serve Robotics in U.S. and Canada)
Geographic scope	Australia, New Zealand, U.S. (Texas, Michigan), U.K.	U.S. (Washington, California, expanded to other states)	U.K. (Leeds and other regions)	U.S., Canada, Japan
Partnerships	Nuro, Starship Technologies, Ford	Acquired Dispatch for in-house development	Starship Technologies	Serve Robotics, Coco Robotics, Mitsubishi Electric, Cartken, Avride
Primary vehicle	DRU (4-wheeled); R2 (mini autonomous van); Ford full-sized autonomous car	Amazon Scout (6-wheeled)	Starship's 6-wheeled ADR	Coco (4-wheeled); Cartken's ADR; Mitsubishi ADR; Serve Robotics' ADR
Delivery type	Pizza orders	Parcel delivery	Groceries	Food orders
Unique features	Separate compartments for hot and cold items; sliding roof; sidewalk travel; multiple experimental ADR types	Scout ambassadors (human operators for monitoring); pandemic-driven demand for contactless delivery	Renewable electricity-powered ADR; low energy consumption; live ADR tracking; supports carbon-neutral city goals	Zero-emissions deliveries (Coco); remote monitoring in Japan; integration with major restaurants like Shake Shack
Sustainability focus	Varies by ADR type; Ford autonomous vehicles not explicitly sustainable	No explicit sustainability focus	Powered by renewable energy; promotes carbon neutrality; reduces short car trips	Zero-emissions focus with Coco; ADR projects span multiple environments
Current status	Continued ADR trials in multiple countries	Discontinued Scout pilots in 2023	Expanding coverage in the U.K.; actively growing postpandemic	Ongoing expansion and pilots in multiple countries (U.S., Canada, Japan); partnerships active in 2024

Even individuals who do not use ADR services directly tend to view the robots favorably, with many describing them as "cute" and expressing a desire to help when robots encounter obstacles, such as getting stuck in snow or traffic lights (Westerlund, 2024).

Jennings and Figliozzi (2020) highlight that ADRs offer substantial time and cost savings while reducing on-road travel per package, thereby enhancing delivery efficiency and resource optimization. Companies such as Domino's, FedEx, Kroger, Walmart, and JD.com have reported significant cost reductions through the use of ADRs, with studies suggesting that ADRs can deliver orders for as little as 6 U.S. cents per mile or average of 40 U.S. cents per trip, which is 20 times cheaper than human delivery costs and six times cheaper than the cost of a consumer's personal trip to the grocery store (Keeney, 2021).

Even though many companies have not publicly disclosed the full economic impact of their ADR trials, the fact that many are expanding their services into new regions suggests they are satisfied with the outcomes. Even companies that have halted their pilots, such as Tesco Express in the United Kingdom, attribute their decisions to external factors, such as company-wide budget cuts or regulatory actions, rather than to unsatisfactory outcomes of their ADR programs. Furthermore, the increasing number of ADR pilots, such as Yogiyo's food delivery trial in South Korea (Kim, 2024), and the growing flow of investments in robotics companies reflect strong interest in robotic delivery systems across industries.

Environmentally, ADRs offer considerable sustainability benefits, particularly when powered by renewable energy. Studies indicate that ADRs have the potential to drastically reduce GHG emissions compared to traditional delivery vehicles (Garus et al., 2024). In fact, the adoption of electricity-powered ADRs instead of gas-powered cars could eliminate 18 million metric tons of CO_2 emissions in the United States by 2030, equivalent to removing 4 million cars from the road (Keeney, 2021). This underscores ADR's potential role in future-focused, eco-efficient urban delivery systems, where both sustainability and cost-effectiveness are paramount considerations (Garus et al., 2024).

2.4.1 Challenges of ADRs for Last-Mile Delivery

According to Ayyildiz and Erdogan (2024), the main challenges associated with adopting ADRs for last-mile delivery are categorized as: (i) political, (ii) economic, (iii) social, (iv) technological, (v) environmental, and (vi) legal. These categories can be analyzed from various perspectives, including those of robotic delivery users, delivery providers, ADR manufacturers, local communities, political decision-makers, and society at large. For example, Koh et al. (2024) found that consumer acceptance of ADRs is influenced by perceived reliability, convenience, security, and privacy associated with robots, while Hossain (2023) highlights delivery cost and perceived safety as additional factors. Moreover, liability issues surrounding the use of autonomous robots continue to feature prominently in political discussions (Westerlund, 2020).

The introduction of ADRs presents significant environmental challenges for robotic delivery providers, including energy consumption, noise, and concerns about durability, recyclability, and robotic waste disposal (Ayyildiz & Erdogan, 2024). Westerlund (2024) emphasizes that sustainability in robotic last-mile delivery remains largely under-addressed, despite ADRs' potential to reduce carbon footprints by operating on green energy. Major challenges related to the production and recycling of robotics components persist, including environmental and ethical concerns surrounding the mining of battery materials, which demand substantial energy and water and often involve unsafe working conditions (Westerlund, 2024). Understanding consumers' perceptions of ADRs' environmental performance is important because it has been found to be associated with greater intentions to adopt robotic delivery (Han et al., 2024).

2.5 Method

To understand how people perceive ADRs from the sustainability perspective, we gathered user-generated comments to a 2023 Reddit posting about Uber Eats' robotic delivery from subreddit UberEATS. The posting was short, asking when Uber Eats started robotic deliveries, and had an image of an ADR attached. Upon cleaning up the data, we created a sample of 440 comments to be used for the analysis. We focused on longer comments because they were anticipated to explain more about people's cognitive perceptions compared to shorter comments that might be restricted to people's emotional perceptions, in other words, sentiments toward the robots.

For the analysis of the data, we applied a Large Language Model, specifically GPT-4, because previous research had used GPT to understand public opinion (Qi et al., 2024) and technology acceptance (Smolinski et al., 2022) by analyzing social media comments from Reddit (Wu et al., 2021). GPT applications like ChatGPT are used on a range of natural language processing tasks, such as language translation, question-answering, text generation, and sentiment analysis (Ray, 2023). Our analysis employed topic analysis to identify themes in a text corpus of user-provided data and to extract exemplary comments for reporting, as recommended by Lee et al. (2024).

As ChatGPT uses deep learning methods to analyze and produce text, it can perceive the context and provide meaningful information, thus making it a helpful tool for collecting and analyzing social media data (Haleem et al., 2022). In terms of analyzing data, ChatGPT can synthesize and summarize large amounts of information, automatically identify emerging trends and patterns in data, and perform automated modeling and prediction on given tasks (Gill & Kaur, 2023). On a hindside, it is known to have accuracy problems, is prone to errors and fabrication of information, and tends to occasionally "hallucinate," in other words, provide plausible sounding but nonsensical or incorrect responses (Currie, 2023; Haleem et al., 2022).

ChatGPT is designed to respond swiftly and thoroughly to "prompts," in other words, instructions input by the user (Haleem et al., 2022). Furthermore, it can pick up new prompts as users feed them, thereby making an analysis easier as the user can ask consecutive questions for clarification, further informing and breaking down the analysis into multiple linked tasks (Haleem et al., 2022; Tarbit et al., 2023). Following the example of Tarbit et al. (2023), a series of 11 chained prompts were entered into ChatGPT to analyze the gathered Reddit comments data; the research questions, main steps, and some exemplary prompts are shown in Table 2.2.

While the prompts associated with our first research question allowed ChatGPT to freely decide on the specific sustainability dimensions, the prompts associated with our second research question were more restrictive, instructing the AI to analyze the data from the perspective of sustainable business in line with Musulin and Strahonja (2023); according to their SD-X-BM framework, value is created to customers only when three strategic dimensions of the business are addressed in terms of sustainability: service design, business model, and user experience.

Table 2.2 Research questions and corresponding AI prompts

Research question	Prompt creation step	AI prompt example
RQ1. What are the main themes in consumers' sustainability perceptions of ADRs	Define context, data, AI's role, writing conventions, and overall task objectives to guide and start the data analysis that draws on multiple linked prompts.	"I will input a corpus of online comments about autonomous robots for food delivery. Please provide the key topics in the corpus with characterizing comments."
—	Set up task parameters and instruct AI to summarize information taken from the previous responses, label specific topics, or provide information on subtopics.	"What do the comments say about sustainability themes related to robots? Provide a 100 words summary and 10 exemplary comments for each theme."
RQ2. How can companies address these themes in their robotic delivery service models?	Give AI a clearly defined task, which will prompt it to transform the output parameters based on the task provided (i.e., a specific business model framework).	"Could you elaborate on each sustainability theme from the perspectives of (i) service design, (ii) business model, and (iii) user experience?"

2.6 The Learning Outcomes

2.6.1 RQ1: What Are the Main Themes in Consumers' Sustainability Perceptions of ADRs?

Our analysis revealed four positive and five negative themes that characterize consumer perceptions of ADRs' sustainability impacts. The positive themes include the societal benefits of (i) techno-centric greening like robotics for delivery, equipped with specific environmental benefits of ADRs, namely (ii) reduction in emissions, (iii) energy efficiency, and (iv) reduced traffic congestion. Conversely, the negative themes include (v) ADR's inferiority compared to drones from both economic and environmental perspectives, (vi) the societal harm of ADRs replacing human labor, and the environmental harms of (vii) creating robotic waste, (viii) increasing maintenance footprint, and (ix) practicing greenwashing, that is, virtue-signaling without necessarily helping the environment. These themes are illustrated in Figure 2.1.

As for positive themes, "techno-centric greening" reflects the favorable attitude toward using technology, such as delivery robotics, to benefit the environment. Examples include comments like, *"Great to see a focus on environmental benefits alongside technological progress"* and *"This feels like a step toward a greener, smarter future."* The theme of "reductions in emissions" included remarks such as *"Electric robots delivering food could be a small but meaningful way to lower our carbon footprint."* Additionally, the "energy efficiency" of ADRs was highlighted in

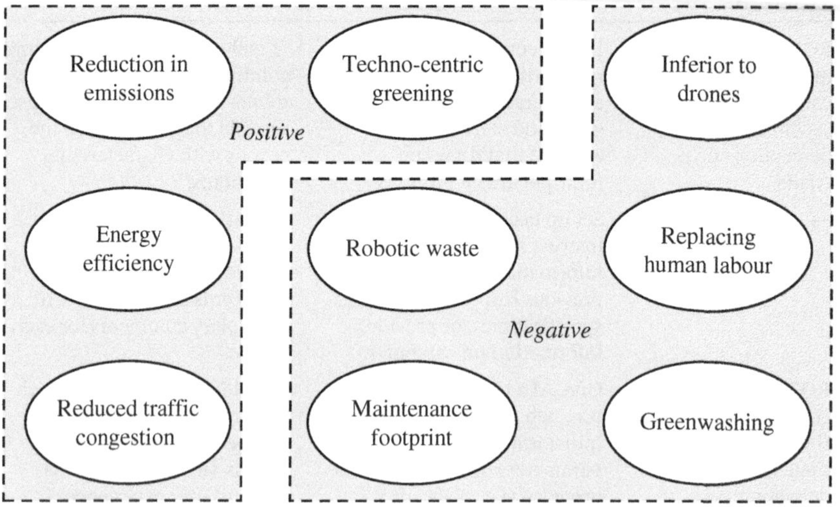

Figure 2.1 Environmental perceptions of ADRs found from social media comments.

various ways, for instance, *"The future of delivery could be energy-efficient robots—no gas, no noise, just clean energy."* Lastly, the theme of "reduced traffic congestion" in urban areas was noted, with comments emphasizing ADRs' size: *"The small size of these bots makes them ideal for reducing road congestion,"* which is advantageous compared to full-sized delivery vehicles, for example, *"They don't clog up the streets like delivery cars do, which is a win for urban living."*

Regarding negative themes, ADRs' "inferiority to drones" was evident. One commenter stated, "*I feel like drones would be more efficient and less wasteful overall,*" while another noted, "*Drones seem better for the environment—no roads, no wheels, just straight to the destination.*" The societal harm caused by ADRs "replacing human labor" was highlighted in comments such as, "*It's ironic that we're replacing humans with robots and calling it green innovation.*" Further, "robotic waste" was addressed with comments pointing to environmental harms during production, such as "*Sure, they save emissions, but what about the resources used to make them in the first place?*" as well as concerns about the end of their service life, exemplified by "*We already have a problem with e-waste—are we just adding to it with these bots?*" These concerns were compounded by a theme of "maintenance footprint," for example, "*Frequent maintenance might erase any environmental benefits they're supposed to offer.*" Finally, the theme of "greenwashing" reflected skepticism about the environmental benefits of ADRs, with comments like "*Are they really better for the environment, or is this just a marketing ploy?*"

2.6.2 RQ2: How Can Companies Address These Themes in Their Robotic Delivery Service Models?

According to the SD-X-BM framework by Musulin and Strahonja (2023), companies must address sustainability in their business across three strategic dimensions: service design, business model, and user experience. Applying this framework to themes in consumers' environmental perceptions, our analysis offers the following recommendations to reinforce positive themes and mitigate negative ones for companies considering the use of ADRs for last-mile delivery.

2.6.2.1 Enforcing Positive Themes

1) **Techno-centric greening**:
   Companies' marketing should emphasize ADR's environmental benefits, like reduced emissions, with carbon tracking per order boosting positive perceptions. Leveraging positive attitudes can drive adoption, especially among eco-conscious customers, while partnerships with environmental organizations enhance credibility and funding opportunities. Users of robotic delivery feel proud of their environmental contribution, leading to stronger brand loyalty, and rewarding users for eco-friendly choices enhances their satisfaction.

2) **Reduction in emissions**:
Companies should prefer routes that minimize emissions by clustering deliveries, with renewable energy-powered charging stations supporting eco-friendly goals. Highlighting environmental benefits can attract eco-conscious customers and investors, while fuel savings support scaling or sustainability-focused marketing. Users appreciate eco-friendly delivery options, especially with transparent emission savings (e.g., "You saved X kg of CO_2").

3) **Energy efficiency**:
Companies should use lightweight robots with efficient motors and renewable energy charging stations to minimize energy use and maximize uptime. Partnering with green energy providers and leveraging off-peak charging can cut costs and reduce grid stress. Highlighting energy-efficient practices like solar-powered hubs builds trust, while exploring regenerative energy technologies further boosts efficiency and brand appeal.

4) **Urban traffic congestion**:
Companies should use ADRs that operate smoothly on sidewalks, with collaborative routing systems ensuring efficient navigation and minimizing urban bottlenecks. Reducing congestion enables premium services in high-traffic areas and allows collaboration with cities on initiatives in exchange for regulatory incentives. Faster delivery times and navigation in urban settings foster a positive perception of the service as efficient and well-organized.

2.6.2.2 Mitigating Negative Themes

5) **Inferior to drones**:
Companies could use drones to complement ADRs in areas with unsuitable sidewalks or vertical delivery needs, thus maximizing efficiency for diverse environments. Drones reduce infrastructure needs but require regulatory investments, yet promoting the hybrid model can appeal to tech-savvy, eco-conscious users. Urban users may prefer robots for reliability, while drones suit rural areas; addressing noise and safety concerns ensures a seamless experience.

6) **Replacing human labor**:
Companies should deploy ADRs where they replace high-emission vehicles, not eco-friendly options; they can use robots for short deliveries and humans for personalized service. Reducing fuel-based vehicle use boosts environmental value, while investing in human-powered delivery alongside automation fosters a mixed workforce. Emphasizing ADRs' environmental advantages over polluting human-operated vehicles prevents alienating users.

7) **Robotic waste**:
Companies should choose ADRs that allow modular repairs and recycling programs to enhance durability and reduce e-waste. The circular economy

approach in manufacturing, with refurbished parts, reduces environmental impact and production costs while appealing to eco-conscious stakeholders. Eco-conscious users support waste-reducing services; transparency about ADRs' life cycle, like recycled materials and repairability, builds loyalty.

8) **Maintenance footprint**:
   Companies should invest in durable designs, regular maintenance, and self-diagnostics to extend the robot lifespan and ensure performance in harsh conditions. Continuous remote software updates and AI-powered predictive maintenance can reduce downtime, repair needs, and replacement costs, boosting return on investment (ROI). Warranties and maintenance contracts enhance reliability and trust, meeting user expectations for dependable, long-lasting technology.
9) **Greenwashing**:
   Companies should ensure transparency in environmental claims by using visible green features like solar panels and providing clear environmental data. Trust can be built through certifications, audits, and sustainability partnerships, bolstering the brand against greenwashing while highlighting ADRs' life cycle benefits. Independent studies on ADRs' sustainability impacts, shared openly, further enhance credibility and address skepticism.

2.7 Plans to Further Develop the Initiative

Garus et al. (2024) emphasize that there is no one-size-fits-all solution to last-mile delivery; the optimal approach depends on a company's delivery system specificity and external factors. Having said that, ADRs offer an efficient and sustainable alternative in urban settings. However, for companies to genuinely benefit from using robotic delivery systems such as SADRs, we need to better address the underresearched area of sustainability. While most of the extant research on ADRs has focused on operational emissions, future research efforts need to consider life cycle emissions and their impacts on nature (Figliozzi, 2020).

Westerlund (2020, 2024) noted that both research and public discussions on smart, autonomous robots have largely neglected the fact that the increased corporate use of cheap and ubiquitous robotics technology, with constant new models and improvements, will become a huge future challenge in terms of disposal and recycling of robotics materials. In line with this argument, we call for more research using a sustainability perspective to robotic delivery, and propose conducting expert interviews and industry surveys, as well as analyzing large social media data to gain insights into environmental concerns and industrial approaches to tackle those concerns.

References

Amazon Scout. (2025, April 22). Wikipedia. Retrieved from https://en.wikipedia.org/w/index.php?title=Amazon_Scout&oldid=1286808851

Ayyildiz, E., & Erdogan, M. (2024). Addressing the challenges of using autonomous robots for last-mile delivery. *Computers & Industrial Engineering, 190*, 110096. https://doi.org/10.1016/j.cie.2024.110096

CBC.ca. (2017). *Domino's Pizza tests driverless car delivery*. https://www.cbc.ca/news/science/dominos-pizza-ford-self-driving-car-1.4266349

Chen, C., Demir, E., Huang, Y., & Qiu, R. (2021). The adoption of self-driving delivery robots in last mile logistics. *Transportation Research Part E: Logistics and Transportation Review, 146*, 102214. https://doi.org/10.1016/j.tre.2020.102214

Co-op. (2023). *Co-op and Starship Technologies expand robot delivery service across Leeds*. https://www.co-operative.coop/media/news-releases/co-op-and-starship-technologies-expand-robot-delivery-service-across-leeds

Currie, G.M. (2023). Academic integrity and artificial intelligence: Is ChatGPT hype, hero or heresy? *Seminars in Nuclear Medicine, 53*(5), 719–730. https://doi.org/10.1053/j.semnuclmed.2023.04.008

De Maio, A., Ohlmann, J. .W., Stoia, S., & Vocaturo, F. (2024). Analysis of in-store crowdshipping in a stochastic dynamic pickup-and-delivery system. *Central European Journal of Operations Research, 33*(3), 1149–1170. https://doi.org/10.1007/s10100-024-00939-8

Domino's. (2024). *DOM – Autonomous delivery vehicle*. https://www.dominos.com.au/about-us/technology/dom

Figliozzi, M.A. (2020). Carbon emissions reductions in last mile and grocery deliveries utilizing air and ground autonomous vehicles. *Transportation Research Part D: Transport and Environment, 85*, 102443. https://doi.org/10.1016/j.trd.2020.102443

Garus, A., Christidis, P., Mourtzouchou, A., Duboz, L., & Ciuffo, B. (2024). Unravelling the last-mile conundrum: A comparative study of autonomous delivery robots, delivery bicycles, and light commercial vehicles in 14 varied European landscapes. *Sustainable Cities and Society, 108*, 105490. https://doi.org/10.1016/j.scs.2024.105490

Gill, S.S., & Kaur, R. (2023). ChatGPT: Vision and challenges. *Internet of Things and Cyber-Physical Systems, 3*, 262–271. https://doi.org/10.1016/j.iotcps.2023.05.004

Ha, N.T., Akbari, M., & Au, B. (2023). Last mile delivery in logistics and supply chain management: A bibliometric analysis and future directions. *Benchmarking: An International Journal, 30*(4), 1137–1170. https://doi.org/10.1108/BIJ-07-2021-0409

Haleem, A., Javaid, M., & Singh, R.P. (2022). An era of ChatGPT as a significant futuristic support tool: A study on features, abilities, and challenges. *BenchCouncil Transactions on Benchmarks, Standards and Evaluations, 2*(4), 100089. https://doi.org/10.1016/j.tbench.2023.100089

Han, J., Seo, S., & Kim, H.J. (2024). Autonomous delivery robots on the rise: How can I cut carbon footprint for restaurant food deliveries? *International Journal of Hospitality Management, 121*, 103804. https://doi.org/10.1016/j.ijhm.2024.103804

Heater, B. (2022). *Amazon scales back Scout delivery robot program*. https://techcrunch.com/2022/10/07/amazon-scales-back-scout-delivery-robot-program/

Hossain, M. (2023). Autonomous delivery robots: A literature review. *IEEE Engineering Management Review, 51*. https://doi.org/10.1109/EMR.2023.3304848

Jaller, M., Pineda, L., Ambrose, H., & Kendall, A. (2021). Empirical analysis of the role of incentives in zero-emission last-mile deliveries in California. *Journal of Cleaner Production, 317*, 128353. https://doi.org/10.1016/J.JCLEPRO.2021.128353

Jennings, D., & Figliozzi, M. (2020). Study of road autonomous delivery robots and their potential effects on freight efficiency and travel. *Transportation Research Record, 2674*(9), 1019–1029. https://doi.org/10.1177/0361198120933633

Keeney, T. (2021). *Robot delivery of groceries could cost 40 cents per trip*. https://www.ark-invest.com/articles/analyst-research/robot-delivery

Kim, J.-W. (2024). Yogiyo to begin autonomous robot delivery tests in August. *The Korea Economic Daily*. https://www.kedglobal.com/robotics/newsView/ked202406210007

Koh, L.Y., Xia, Z., & Yuen, K.F. (2024). Consumer acceptance of the autonomous robot in last-mile delivery: A combined perspective of resource-matching, perceived risk and value theories. *Transportation Research Part A: Policy and Practice, 182*, 104008. https://doi.org/10.1016/j.tra.2024.104008

Lee, V., van der Lubbe, S., Goh, L., & Valderas, J. (2024). Harnessing ChatGPT for thematic analysis: Are we ready? *Journal of Medical Internet Research, 26*, e54974. https://doi.org/10.2196/54974

Musulin, J., & Strahonja, V. (2023). User experience, business models, and service design in concert: Towards a general methodological framework for value proposition enhancement. *Sustainability, 15*, 12509. https://doi.org/10.3390/su151612509

Ostermeier, M., Heimfarth, A., & Hübner, A. (2022). Cost-optimal truck-and-robot routing for last-mile delivery. *Networks, 79*(3), 364–389. https://doi.org/10.1002/net.22030

Qi, W., Pan, J., Lyu, H., & Luo, J. (2024). Excitements and concerns in the post-ChatGPT era: Deciphering public perception of AI through social media analysis. *Telematics and Informatics, 92*, 102158. https://doi.org/10.1016/j.tele.2024.10.158

Ray, P.P. (2023). ChatGPT: A comprehensive review on background, applications, key challenges, bias, ethics, limitations and future scope. *Internet of Things and Cyber-Physical Systems, 3*, 121–154. https://doi.org/10.1016/j.iotcps.2023.04.003

Salgado, B.A. (2024). *Avride delivery robots coming to Austin through partnership with Uber Eats; what to expect*. https://www.msn.com/en-us/money/other/avride-delivery-robots-coming-to-austin-through-partnership-with-uber-eats-what-to-expect/ar-AA1uatev?ocid=socialshare

Serve Robotics. (2023). *Serve unveils commercial deal with Uber to enable scaling of robotic delivery*. https://www.serverobotics.com/uber-scaling

Smolinski, P. R., Januszewicz, J., & Winiarski, J. (2022). *Scaling technology acceptance analysis with large language model (LLM) annotation systems*. Proceedings of the 32nd International Conference on Information Systems Development (ISD 2024) (26, 28 August 2024). Gdansk, Poland.

Tarbit, J., Wirtz, J., Kunz, W., & Hartley, N. (2023). Interpretation of corporate digital responsibility risks and concerns by automated service technologies: An AI co-created article. *ROBONOMICS: The Journal of the Automated Economy*, *4*, 52.

The Robot Report Staff. (2024). *Uber Eats rolls out Coco's sidewalk delivery robots in LA*. https://www.therobotreport.com/uber-eats-rolls-out-coco-sidewalk-delivery-robots-in-la/

Uber. (2024). *Bringing the future to today: Making robot delivery a reality*. https://www.uber.com/en-CA/blog/so-funakoshi/

Westerlund, M. (2020). The ethical dimensions of public opinion on smart robots. *Technology Innovation Management Review*, *10*(2), 25–36. https://doi.org/10.22215/timreview/1326

Westerlund, M. (2024). Social acceptance of autonomous food delivery robots: An analysis of public commentaries. *ROBONOMICS: The Journal of the Automated Economy*, *5*, 57.

Wu, W., Lyu, H., & Luo, J. (2021). Characterizing discourse about COVID-19 vaccines: A Reddit version of the pandemic story. *Health Data Science*, *2021*, 9837856. https://doi.org/10.34133/2021/9837856

Wu, M., Lin, A.S.Q., & Yuen, K.F. (2024). The effects of motivated consumer innovativeness on consumer acceptance of autonomous delivery robots. *Journal of Retailing and Consumer Services*, *81*, 104030. https://doi.org/10.1016/j.jretconser.2024.104030

3

Assessing the Impact of an App-Based Scheduling Platform on the Overall CO_2 Emissions and Travel Time of Minibus Taxis in Four Cities (Municipalities) in Uganda

Gabriel Okello[1,2], Vianney Tumwesige[3], Derrick Joseph Njuba[3], Richard Nelson Sserunjogi[4], and Tom Courtright[5]

[1] *Institute for Sustainability Leadership, University of Cambridge, Cambridge, UK*
[2] *Department of Research and Innovation, African Centre for Clean Air, Kampala, Uganda*
[3] *Operations Department, Green Heat (Uganda) Limited, Wakiso, Uganda*
[4] *AirQo, Department of Computer Science, Makerere University, Kampala, Uganda*
[5] *Centre for Transport Studies, University of Cape Town, Cape Town, South Africa*

3.1 Introduction

Privately owned minibus paratransit forms a major part of the mobility solution in the Global South, especially in sub-Saharan Africa. Some cities across sub-Saharan Africa, including Accra (Agyemang, 2015; Poku-Boansi & Marsden, 2018), Cape Town (Schalekamp, 2017), Dar es Salaam (Chengula and Kombe, 2017), Johannesburg (Asimeng and Heinrichs, 2021), Lagos (Asimeng & Heinrichs, 2021), Nairobi (Klopp and Cavoli, 2019), and so on, have sought to replace minibus paratransit with contracted scheduled services, typically bus rapid transit (BRT), but have had limited success. The biggest challenge has been cost and difficulty of transforming the existing system.

The current minibus paratransit operations include a collection of independent operators that provide much of the on-demand public transport services within and outside cities. The routes and services are determined by drivers in response to real-time passenger demand, rather than by contractual obligation to a regulatory authority. Other names for minibus taxis elsewhere include trotro (Ghana), car rapides (Senegal), and matatu (Kenya) (Behrens et al., 2016; Cervero, 2000; Golub et al., 2012; Kumar and Diou, 2010; Saddier and Johnson, 2018).

Green Innovations in Supply Chain Management: Case Studies and Applications, First Edition. Edited by Marcia Mkansi, Godfrey Mugurusi, Aaron L. Nsakanda, Temidayo Akenroye, and Frank Tietze.

Published 2026 by John Wiley & Sons, Inc.

Matatus, as they are commonly known in Kenya and Uganda, transport up to 70% of the total urban travelers and dominate most social and economic aspects of urban mobility (Behrens et al., 2015; Welle, 2019). They are part of a broader paratransit system that operates with little or no regulation in many developing cities of Africa and the Global South (Behrens et al., 2015), including Kampala. The matatus operate a flexible and semi-adaptive mode of operation, with stops, schedules, fares, and routes primarily determined by drivers responding to perceived demand (Lucas, 2019). Matatu drivers often plan their routes according to the occupancy status of the matatu and anticipated demand, and adjust based on traffic, weather, and other factors (Gauthier and Weinstock, 2010). Matatus usually accommodate 14 passengers excluding the driver and a conductor. Like in many countries, decision-making is divided between the owner and a driver. The owner provides the vehicle with an operating license that dictates the specific route on which the vehicle is permitted, as well as which terminals can be used for loading (Behrens et al., 2016; McCormick et al., 2015). Drivers determine the operational specifics of the service, deciding when to provide service. Drivers collect fare revenue and either pay owners a rental fee, or "target," each day or split the money with the vehicle owner on a commission basis, where typically 70% accrues to owners and 30% to drivers (McCormick et al., 2015). The conductor is usually paid from the driver's earnings and is responsible for collecting fares on board. Matatus serve crucial demands and needs for mobility and accessibility without government subsidy. However, they also have negative characteristics, as summarized in Table 3.1 (Plano et al., 2020).

Table 3.1 Summary of positive and negative elements of paratransit services

Positive	Negative
• Short access distance/high coverage (Hitge and Vanderschuren, 2015) • Frequent (Cervero, 2000) • Greater security in smaller vehicle (Cervero, 2000) • Customer service, such as older passengers assisted with boarding/alighting (own observations) • Demand responsive (McCormick et al., 2015) • Job creation for low-skilled young men (Schwanen, 2018) • Subsidy free (Behrens et al., 2016)	• Poorly maintained vehicles (Cervero, 2000) • Poor off-peak frequency (McCormick et al., 2015) • Aggressive driving stemming from insecurity of pay (McCormick et al., 2015) • Overloading of passengers (Cervero, 2000) • Passengers unable to board full vehicles near terminals (Behrens et al., 2016) • Informal driver employment results in unpredictable income, job insecurity, and no benefits (Behrens et al., 2016)

Research has shown that traffic-related air pollution (TRAP) contributes to ambient air pollution from both the combustion of fossil fuels in the vehicle fleet and non-exhaust emissions (Boogaard et al., 2022). Globally, 25% of urban ambient particulate matter ($PM_{2.5}$) is caused by traffic (Karagulian et al., 2015). This challenge is exacerbated by the fact that matatus are mainly imported second-hand vehicles (usually over 15 years old from the date of manufacture) and their mode of operation involves stopping and starting along the route, with significant amounts of idling time to look for passengers and to pick up bystanders (Booysen et al., 2022). Prolonged travel times caused by traffic congestion, often in sprawling and rapidly urbanizing cities, lead to increased fuel consumption and emissions (Smit et al., 2008). Emissions from TRAP have been linked to environmental, economic, and health issues (Dalal et al., 2011; Landrigan et al., 2018; Croitoru et al., 2020; World Health Organization, 2018).

Uganda's transportation system emits 2 Mt of CO_2 per year, which is 38% of Uganda's total carbon emissions (Crippa et al., 2019), and grappling with high levels of ambient air pollution. One of the cost-effective ways to address TRAP and related challenges like time wastage on many stoppages could be by operating transport routines with clear schedules, thus reducing stoppages and saving time and reducing air pollution as a result of idling.

3.2 Literature Review: Uganda's Urban Transport and Technological Advances

Uganda's urban transport faces significant challenges, including severe traffic congestion that reduces productivity (Stewart, 2024). Poor road networks worsen congestion, and the lack of integration between transport modes and disjointed planning undermine the system's efficiency, disrupting the smooth flow of people and goods, thus affecting travel times.

Conversely, technological advancements, such as smartphone transit apps, app-based ride-sharing, and public transit platforms, have emerged as a cost-effective solution to reduce individual travel times and promote public transport usage (Bian et al., 2023). Transit agencies and private developers have launched numerous apps offering fare payment, trip planning, and real-time vehicle updates (Gooze et al., 2013; McCord et al., 2015). With the widespread adoption of smartphones, these apps have become vital for delivering timely transit information. However, the implementation of technology apps for public transport is still limited in many African cities like Kampala.

The adoption of smartphone transit apps is growing rapidly in Uganda, particularly among motorcycle taxis (boda bodas) and four-seater taxis, with platforms like Uber, SafeBoda, Ride Now App, Lolo Drive, Spesho Taxi, and Bolt dominating the market. Public transport vehicles, such as 14-seater minibuses and larger buses, are served by fewer apps like Easy Matatu (EM) and KaCyber. However, there is limited understanding of the overall impact of these app-enabled platforms on travel time, traffic congestion, and CO_2 emissions in urban areas.

This chapter addresses this knowledge gap by evaluating the impact of an app-based scheduling platform on CO_2 emissions and travel time for minibus taxis in four Ugandan cities, while analyzing key trends in app usage. This study offers insights to guide future research, urban governance, and planning, aligning digitalization and mobility innovations with sustainable urban development goals. Against this backdrop, we applied sequential quantitative methods to assess the impact of using an app-based platform (mobile app) that enables commuters to book seats on scheduled express rides that are operated by trained matatu drivers. Through the mobile app commuters traveling to the same destination are efficiently pooled together, thus eliminating the need to stop frequently along the route. In this study, we leveraged the availability of the app-based platform called EM app and quantitative assessment to:

1) Investigate if the use of online-based app contributed to reduction of CO_2 emissions.
2) Investigate if the use of online-based platform reduced stoppages along the travel route and/or fuel savings for minibus taxis using the EM app.
3) Assess if there was reduced commute time by using the EM app.

3.3 Materials and Methods

3.3.1 Study Design and Setting

We undertook a cross-sectional survey between October 2022 and January 2023 to examine the impact of an online booking platform for minibus taxis on CO_2 emissions, travel time, and overall fuel cost for matatus in Kampala, Mukono, Wakiso, and Entebbe districts in Uganda. These districts represent some of the fastest-growing cities in Uganda, owing to the rapid population growth, rural-urban migration, and the associated urbanization. For example, Kampala, which was originally designed to host 300,000 people (UN-Habitat, 2016), has rapidly expanded over the past 40 years and now hosts approximately 1.75 million people living in the city, with the population considerably increasing to around 4.5 million at certain periods of the day due to influx of commuters and workforce (Kampala Capital City Authority, 2019). With an annual demographic growth rate of 3.9%, the planning functions and infrastructure of the city are struggling to keep pace (Kampala Capital City Authority, 2019). The increase in the population of Kampala city has led to very high traffic congestion and growing informal settlements.

3.3.2 Data Collection

3.3.2.1 Identifying the Routes

We held two virtual and two in-person meetings with EM representatives to obtain the routes and operation times (and days) of EM taxis, get training on how to use the EM app, and introductions to the EM drivers. This was an iterative process where more themes, for example, data collection using global positioning system (GPS), were discussed, reviewed, and added.

Only routes that had matatus using the online platform were included in the study. Routes that did not have matatus using the online platform were excluded. The number of matatus using the online platform on each particular route was obtained (Table 3.2). The start and end points of each route were obtained from the various routes as shown in Table 3.2. We defined matatus using the online platform as EM taxis and those not using the platform as non-Easy Matatu (NEM) taxis. Under the EM taxis we also defined two subcategories including online EM and offline EM, representing matatus that were registered on the online platform and using the platform at the time of the research and those that were registered on the platform but for some reason were offline during the time for conducting the research, respectively.

The routes selected for the study were then divided into three categories (Figure 3.1) according to distance from the city center/city square. The routes included categories 1, 2, and 3.

a) **Category 1**: 0 to 15 km
b) **Category 2**: 0 to 30 km
c) **Category 3**: 0 to 45 km

Table 3.2 The routes details and number of minibuses using the EM app

Route	Starting point	Ending point	No. of EM	Status*
Entebbe	Total Kitoro	Total city center	12	Active
Najjera	Shell Kiira	City square	7	Active
Namugongo	St Michael Sonde	City square	2	Active
Gayaza	Stabex Manyangwa	City square	2	Active
Mukono	Wantoni	City square	1	Active
Matugga	Gaz Matugga	City square	2	Active
Bweyogerere	Hertwin Bweyogerere	City square	1	Inactive
Luzira	Portbell	City square	1	Closed
Toyota (Namanve)	1st Street and Shell	Toyota Namanve	2	Inactive

* Status of the route during the research period.

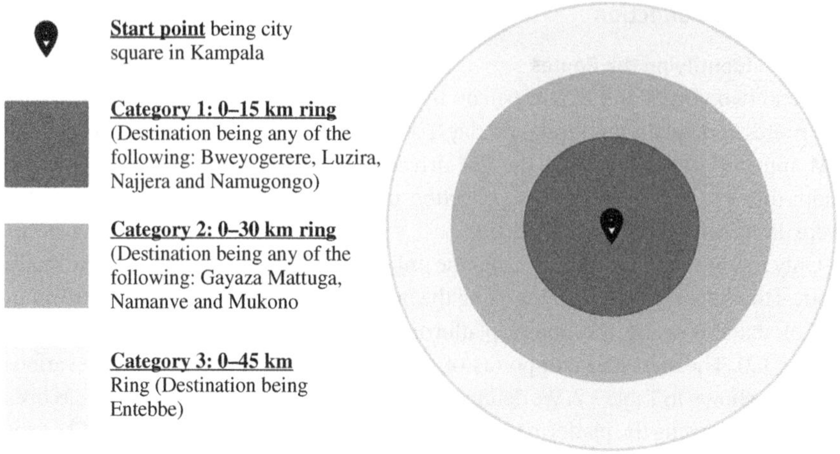

Figure 3.1 Study routes defined by distance from start to finish.

3.3.3 Sample Size

We conducted data collection on a total of 36 matatus, with 18 using the online EM app and 18 NEM app users selected from three defined route categories from 24 October 2022 to 11 January 2023. The pragmatic sample size was based on the resources and number of EM available for the route categories. The sample size included minibus taxis using the EM app and those not using the app on the three route categories, operating on the same routes and similar time schedules. The number of trips by participating matatus (those with EM app and those without) was collected per route with detailed number of trips, passengers per trip, and distance covered (Supplementary File SF1).

3.3.3.1 Transit Data Collection

We developed a data collection tool (Figure 3.2) from a comprehensive review of the published research literature and expert consultation. Feedback from the experts was also incorporated into the tool prior to data collection. The data collected by the tool included: fuel consumption, year of manufacture of the matatu, number of stops, idling time using stopwatches (waiting or picking/dropping passengers), and overall time taken on the route. The tool was used to collect data for both EM (online and offline) and NEM.

Eight research assistants were recruited, trained, and assigned to the identified routes. After the training, a pilot was conducted for two weeks to test the rigor of the tool and assess whether the methodology and the data collection tool required further refining. We refined the tool prior to official commencement of the study.

Vehicle Registration Plate		START LOCATION (Lat/Long):				START TIME:	
		END LOCATION (Lat/Long):				END TIME:	
		TOTAL TRIP TIME:					
Route Name				DATE			
Route Code;		Number of Passengers (Start Point)			Number of Passengers (End Point)		
Fuel Type		Is there a Route Diversion		Name of the Driver:	Peak Hours/	GPS Tracking Recorded	
Engine Size (c.c.)		Yes	No		Off-peak Hours	Yes	
Vehicle model						No	
Fuel at the start							
Refuelling							
Stop/Rest areas (Use any common feature in the area).		Stop time	Start time	Idling Time (minutes)	Number of passengers dropped off	No. of passengers boarded	Any other stoppages besides*
Start Car Mileage (KM)			End, Car Mileage (KM)		Total KM Travelled		
Driver's Signature		Supervisor's Signature		Comments			
Research Assistant Signature:		GPS Data Reports Submitted	Yes				
			No				

Peak hours_07:00 hours–10:00 hours/16:00 hours–19:00 hours **Off-peak hours**_10:00 hours–16:00 hours

*Additional "activities" include police encounters, riots, halting traffic due to movements of Very Important Persons (VIPs) etc.

Figure 3.2 Data collection tool.

More parameters were added in the template during refinement. These included fuel cost per liter, the fuel consumption per trip, vehicle model, and the number of passengers boarded and dropped off.

For data collection two research assistants were assigned per route. One research assistant was responsible for collecting data from the EM and the other research assistant for the NEM. The research assistant responsible for collecting data from the EM booked the seat a day earlier. Both research assistants collected data on a particular route on the same day from the same start point at about the same time.

Data collection was carried out for six out of the seven days in the week (from Monday to Saturday). This is because all matatus were offline on Sunday and so it would be difficult to get comparative data between EM and NEM on the same route. The overall number of trips made over the study period was documented; this period covered peak and off-peak hours on any given study day. The distance covered per route was recorded by GPS devices across all the classified matatu groups. Both fuel type and engine size for each participating matatu were documented.

Research assistants recorded the fuel consumption of the matatus, number of stops, matatu idling time using stopwatches (waiting or picking/dropping passengers). Start and stop time durations were cross-referenced with GPS data for each trip. To estimate the delays occurring at each trip per route, idling time (in minutes) was recorded as the matatus picked up passengers and during traffic jam. The overall idling time was expressed in hours. The research assistants recorded the various reasons for idling, for example, passenger pickup, passenger drop-off, traffic congestion, etc. This was compared to the total time per trip. The research assistants were paid for each route for an EM and NEM operating at similar times on the same day and same route. This was done to obtain a comparative analysis of the routes, the days and time of operation. Research assistants also recorded when the routes for the EM and NEM changed.

For internal validity, we compared stoppages (both using the EM app and those not using the app) along the different route categories to explore if the various route categories based on distance from the city center could have affected stoppages.

Aggregate data generated from 36 matatus were processed to produce summary statistics to enable subsequent assessment and analysis.

3.4 Data Analysis

The data collected from EM and NEM were cleaned, aggregated, processed, and adjusted for local confounders (rain disruptions, public holidays, traffic stops, special ceremonies like campaigns along any route) using Excel.

We used the Intergovernmental Panel on Climate Change (IPCC) guidelines for calculating the carbon footprint due to idling of vehicle emissions (Eggleston et al., 2006). An oxidation factor was applied to the carbon content to account for a small portion of the fuel that is not oxidized into CO_2. The oxidation factor was 0.99 in accordance with the IPCC for all oil and oil products. This translates into 99% of the carbon in the fuel oxidized, while 1% remains unoxidized. However, the oxidation efficiency was affected by the age and the vehicle engine factored in. To calculate the CO_2 emissions, the carbon content of the type of fuel used was multiplied by the ratio of the molecular weight (m.w.) of CO_2 (m.w. 44) to the molecular weight of carbon (m.w. 12): 44/12.

CO_2 emissions for both EM and NEM were therefore calculated by multiplying estimated fuel used with a default CO_2 emission factor. The approach is represented in equation (3.1).

$$\text{Emission} = \sum_{a} \left(FC_a \times EF_a \right) \tag{3.1}$$

where (Barth & Boriboonsomsin, 2008)

- Emission = Emissions of CO_2 (kg),
- FC = fuel consumed (TJ),
- EF_a = emission factor (kg/TJ). This is equal to the carbon content of the fuel multiplied by 44/12,
- a = type of fuel (e.g., diesel, petrol),
- CO_2 emissions from a gallon of diesel = 2778 grams × 0.99 × (44/12) = 10,084 grams = 10.1 kg/gallon = 22.2 pounds/gallon (Eggleston et al., 2006).

We explored the frequency distribution of measured idling, stoppages, and trips by time of the day (peak and off-peak time) in order to assess the reduction in emissions according to time of the day, day of the week, and route category.

Stoppages and idling frequency distributions for the matatus using the EM app and those not using the EM app were also analyzed for correlations according to time of the day, day of the week, and route category.

We summed up the overall number of stoppages for loading and offloading passengers along the travel route for matatus using the EM app and those that were not using the EM app to assess if using the EM app reduced the number of stoppages. We also explored if there were any fuel savings due to reduced stoppages along the route and if the commute time was reduced by using the EM app.

Data collected by research assistants across all designated routes and parameter estimates were used to model future impacts of the EM interventions.

3.5 Results

Analysis was complicated by the fact that many of the matatus using the EM app would turn the app off during the day to pick up regular customers, resulting in three operational modes of the matatus: NEM, off-app EM, and on-app EM.

3.5.1 Difference in CO_2 Emissions Between Matatus Using the EM App and Those Not Using the App

During the research period, we found that matatus using the EM app emitted an average of 13.07 kg of CO_2 per trip, whereas matatus not using the app emitted an average of 13.61 kg of CO_2 per trip. An independent t-test ($p > 0.05$) conducted indicated that there was no statistically significant difference between average CO_2 emissions per trip between matatus using the EM app and those not using the app.

Table 3.3 shows the difference in CO_2 emissions between matatus using the EM app and those not using the app.

We calculated the CO_2 emissions due to idling using equation (3.1) as shown in Table 3.3.

3.5.2 Stoppages Along the Travel Route for Matatus Using the EM App and Those Not Using the EM App

According to Table 3.4, Matatus that did not use the app had an average of 17 stops per trip, while those using the app had an average of 14.7 stops per trip. This disparity was most significant in drop-off and pickup stops, as NEM had 55% more stops than their EM counterparts, thereby providing greater convenience to passengers but resulting in longer onboard trip time. However, it should be noted that on-app EM were primarily used during rush hours, which meant they had more stops due to traffic jams, averaging around 5.5 stops per trip compared to NEM, which operated throughout the day and averaged 3.8 stops per trip.

Table 3.3 CO_2 emissions per trip for matatus

	Mean	Standard deviation	Minimum	Maximum
	CO_2 emissions in kg			
EM ($n = 59$)	13.07	3.78	7.15	20.69
NEM ($n = 142$)	13.61	4.28	7.99	24.68

Table 3.4 Average stops per trip for EM and NEM

App usage	Traffic jam stops	Traffic light stops	Refueling stops	Drop-offs and pickups	Total stops
NEM	3.8	1.4	0.4	11.5	17.2
Off-App EM	1.3	1.2	0.5	9.7	12.9
On-App EM	5.5	1.3	0.3	7.4	14.8

Table 3.5 Stops per trip distance category

App usage	0–15 km	15–30 km	30–45 km
NEM	16.3	15.7	20.1
Off-App EM	11.1	10.5	18.1
On-App EM	18.7	13.4	16.1
Grand total	**13.9**	**15.2**	**19.2**

Contrary to expectations, longer trips (30–45 km) using the EM app had fewer stops than shorter trips, unlike NEM trips, where longer journeys had more stops (Table 3.5).

EM trips on mid-to-long distances had fewer pickup and drop-off stops compared to regular matatus.

For example, in the long-distance Entebbe-Kampala route, on-app matatus moved at an average of 34 km/h compared to only 26.7 km/h for offline matatus (Table 3.6).

3.5.3 Commute Time Between Matatus Using the EM App and Those Not Using the EM App

On-app EM had a higher average speed of 22 km/h compared to NEM at 19.9 km/h (Table 3.6). This difference was even more pronounced on specific routes, such as the long-distance Entebbe-Kampala, where on-app EM moved at an average speed of 34 km/h, while NEM only managed 26.7 km/h.

EM trips averaged 1 hour and 13 minutes shorter than NEM trips at 1 hour and 25 minutes (Table 3.6). Despite covering slightly longer distances (25.8 versus 24.1 km), EM trips were faster, even on the same routes.

Table 3.6 Overall average time spent on trips

		Average distance per trip (km)	Average speed per trip (km/h)	Average time per trip
Entebbe-Kampala	On-App EM	40.1	34.0	1:17:00
	NEM	35.1	26.7	1:23:13
All trips	On-App EM	25.8	22.0	1:13:58
	NEM	24.1	19.9	1:25:00

Table 3.7 Average trip time per trip category

App usage	0–15 km	15–30 km	30–45 km
NEM	1:00:25	1:12:69	1:28:54
Off-App EM	0:55:30	1:09:30	1:37:12
On-App EM	1:03:40	1:13:49	1:04:15

We analyzed trip durations by route category and found that EM saved passengers up to 30 minutes on the longest trips, even on regular routes. However, the time savings were less significant for shorter trips (Table 3.7).

3.5.4 Fuel Savings Due to Reduced Stoppages Along the Route

On-app EM saved on fuel costs, spending 798 UGX per km compared to 837 UGX and 896 UGX for NEM and off-app trips, respectively. With higher speeds and fewer stops, drivers achieved greater fuel efficiency, traveling an extra 0.5 km/l of petrol (Table 3.8).

Table 3.8 Fuel saving between matatus using the app and matatus not using the app

App usage	Total fuel purchased (liters)	Total distance traveled (km)	km/liter	UGX D/liter
NEM	3.7	24.1	6.5	837
Off-App EM	3.2	19.7	6.2	896
On App EM	3.7	25.8	7.0	798
Average	**3.6**	**23.7**	6.5	837

3.6 Discussion

Our study is the first to investigate the impact of mobile applications on reducing transport emissions in East Africa. Similar research in India by the International Council on Clean Transportation (ICCT) found that ride-hailing apps like Uber and Ola reduced carbon emissions by up to 69% compared to traditional transport (Rokadiya et al., 2019), while a study conducted in China reported up to 52% reduction with apps like Didi Chuxing compared to private car use (Zhou et al., 2021). Consistent with these findings, our study showed reduced CO_2 emissions per trip for matatus using the EM app compared to those not using the app. While an independent *t*-test showed no significant difference in average CO_2 emissions per trip between matatus using the EM app and those not using it, there were other benefits shown from matatus using the EM app such as reduced travel time, fewer stops, and savings on fuel.

Our results also showed that matatus using the app had 16 stops per trip compared to 20 stops for NEM. This variance was largest in drop-off and pickups, as NEM had 55% more drop-offs and pickup stops than EM, delivering significant benefits to passengers. Longer on-app EM trips (30–45 km) had fewer stops than shorter trips in categories 1 and 2, unlike NEM, which had more stops on longer trips. Additionally, EM saved on fuel costs per kilometer due to higher speeds and fewer stops.

This study demonstrates that transit apps, app-based ride-sharing, and public transit platforms deliver several benefits for users, owners, and passengers, including reduced CO_2 emissions, fewer stops, shorter trip times, and higher speeds. This, in turn, leads to fuel efficiency, fuel savings for drivers, and quicker trips for passengers. Similar studies in other parts of the world have shown similar benefits for app-based ride-sharing services (Rokadiya et al., 2019).

However, further investigation is needed to understand the reasons why drivers may turn off the app during off-peak hours, which may reduce the benefits of the EM app. A qualitative study could provide insights into this issue and help identify strategies to encourage greater app usage among drivers during off-peak periods.

A recent study by Okello and colleagues (2023) noted very few strategies exploring transport-related air pollution in Africa, yet there has been a massive increase of motorized transport comprising pre-owned (second-hand) vehicles and two and three wheelers in Africa.

3.6.1 Strengths and Limitations

A limitation of our study was the inability to measure tailpipe emissions from matatus using the EM app and those that did not, as the company with the capacity to carry out these measurements had not resumed operations. However, this did not significantly affect the model, as sufficient data on fuel type, number of stops, and stop categories were available for emissions estimation. Additionally, the operation of the EM app was more complex than anticipated, with some drivers turning off the app during the day, resulting in a smaller sample of on-app EM trips. Future research could explore these factors through qualitative methods, such as focus group discussions with matatu drivers. Factors surrounding these actions could be investigated using qualitative and focused group discussions with the matatu drivers.

3.7 Conclusion and Recommendations

Some cities across sub-Saharan Africa have sought to replace minibus paratransit with contracted scheduled services, typically BRT, but have had limited success. The biggest challenge has been cost and difficulty to transform the existing system.

Our study has demonstrated that using an app-based platform that enables commuters to book seats on schedule contributed to reduced air pollution, travel time, and fuel usage by reducing stoppages.

Smartphone transit apps, app-based ride-sharing, and public transit platforms for booking and route scheduling have the potential to reduce transport-related emissions from idling, reduce travel time, and increase profit for matatu drivers. More research is needed for a longer period. Insights from this study could guide future research, urban governance, and planning, aligning digitalization and mobility innovations with sustainable urban development goals for cities. City authorities such as Kampala Capital City Authority could accelerate the adoption for technologies to prioritize scheduled and express routes over traditional pickup and drop-off routes. The city authorities should also provide incentives for special privileges for express taxis, for example, stickers, special pickup/drop-off zones, and regulate taxis from picking up passengers randomly along express routes.

Funding

Gabriel Okello is a senior research associate and Kings Global Sustainability Fellow leading the Air Quality and NCDs fellowship supported by a philanthropic gift from AstraZeneca, Grant No G111521. The views expressed in this chapter are those of the author(s) and not necessarily those of AstraZeneca.

Declaration of Competing Interest

The authors declare that they have no known competing financial interests or personal relationships that could have appeared to influence the work reported in this chapter.

References

Agyemang, E. (2015). The bus rapid transit system in the Greater Accra Metropolitan Area, Ghana: Looking back to look forward. *Norsk Geografisk Tidsskrift – Norwegian Journal of Geography, 69*, 28–37.

Asimeng, E.T., & Heinrichs, D. (2021). Why do paratransit operators resist participation in bus rapid transit? Case evidence from Bogota, Mexico City, Johannesburg and Lagos. *Transport Reviews, 41*, 115–135.

Barth, M., & Boriboonsomsin, K. (2008). Real-world carbon dioxide impacts of traffic congestion. *Transportation Research Record, 2058*, 163–171.

Behrens, R., Ferro, P.S., & Golub, A. (2015). International case studies of hybrid public transport system regulation and complementarity. In Behrens, R., McCormick, D., & Mfinanga, D. (Eds.), *Paratransit in African cities* (pp. 237–259). Routledge.

Behrens, R., McCormick, D., & Mfinanga, D. (2016). *Paratransit in African cities.* New York, NY: Routledge.

Bian, J., Li, W., & Lee, C. (2023). Current practices and emerging trends of transit apps for fixed-route bus services in the US. *Journal of Public Transportation, 25*, 100052. https://doi.org/10.1016/j.jpubtr.2023.100052

Boogaard, H., Patton, A., Atkinson, R., Brook, J., Chang, H., Crouse, D., Fussell, J., Hoek, G., Hoffmann, B., Kappeler, R., Joss, M.K., Ondras, M., Sagiv, S.K., Samoli, E., Shaikh, R., Smargiassi, A., Szpiro, A.A., Van Vliet, E.D.S., Vienneau, D., ... Forastiere, F. (2022). Long-term exposure to traffic-related air pollution and selected health outcomes: A systematic review and meta-analysis. *Environment International, 164*, 107262.

Booysen, M., Abraham, C., Rix, A., & Ndibatya, I. (2022). Walking on sunshine: Pairing electric vehicles with solar energy for sustainable informal public transport in Uganda. *Energy Research and Social Science, 85*, 102403.

Cervero, R. (2000). *Informal transport in the developing world.* UN-HABITAT.

Chengula, D., & Kombe, K. (2017). Assessment of the effectiveness of Dar Es Salaam bus rapid transit (DBRT) system in Tanzania. *International Journal of Sciences, 36*, 10–30.

Crippa, M., Oreggioni, G., Guizzardi, D., Muntean, M., Schaaf, E., Lo Vullo, E., Solazzo, E., Monforti-Ferrario, F., Olivier, J.G., & Vignati, E. (2019). *Fossil CO2 and GHG emissions of all world countries.* Luxembourg: Publications Office of the European Union.

Croitoru, L., Chang, J., & Christina; Kelly, Andrew. (2020). *The cost of air pollution in lagos.* World Bank. http://hdl.handle.net/10986/33038 License: CC BY 3.0 IGO

Dalal, S., Beunza, J.J., Volmink, J., Adebamowo, C., Bajunirwe, F., Njelekela, M., Mozaffarian, D., Fawzi, W., Willett, W., Adami, H.-O., & Holmes, M.D. (2011). Non-communicable diseases in sub-Saharan Africa: What we know now. *International Journal of Epidemiology, 40*, 885–901.

Eggleston, H.S., Buendia, L., Miwa, K., Ngara, T., & Tanabe, K. (2006). *IPCC guidelines for national greenhouse gas inventories.*

Gauthier, A., & Weinstock, A. (2010). Africa transforming paratransit into BRT. *Built Environment, 36*, 317–327.

Golub, A., Behrens, R., & Ferro, P. S. (2012). *Planned and paratransit service integration through trunk and feeder arrangements: An international review.* 31st Annual Southern African Transport Conference 2012.

Gooze, A., Watkins, K.E., & Borning, A. (2013). Benefits of real-time transit information and impacts of data accuracy on rider experience. *Transportation Research Record, 2351*, 95–103. https://doi.org/10.3141/2351-11

UN Habitat. (2016). *UN-habitat country programme document 2016 – 2021 – Uganda | UN-habitat.* https://unhabitat.org/un-habitat-country-programme-document-2016-2021-uganda.

Hitge, G., & Vanderschuren, M. (2015). Comparison of travel time between private car and public transport in Cape Town. *Journal of the South African Institute of Civil Engineering, 57*, 35–43.

Kampala Capital City Authority (2019). *Statistical abstract for Kampala City.* Kampala Capital City Authority.

Karagulian, F., Belis, C.A., Dora, C.F.C., Prüss-Ustün, A.M., Bonjour, S., Adair-Rohani, H., & Amann, M. (2015). Contributions to cities' ambient particulate matter (PM): A systematic review of local source contributions at global level. *Atmospheric Environment, 120*, 475–483.

Klopp, J.M., & Cavoli, C. (2019). Mapping minibuses in Maputo and Nairobi: Engaging paratransit in transportation planning in African cities. *Transport Reviews, 39*, 657–676.

Kumar, A., & Diou, C. (2010). *The Dakar bus renewal scheme: Before and after.* Sub-Saharan Africa Transport Policy Program (SSATP) discussion paper, no. 11 Washington, DC: World Bank.

Landrigan, P.J., Fuller, R., Acosta, N.J., Adeyi, O., Arnold, R., Baldé, A.B., Bertollini, R., Bose-O'Reilly, S., Boufford, J.I., Breysse, P.N., Chiles, T., Mahidol, C., Coll-Seck, A.M., Cropper, M.L., Fobil, J., Fuster, V., Greenstone, M., Haines, A., … Zhong, M. (2018). The Lancet Commission on Pollution and Health. *The Lancet, 391*, 462–512.

Lucas, K. (2019). A New evolution for transport-related social exclusion research? *Journal of Transport Geography, 81*, 102529.

McCord, M.R., Mishalani, R.G., & Ettefagh, M. (2015). Effect of real-time passenger information systems on perceptions of transit's favorable environmental and traffic reduction roles. *Transportation Research Record, 2538*, 102–109. https://doi.org/10.3141/2538-12

McCormick, D., Schalekamp, H., & Mfinanga, D. (2015). The nature of paratransit operations. In *Paratransit in African cities* (pp. 75–94). Routledge.

Okello, G., Nantanda, R., Awokola, B., Thondoo, M., Okure, D., Tatah, L., Bainomugisha, E., & Oni, T. (2023). Air quality management strategies in Africa: A scoping review of the content, context, co-benefits and unintended consequences. *Environment International, 171*, 107709. https://doi.org/10.1016/j.envint.2022.107709

Plano, C., Behrens, R., & Zuidgeest, M. (2020). Towards evening paratransit services to complement scheduled public transport in Cape Town: A driver attitudinal survey of alternative policy interventions. *Transportation Research Part A: Policy and Practice, 132*, 273–289.

Poku-Boansi, M., & Marsden, G. (2018). Bus rapid transit systems as a governance reform project. *Journal of Transport Geography, 70*, 193–202. https://doi.org/10.1016/j.jtrangeo.2018.06.005

Rokadiya, S., Wappelhorst, S., Pavlenko, N., & Bandivadekar, A. (2019). *Near-term incentives for electrifying ride-hailing vehicles in India*. Working Paper 2019-21.

Saddier, S., & Johnson, A. (2018). *Understanding the Operational Characteristics of Paratransit Services in Accra, Ghana: A Case Study*. 37th Annual Southern African Transport Conference 2018.

Schalekamp, H. (2017). Lessons from building paratransit operators' capacity to be partners in Cape Town's public transport reform process. *Transportation Research Part A: Policy and Practice, 104*, 58–66. https://doi.org/10.1016/j.tra.2017.08.002

Schwanen, T. (2018). Towards decolonised knowledge about transport. *Palgrave Communications, 4*, 1–6.

Smit, R., Brown, A.L., & Chan, Y.C. (2008). Do air pollution emissions and fuel consumption models for roadways include the effects of congestion in the roadway traffic flow? *Environmental Modelling & Software, 23*, 1262–1270. https://doi.org/10.1016/j.envsoft.2008.03.001

Stewart, M.S. (2024). *Assessing the sustainability of urban transport infrastructure in Uganda: Challenges, opportunities and strategies.*

Welle, B. (2019). *Paratransit in African cities: Operations, regulation, and reform*. Routledge.

World Health Organization. (2018). *9 out of 10 people worldwide breathe polluted air, but more countries are taking action*. https://www.who.int/news-room/detail/02-05-2018-9-out-of-10-people-worldwide-breathe-polluted-air-but-more-countries-are-taking-action.

Zhou, S., Chen, J., Wu, Z., & Qiu, Y. (2021). Electrification of online ride-hailing vehicles in China: Intention modelling and market prediction. *Energies, 14*, 7380.

4

Mobile Application-Enabled Logistics Crowdsourcing and the Triple Bottom Line in South Africa's E-Grocery Industry

Marcia Mkansi[1]*, Aaron L. Nsakanda*[2]*, Godfrey Mugurusi*[3]*, and Masilela Phumlani*[1]

[1] *College of Economic and Management Sciences, University of South Africa, Pretoria, South Africa*
[2] *Sprott School of Business, Carleton University, Ottawa, Ontario, Canada*
[3] *Department of Industrial Economics and Technology Management in Gjøvik, Norwegian University of Science and Technology, Trondheim, Norway*

4.1 Introduction

The rapid growth of e-commerce has transformed logistics operations worldwide, with the e-grocery sector emerging as a key driver of digital retail expansion. In South Africa, the rise of online grocery shopping presents both opportunities and challenges for last-mile delivery (Mkansi & Nsakanda, 2023). Traditional logistics models struggle to meet the increasing demand efficiently, while addressing economic, social, and environmental sustainability concerns—the three pillars of the triple bottom line. In these models, the movement of goods and services is frequently hindered by traffic congestion, multiple stops, and logistical inefficiencies, all of which contribute to excessive fuel consumption and increased CO_2 emissions per vehicle-kilometer. These inefficiencies not only escalate operational costs but also impact on-time delivery performance, which is a crucial factor in customer satisfaction and business competitiveness. Moreover, logistics providers are under increasing pressure to manage their environmental footprint and ensure compliance with evolving sustainability standards (Evangelista et al., 2018). This dual challenge of cost-effectiveness and sustainability underscores the need for more efficient logistics strategies.

Green Innovations in Supply Chain Management: Case Studies and Applications, First Edition. Edited by Marcia Mkansi, Godfrey Mugurusi, Aaron L. Nsakanda, Temidayo Akenroye, and Frank Tietze.

Published 2026 by John Wiley & Sons, Inc.

As a result, innovative approaches such as crowdsourcing-based logistics, enabled by mobile applications, are increasingly being viewed as attractive solutions to improve delivery efficiency, reduce environmental impact, and create a more resilient and sustainable e-grocery supply chain (Srivastava et al., 2022). Passenger transport, supply, and distribution crowdsourcing have seen huge growth over the past decade, with services such as Uber, taxify, inDriver, and Piggy Baggy leading the way. However, according to previous research (Kafle et al., 2017), the existing supply and distribution crowdsourcing services add more cars on the road, which ultimately give rise to CO_2 emissions and congestion. Additionally, these existing services cannot always guarantee on-time delivery and, in some cases, encourage people to start businesses due to the margins associated with their models. Therefore, while crowdsourcing-based logistics presents promising opportunities, it also raises concerns about its long-term sustainability and effectiveness in reducing urban congestion and emissions. Recent studies highlight the necessity of adopting sustainable practices to mitigate potential increases in congestion and CO_2 emissions (e.g., Hu et al., 2024).

This chapter explores a novel approach to addressing these challenges using a mobile application innovation that helps reduce road-related CO_2 emissions while promoting a green crowdsourcing model. It highlights how communities and businesses in developing contexts can leverage digital technologies to advance sustainability. In South Africa, where urban congestion, high logistics costs, and socioeconomic disparities present significant hurdles, mobile application-enabled crowdsourcing could offer a viable solution to enhance logistics sustainability (Govender & Rampersad, 2016). However, successful implementation of crowdsourced logistics in the South African e-grocery industry is contingent on overcoming key challenges. In particular, understanding stakeholder perceptions is critical in determining the adoption and scalability of these innovations, as well as designing strategies that optimize the economic, environmental, and social benefits (Buldeo et al., 2019). Accordingly, this study is guided by two research questions, one exploratory in nature and the other confirmatory:

a) How do stakeholders in South Africa's e-grocery sector, particularly those considered potential users, perceive the social, economic, and environmental impacts of a green crowdsourcing mobile application?
b) How can simulation modeling validate stakeholder perceptions of the green crowdsourcing mobile application's impact on the triple bottom line in South Africa's e-grocery industry?

The exploratory research question seeks to uncover and understand stakeholder perceptions using a qualitative approach to explore attitudes, motivations, and experiences. The confirmatory research question focuses on measurable impacts related to the triple bottom line and aims to quantify the insights obtained from stakeholders through simulation modeling. By leveraging stakeholder insights and simulation modeling, we assess how this platform can contribute to the triple bottom line—enhancing financial viability, reducing carbon emissions, and fostering social benefits. The findings provide actionable recommendations for policymakers and industry leaders seeking to leverage digital innovations for sustainable logistics development in South Africa.

The remainder of the chapter is organized as follows: Section 4.2 presents a review of the relevant literature. Section 4.3 describes the research methodology, including the study design used to assess stakeholder perceptions and simulate the impact of the proposed mobile application. Section 4.4 outlines the implementation of the methodology and discusses the key findings. Finally, Section 4.5 concludes the chapter with a summary of the main insights and their implications.

4.2 Literature Review

The growing urgency to reduce carbon emissions in the transport sector has intensified the search for innovative, sustainable delivery solutions. This literature review is structured into three key sections that collectively frame the foundation for this study. First, it examines the transport-based carbon emissions challenge, outlining the environmental impact of conventional logistics and mobility patterns. Second, it explores digitally enabled crowdsourcing delivery models and their implications for sustainability, drawing on the triple bottom line framework—economic, environmental, and social benefits. Finally, it reviews analytical research on crowdsourced last-mile delivery, with a particular focus on simulation modeling approaches.

4.2.1 Digitally Enabled Crowdsourcing Delivery Models and the Triple Bottom Line

Several national and international crowd logistics types exist, encompassing services related to passenger transport, storage, goods delivery, and shipping, among others (Frehe et al., 2017). In the case of storage-based crowd logistics, individuals rent out personal spaces such as basements, spare rooms, or entire homes. Shipping-based crowd logistics typically rely on existing platform carriers, effectively piggybacking on their infrastructure. Notable examples include Maritime Carrier Services and Morgan Cargo, which offer air and sea freight logistics solutions. Table 4.1 provides a summary of the different types of crowd logistics.

Table 4.1 Types of crowd logistics

Types	Description	Example(s)
Passenger	Responsible for transporting people from one location to another	Uber, Taxify
Local delivery service	Utilizes motor vehicles, motorcycles, bicycles, or people walking	Amazon crowd logistics, DHL
Goods	Delivery of packaged goods to clients	EOM, PostNet, Uber Eats
Storage	Achieved by renting storage from multiple individuals such as a basement, a garage, or rooms	Boxbee, Storemates, Stash, Makespace, Roost, storenextdoor
Shipping	Use of a platform to purchase goods from abroad and getting a person from a platform to be the carrier	Maritime Carrier Services, Morgan Cargo air and sea freight logistics

While various types of crowd logistics exist, the majority are part of business-to-customer, primarily focusing on the delivery of packaged goods and foods. Ciobotaru and Chankov (2021) present a taxonomy that categorizes crowd-sourced delivery business models into six distinct clusters. In addition to the two clusters that involve shipping goods from other countries using travelers as couriers, the remaining clusters create value by utilizing commuters or non-professional drivers for deliveries. The authors argue that business models that leverage commuters and their underutilized assets offer the greatest potential for more sustainable last-mile deliveries. Other services that target nonprofessional drivers often require additional resources, such as more vehicle trips and increased fuel consumption, which can lead to higher pollution levels compared to traditional last-mile couriers.

Alharbi et al. (2022) present three crowd-based last-mile business models commonly used in emerging economies (Business-to-Customers, Customer-to-Customer, and Business-to-Business). Drawing on semi-structured interviews, they assessed the economic, environmental, and social benefits of each model using the criteria summarized in Table 4.2, along with their implementation challenges and limitations.

Rześny-Cieplińska and Szmelter-Jarosz (2019) assessed crowd logistics solutions from the perspectives of various stakeholders, using 20 sustainability criteria identified in the literature and categorized into social, economic, and environmental dimensions. The importance of each criterion for the groups of stakeholders was then evaluated using the Analytical Hierarchy Process approach. Table 4.3 provides a summary of sustainability dimensions that are the most pertinent for last-mile deliveries.

Table 4.2 Summary of the sustainability benefits from crowd logistics

Sustainability benefits		
Environmental	**Social**	**Economic**
• Reduction in traffic and congestion • Reduction in traffic emissions • Reduction in CO_2	• Private interactions • Supporting community	• Greater variety of goods • Faster delivery • More flexibility • More convenient • Better priced • More personal • Less missed deliveries

Source: Alharbi et al. (2022)/MDPI/CC BY 4.0.

Table 4.3 Summary of the environmental criteria pertinent for last-mile deliveries

Sustainability dimensions		
Environmental	**Social**	**Economic**
• Reduction in traffic and congestion • Reducing noise • Reduction in CO_2 emissions	• Connecting individual providers and consumers • Voluntary character • Tracking, transparency • Simplicity and trust • Indicating the country's specifics and ethics in business model	• Access to adequate IT infrastructure • Attractive revenue model • Time of delivery • Strategy of cooperation

Source: Rześny-Cieplińska and Szmelter-Jarosz (2019)/MDPI/CC BY 4.0.

Mohri et al. (2023) offered a review of crowdsourced delivery systems through the lens of sustainability's three main pillars: environmental, economic, and social. They analyzed existing literature and argue that, by leveraging the underutilized crowd and associated resources, these systems have the potential to reduce the environmental impacts of urban deliveries, to support low-income participants, and foster community development with new social constructs in the market. These findings are also supported by other scholars, including Rai et al. (2017), Michel et al. (2023), and Samad et al. (2023).

Building on these supporting studies, it is evident that crowdsourced delivery systems hold significant promise in advancing sustainability across environmental, economic, and social dimensions. However, while the potential benefits are well-theorized, there remains a critical need for empirical research that directly compares the sustainability impacts of companies adopting crowdsourced delivery models versus those relying on traditional delivery methods. Such comparative analysis would provide valuable insights into real-world outcomes, helping businesses, policymakers, and stakeholders better understand the trade-offs, challenges, and opportunities associated with integrating crowdsourced logistics into urban delivery networks.

4.2.2 Analytical Research on Crowdsourcing Last-Mile Delivery: Simulation Modeling

Analytical approaches to crowdsourced last-mile delivery have been widely used to analyze, design, and improve the performance of this emerging logistics system. This section focuses specifically on simulation methodologies. While optimization methods are effective in identifying optimal solutions under defined constraints, simulation modeling is particularly well-suited to capturing the high levels of uncertainty, variability, and human behavior inherent in crowdsourced delivery. It also provides greater flexibility for testing a wide range of operational scenarios. Comprehensive reviews of optimization-based approaches to crowdsourced last-mile delivery are provided by Alnaggar et al. (2021) and Shahin et al. (2024).

Chen and Chankov (2017) proposed an agent-based simulation model to assess the performance of a crowdshipping company that relies on crowd couriers to deliver packages from a distribution center to its final destination along their daily commute. The study measured the performance using two metrics: the delivery service level (measured as the percent of the available packages that are delivered by the crowd couriers) and the detour time (the maximum extra time accepted by crowd couriers). The findings revealed that the maximum detour distance that crowd couriers are willing to accept significantly affects both the service level and the number of packages successfully delivered. This study was conducted in the context of an intra-city courier company providing delivery services in an urban area.

Castillo et al. (2018) developed a discrete-event simulation model to evaluate the logistics effectiveness of a crowdsourced vehicle fleet to a dedicated fleet for same-day delivery from a centralized distribution center within an urban customer network. The study measured effectiveness according to two key metrics: the delivery rate and the total number of deliveries performed. The findings indicated that tight time windows for deliveries have a greater negative impact on the on-time delivery performance of a crowdsourced fleet of vehicle compared to a dedicated fleet. Similar findings were also observed for the number of deliveries completed. Higher daily demand variability had a more significant negative impact on the performance of crowdsourced fleet of vehicles compared to a dedicated fleet. This study was conducted in the context of a courier company providing same-day delivery services in an urban area.

Guo et al. (2019) conducted a discrete event simulation-based case study to explore the feasibility of integrating crowdsourced delivery with the conventional delivery methods to address the challenges of e-grocery last-mile logistics. The study focused on traditional (offline) customers of an e-grocery store who acted as crowd couriers, voluntarily delivering grocery packages to their neighbors in a highly urbanized setting. Various scenarios were analyzed, including increased participation of offline customers in crowdsourcing, growth in online orders, and changes in delivery time windows. The study measured performance using two key metrics: the matching rate (the percentage of grocery parcels delivered by offline crowd couriers) and total delivery costs. The findings indicated that integrating crowdsourced delivery as a supplement to the conventional delivery network can help reduce last-mile logistics costs. The offline participation rate is a critical factor in ensuring the feasibility of transitioning to this hybrid last-mile delivery model.

Simoni et al. (2020) used simulation to evaluate the sustainability impacts of adopting a crowdsourced last-mile delivery model. The study considered various factors, including the mode of transport used (public transit vs private cars), the level of supply-and-demand matching for delivery services, the different levels of the detour distance (i.e., deviation from original trips), and the different traffic conditions. Performance was measured using three key metrics: greenhouse gas emissions (CO_2 emissions), changes in air pollutants (CO, PM_{10}, and NO_x), and variations in traffic delays. The findings revealed the impact of adopting a crowdsourced delivery model on emissions and traffic congestion varies significantly based on specific implementation features. The study was conducted in the context of city logistics in an urban area.

Castillo et al. (2022) developed a multimethod simulation approach that combines agent-based and discrete event simulations to examine hybrid last-mile delivery systems that integrate privately owned delivery fleets with crowdsourced drivers. Utilizing real-world delivery data from a major U.S. retail pharmacy, the authors analyzed how factors such as driver autonomy, compensation, fleet size, fleet mix, and demand intensity affect the cost and service performance of these hybrid systems. The findings reveal that both low and high compensation levels can result in higher delivery costs compared to median compensation, and that changes in compensation significantly affect delivery speed. These results underscore the importance of carefully designing hybrid delivery systems, considering the interplay between compensation strategies, fleet composition, and demand patterns to achieve both cost efficiency and service quality.

Building on the work of Simoni et al. (2020), this chapter examines both the sustainability and economic impacts of crowdsourced last-mile delivery. The study is set in a developing country with a distinctive geographical landscape that includes urban, township, and rural areas, as well as widely dispersed delivery distances. Additionally, similar to Guo et al. (2019), this research focuses on the application of the crowdsourced delivery model within a specific industry—the e-grocery sector.

4.3 Research Methodology

In this study, a mixed-methods approach is used to evaluate the sustainability impact of a digitally enabled crowdsourcing app in the last-mile e-grocery delivery. The methodology is divided into three key phases: a digital solution development, a qualitative insights, and a quantitative impact analysis.

The initial phase consisted of the development of a prototype of a digitally enabled crowdsourcing app following the design science approach similar to the one discussed by Mkansi et al. (2024). The app integrates key functionalities such as real-time order matching, route optimization, dynamic pricing, and sustainability tracking. A qualitative phase followed in which the application was deployed in a controlled pilot project to assess its operational feasibility, user acceptance, and social and economic impacts. Semi-structured interviews were conducted with randomly selected customers of a local major distributor of fast-moving consumer goods. The participants included individuals who consented to act as piggyback couriers and those who purchased e-grocery online. Finally, a quantitative analysis was carried out using the major local distributor of fast-moving consumer goods serving as the case study. A simulation-based analysis was conducted to compare the overall environmental and economic impacts of both the traditional e-grocery and the crowdsourced delivery models.

4.3.1 Prototype Development Using a Design Science Approach

The study undertook the development of a mobile app artifact (Dilivari), to be used for the delivery of last-mile products across different areas to meet the needs of the different stakeholders in urban and rural areas. Some of the core features that were deduced through problem relevance, design as a process, and design artifact using a waterfall methodology produced the unified user interface (UI). The app is encapsulated. This is accomplished by hosting the Dilivari mobile app in the Microsoft Azure shell flyout. According to Wulf and Blohm (2020) the unified UI uses responsive web design principles to provide an optimal viewing and interaction experience for any screen size, device, or orientation. The Dilivari app went through design and evaluation as a quality assurance mechanism for testing the artifact before deployment for use by the stakeholders. Design and evaluation involved conducting the quality assurance mechanism to ensure the artifact meets the requirements. The evaluation that was conducted undertook functional and compatibility testing, performance testing, middle layer testing, and confidentiality layer testing. Functional testing was conducted to determine the extent to which the app met requirements, evolving technology, and customer expectations (Akanmu et al., 2013). Knott (2015) highlights that performance testing used for evaluating all layers of the app from the front end, the middle layer, to the back end of the technology stack is required. Middle layer testing was conducted to determine whether the middle layer linked very well between the UI and the back end—the database. Finally, Paul and Aithal (2019) emphasize the importance of security testing highlighting the challenge imposed by hackers, most of whom are unethical. The app performed to an optimal level across computer experimentation. The green crowdsourcing app development was unique as opposed to the traditional outsourcing strategies that are characterized by contracts between two parties.

The Dilivari app architecture (see Figure 4.1) was developed providing the UI, user communication layer, information retrieval layer, and database layer to provide a blueprint of the system that defined the structure of the system. The development process adhered to major principles for the implementation of architecture, that is, separating concerns, minimizing interdependencies, and isolating fundamental components for the modularization of the app into smaller functions (da Costa et al., 2019).

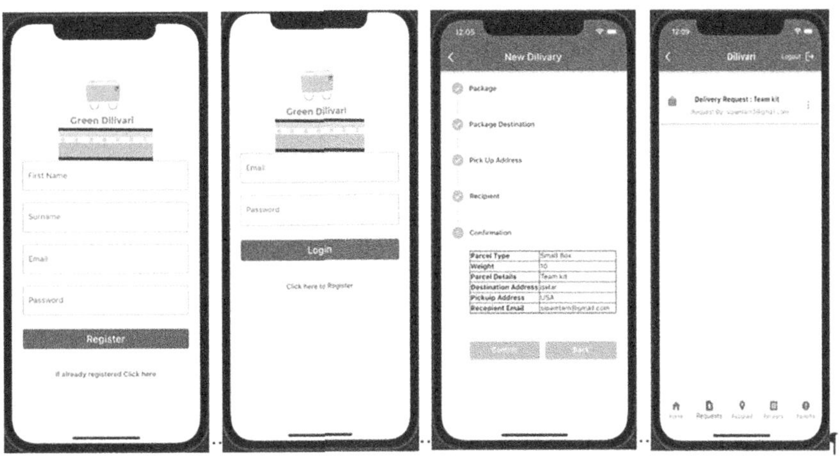

Figure 4.1 Dilivari application (available for download from Apple Store and Google Play Store).

4.3.2 Qualitative Pilot Study and Data Collection

We piloted Dilivari in a real-life setting to evaluate its acceptance among potential users and to assess its social and economic impacts. Participants in the pilot study were customers of a major local distributor of fast-moving consumer goods. Over two days, data was collected from randomly selected customers who were invited to participate as piggyback couriers. They were asked whether they would be willing to pick up an order and deliver it along their planned route to their final destination, and to explain their reasons for accepting or declining. A total of 60 potential participants were approached, resulting in a response rate of 21.7%, as summarized in Table 4.4.

We also conducted semi-structured interviews with e-grocery customers who received their orders via a piggyback driver rather than a traditional courier service. They were asked to evaluate the service based on delivery time and to share any concerns or reluctance about being served by a piggyback driver. A total of 10 e-grocery customers participated in the pilot. The interviewers either acted as piggyback drivers themselves or accompanied the drivers to gather the interview data.

4.3.3 Quantitative Case Study and Simulation Analysis

The final phase involves a quantitative comparison of the sustainability and economic impacts of traditional e-grocery logistics versus the crowdsourced delivery model enabled by the app. A simulation-based analysis is conducted to evaluate the overall environmental and economic impact of both approaches.

Data for simulation model development and testing was collected from a major local distributor of fast-moving consumer goods. The distributor provided secondary data from its e-grocery operations, covering a period of 250 days in the year 2022. We also collected data from the courier company that provides delivery services to the distributor, focusing on vehicle types, fuel consumption, and the distribution of deliveries between point-to-point (PTP) and multi-stop

Table 4.4 Breakdown of participants to the study

Participants	Participants who declined	Participants who accepted	Total
Males alone	9	6	15
Females alone	15	—	15
Family without accompanying kids	10	5	15
Family with accompanying kids	13	2	15
TOTAL	47	13	60

(MTS) routes. Finally, we collected data from two car dealers to identify the most common vehicles used in South Africa, with a focus on vehicle types, market distribution, and fuel consumption.

While the major local distributor serves e-grocery buyers across all provinces in South Africa, this study is limited to Gauteng province. Over 60% of the total orders received are from this province, which encompasses the city of Pretoria, where the company is based.

Tables 4.5–4.8 provide the descriptive statistics of the various data collected from various sources.

Table 4.5 E-grocery orders information

	Minimum	Average	Maximum
Online orders per day	1	5	9
Order originating distance (kilometers)	2.8	51.6	131

Source: Compiled by the authors from data provided by the distributor company.

Table 4.6 Type of vehicles information

Vehicle type	Usage proportion (%)	Fuel usage proportion (%)		Fuel consumption (L/100 km)	
		Petrol	Diesel	Petrol	Diesel
1	10	50	50	6	4.7
2	90			8.5	6.5

Source: Compiled by the authors from data provided by the courier service provider company.

Table 4.7 Type of vehicles information

Vehicle type	Usage proportion (%)	Fuel usage proportion (%)		Fuel consumption (L/100 km)	
		Petrol	Diesel	Petrol	Diesel
Small size	20	60	40	6	4.2
Medium size	20			7	5.3
Large size	60			8.5	5.8

Source: Compiled by the authors from data provided by two car dealer companies.

Table 4.8 Delivery mode information

Delivery mode	Distribution (%)
Point-to-point	40
Multi-stop routes	60

Source: Compiled by the authors from data provided by the courier service provider company.

4.4 Implementation and Results

We present in this section the outcomes obtained from deploying the Dilivari app in a real-world setting as well as the simulated results of the broader economic and environmental implications of the crowdsourced delivery model.

4.4.1 Insights from User Interviews and Social Impact Assessment

Although the sample size was limited, the semi-structured interviews provided valuable insights into user perceptions of the piggyback delivery model.

Among those invited to act as piggyback drivers, several participants expressed concerns around trust and security, particularly the risk of being scammed or held accountable for damaged or missing goods. Others cited uncertainty about how payments would be processed and potential complications with the South African Revenue Service (SARS), especially around declaring informal income.

On the other hand, those who agreed to act as piggyback drivers were primarily motivated by the potential for financial gain, a sense of social responsibility ("ubuntu"), and environmental awareness.

Among e-grocery customers who received their orders via piggyback drivers, concerns also centered on trust and privacy. Some were initially hesitant about a stranger handling their groceries. Despite these concerns, most of these users were ultimately pleased with the service—especially the faster delivery times compared to traditional couriers. Table 4.9 presents a summary of the key findings.

These findings, while based on a small sample, highlight key themes that can inform the refinement of the crowdsourced delivery model and guide future, larger-scale implementations. These findings are in agreement with previous studies (Vallas & Schor, 2020). Carbone et al. (2017) established that crowdsourcing involves ordinary people taking on critical tasks in distribution becoming critical resources in the supply chain management. Gig work involves the use of independent contractors (Vallas & Schor, 2020), online platforms, and temporary workers (Donovan et al., 2016). This is critical toward revitalization of the livelihood of the people who

Table 4.9 Summary of key findings by groups of respondents

Respondent group	Accepted to act as piggyback drivers
Key motivations	• Financial gain • Helping others (social responsibility) • Supporting environmental sustainability
Main concerns	None prominently reported in this group
Illustrative quote	"It's a small way I can make some extra cash while helping someone and reducing delivery trips." "If I'm already going that way, why not help and do some good for the planet too?"
Respondent group	Declined to act as piggyback drivers
Key motivations	• N/A
Main concerns	• Trust and security concerns • Uncertainty about payment process • Fear of tax issues with South Africa Revenue Service (SARS)
Illustrative quote	"I'd need to know exactly how I get paid and whether this might get me into trouble with SARS"
Respondent group	Received goods via piggyback drivers
Key motivations	• Faster delivery time • Satisfactory service quality
Main concerns	• Initial hesitation about trust and privacy • Discomfort with strangers delivering their orders
Illustrative quote	"It felt odd not knowing who exactly had my order, but it arrived fine and on time." "I'd use it again." "I got my groceries quicker than usual, and everything was in good shape—so I'd use it again."

were impacted by the economic downturn in South Africa. Furthermore, people already employed can earn some income by delivering groceries while traveling to or from workplaces. Crowdsourcing creates a wide network of independent workers who can supplement their incomes through gig work (Zalia et al., 2021).

Empirically, our results align with several triple bottom line criteria defined in previous studies (Alharbi et al., 2022; Rześny-Cieplińska & Szmelter-Jarosz, 2019), indicating that Dilivari has the potential to generate meaningful social, environmental, and economic value. Socially, participants expressed a willingness to support others in their community by acting as piggyback drivers. Environmentally, the application could reduce the need for dedicated delivery trips by leveraging existing travel routes. Economically, customers valued the faster and more convenient service, while some saw financial and personal value in participating.

4.4.2 Environmental and Economic Impact Comparison

In this study, simulation modeling was used to evaluate the two scenarios shown in Figures 4.2 and 4.3. The first scenario compares two delivery methods: PTP delivery using the current courier service model versus delivery through a piggyback driver. This scenario specifically examines 40% of current online orders that are fulfilled using the PTP delivery mode.

The second scenario evaluates the fulfillment of all current online orders using two different delivery methods: (i) Current Service Model where 40% of orders are delivered using the PTP method, while the remaining 60% are delivered via MTS routes; and (ii) Piggyback Model where 100% of online orders are delivered by piggyback drivers.

The simulation model was built using Monte Carlo methods within Arena software, adhering to best practices for creating valid and credible simulation models (Law & McComas, 2001). This includes input modeling and analysis, model development and implementation, as well as model verification and validation.

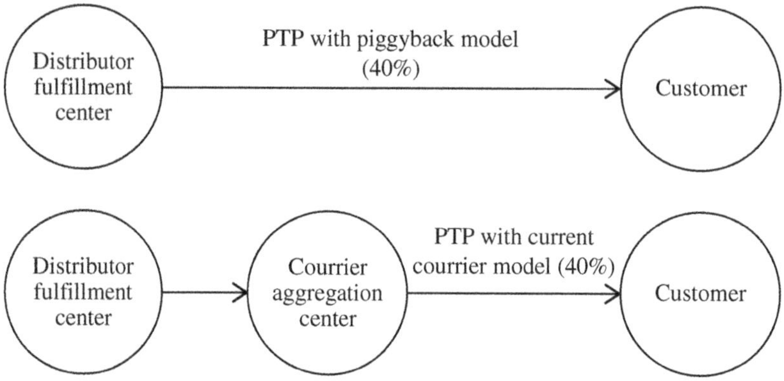

Figure 4.2 Scenario 1–Exclusive PTP delivery model.

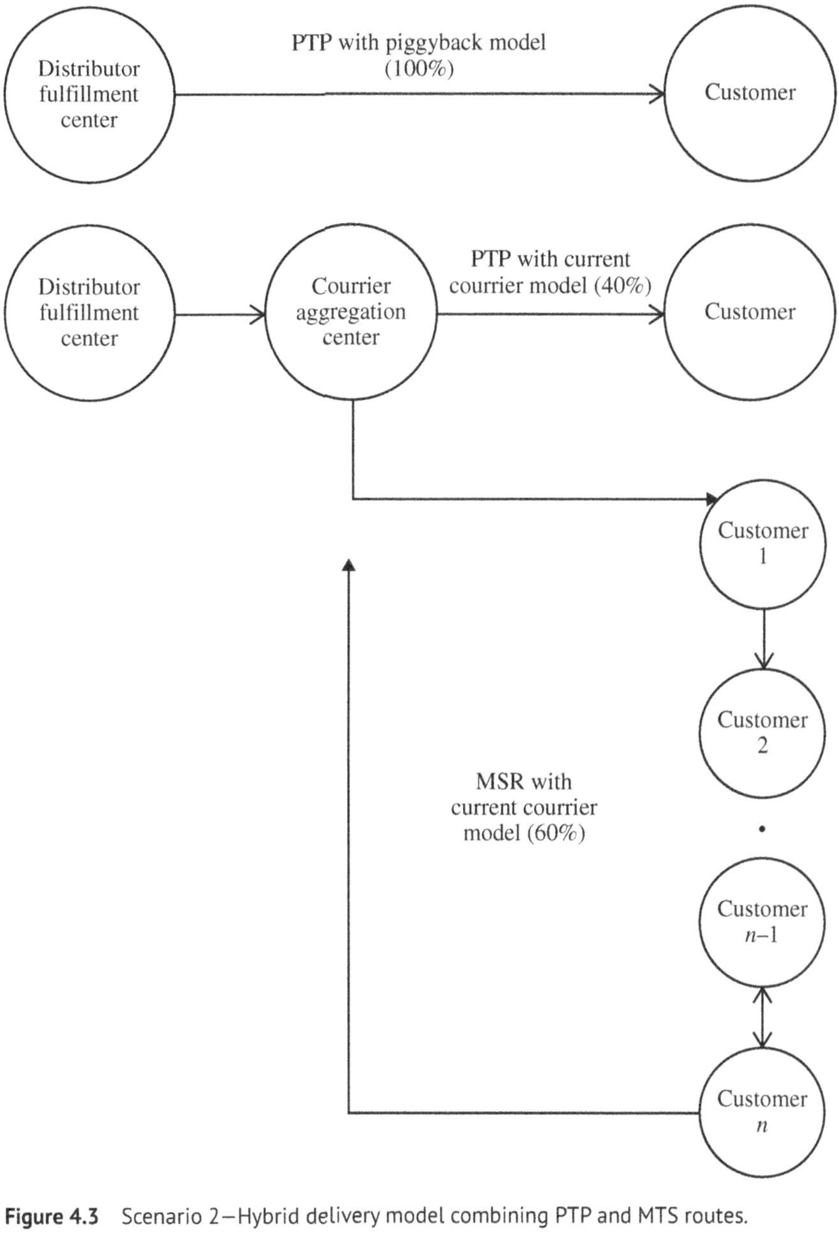

Figure 4.3 Scenario 2—Hybrid delivery model combining PTP and MTS routes.

The output measures considered are the carbon emissions (measured in kg) and the total distance reduction (in percentage). The carbon footprint for each scenario is calculated as follows:

$$T = \sum_{i=1}^{N} E_i$$

where:

$E_i = (d_i \times f_{jk} \times v_k), \forall j, k$	Carbon emissions emitted with the delivery of order I (kg)
d_i	Distance traveled with the delivery of order i (km)
f_{jk}	Per liter consumption of vehicle type j using fuel type k (L/km)
v_k	CO_2 emission factor for fuel type k (kg/L)
T	Total carbon emissions (kg)
N	Total number of orders

The emissions factors for various fuel types, published by Environment Canada, were used in the total emissions calculations, in particular 2.29 kg/L for gasoline and 2.66 kg/L for diesel (Natural Resources Canada, 2014).

4.4.3 Findings and Discussion

Tables 4.10 and 4.11 summarize our findings for each scenario analysis under consideration. In each scenario, we compare the current model, which relies on delivering through a carrier, with an alternative model that consists of delivering

Table 4.10 Exclusive PTP delivery model output measures (n = 100, 95% confidence level)

				t-paired	
	Minimum	Average	Maximum	Half-length	Confidence interval
CO_2 emissions reduction (kg/order)	6.274	6.806	7.389	0.055	(6.75, 6.86)*
CO_2 emissions reduction (in percentage)	76.10	77.90	79.30	0.001	(77.80, 78.00)*
Distance reduction (in percentage)	75.2	76.70	78.20	0.001	(76.6, 76.80)*

* A significant difference.

Table 4.11 Hybrid delivery model output measures ($n = 100$, 95% confidence level)

				t-paired	
	Minimum	**Average**	**Maximum**	**Half-length**	**Confidence interval**
CO_2 emissions reduction (kg/order)	2.934	3.364	3.770	0.032	(3.333, 3.396)*
CO_2 emissions reduction (in percentage)	60.52	63.47	65.84	0.002	(63.26, 63.68)*
Distance reduction (in percentage)	57.98	61.37	63.43	0.002	(61.17, 61.57)*

* A significant difference.

through a piggyback driver using the crowdsourcing mobile application. In both scenarios, the paired *t*-test analyses indicate a significant difference in CO_2 emissions per order (kg) and delivery distance reductions (km) between the two delivery methods. Delivery through a piggyback driver reduced the total distance traveled by an average of 76.7% when using an exclusive PTP model (Scenario 1). As a result, CO_2 emissions per order decreased by an average of 6.806 kg, representing a 77.9% reduction in the carbon footprint compared to the current delivery model.

Similar findings were observed in the analysis of the second scenario. The difference between the two scenarios narrowed due to some orders being delivered via MTS routes or multiple runs in the second scenario. However, delivery through a piggyback driver still produced better results, reducing the total distance traveled by an average of 61.37% compared to using the current hybrid model. Consequently, CO_2 emissions per order decreased by an average of 3.36 kg, representing a 61.37% reduction in the carbon footprint compared to the current hybrid delivery model.

Our findings highlight the role of green crowdsourcing applications in reducing carbon emissions and quantify this impact through reductions of both distance traveled and CO_2 emissions. This supports previous work that pointed out that these applications leverage unused capacity in already planned journeys, therefore contributing toward environmental sustainability (Rai et al., 2017; Zalia et al., 2021).

In terms of impacts on people, the piggyback delivery model eliminates the aggregation step, thereby improving delivery speed, particularly for long distance deliveries. Additionally, as the cost of delivery through a piggyback driver is lower than using a courier, the PTP piggyback model offers notable benefits. It reduces delivery costs, allowing people (buyers) to save money. It enables ordinary people (piggyback drivers) to earn additional income by delivering groceries while commuting to or from work. This contributes to creating a broad network of independent workers who can supplement their earnings through the gig economy (Zalia et al., 2021), supporting the livelihood of those affected by South Africa's economic downturn, particularly residents of townships. However, this shift could also impact traditional courier jobs, potentially leading to reduced demand for conventional delivery services.

4.5 Conclusions and Future Directions

The study established that the green crowdsourcing application impacts people, profitability, and environment especially in developing countries and provides an excellent ground to understand the mechanisms employed under green crowdsourcing to overcome challenges of traditional logistics and existing crowdsourcing applications. This perspective, we believe, provides grounds to develop the most appropriate policies to spur crowdsourcing in developing countries. From a research perspective, this study presents many opportunities for theory testing, theory elaboration, and theory extension of crowdsourcing in Africa and beyond.

Despite offering valuable insights, this study has several limitations that should be acknowledged. First, the exploratory phase relied on a relatively small sample size of participants to assess stakeholder perceptions, which may limit the generalizability of the findings across South Africa's broader e-grocery sector. While the qualitative data provided rich, context-specific insights, a larger and more diverse sample would strengthen the robustness of the conclusions. Second, the confirmatory phase assumes consistent availability of participants for last-mile deliveries—a simplification that does not fully reflect real-world dynamics. As highlighted by Savelsbergh and Ulmer (2024), crowd availability is a well-known challenge in crowdsourced logistics, with fluctuations in participation potentially affecting the reliability and scalability of such models.

These limitations point to the need for further research using larger datasets and dynamic modeling approaches that account for variability in crowd participation. Future research could also broaden the study by considering various scenarios through simulation and testing them empirically in real-life settings.

References

Akanmu, A., Anumba, C., & Messner, J. (2013). Scenarios for cyber-physical systems integration in construction. *Journal of Information Technology in Construction (ITcon)*, *18*(12), 240–260.

Alharbi, A., Cantarelli, C., & Brint, A. (2022). Crowd models for last mile delivery in an emerging economy. *Sustainability*, *14*(3), 1401. https://doi.org/10.3390/su14031401

Alnaggar, A., Gzara, F., & Bookbinder, J.H. (2021). Crowdsourced delivery: A review of platforms and academic literature. *Omega*, *98*, 102139. https://doi.org/10.1016/j.omega.2019.10.139

Buldeo Rai, H., Verlinde, S., & Macharis, C. (2019). The impact of logistics solutions on the delivery of sustainable urban freight transport. *Research in Transportation Economics*, *73*, 67–77.

Carbone, V., Rouquet, A., & Roussat, C. (2017). The rise of crowd logistics: A new way to co-create logistics value. *Journal of Business Logistics*, *38*(4), 238–252. https://doi.org/10.1111/jbl.12164

Castillo, V.E., Bell, J.E., Rose, W.J., & Rodrigues, A.M. (2018). Crowdsourcing last mile delivery: Strategic implications and future research directions. *Journal of Business Logistics*, *39*(1), 7–25.

Castillo, V.E., Bell, J.E., Mollenkopf, D.A., & Stank, T.P. (2022). Hybrid last mile delivery fleets with crowdsourcing: A systems view of managing the cost-service trade-off. *Journal of Business Logistics*, *43*(1), 36–61. https://doi.org/10.1111/jbl.12288

Chen, P., & Chankov, S. M. (2017, December 10–13). Crowdsourced delivery for last-mile distribution: An agent-based modelling and simulation approach. *Proceedings of the 2017 IEEE International Conference on Industrial Engineering and Engineering Management (IEEM)* (pp. 1271–1275). Singapore.

Ciobotaru, A., & Chankov, S. (2021). Towards a taxonomy of crowdsourced delivery business models. *Journal of Business Logistics*, *42*(4), 501–523. https://doi.org/10.1111/jbl.12281

da Costa Junior, J., Diehl, J.C., & Snelders, D. (2019). A framework for a systems design approach to complex societal problems. *Design Science*, *5*, e2.

Donovan, S. A., Bradley, D. H., & Shimabukuru, J. O. (2016). *What does the gig economy mean for workers?*

Evangelista, P., Santoro, L., & Thomas, A. (2018). Environmental sustainability in third-party logistics service providers: A systematic literature review from 2000–2016. *Sustainability*, *10*(5), 1627. https://doi.org/10.3390/su10051627

Frehe, V., Mehmann, J., & Teuteberg, F. (2017). Understanding and assessing crowd logistics business models – Using everyday people for last mile delivery. *Journal of Business & Industrial Marketing*, *32*(1), 75–97. https://doi.org/10.1108/JBIM-10-2015-0182

Govender, J.P., & Rampersad, R. (2016). The adoption of e-commerce in South African supermarkets: A comparative study. *Problems and Perspectives in Management*, *14*(2), 194–202.

Guo, X., Lujan Jaramillo, Y.J., Bloemhof-Ruwaard, J., & Claassen, G.D.H. (2019). On integrating crowdsourced delivery in last-mile logistics: A simulation study to quantify its feasibility. *Journal of Cleaner Production*, *241*, 118365.

Hu, S., Shu, S., Chen, Z.R., Shao, Y., Na, X., Xie, C., Stettler, M., & Lee, D. (2024). Sustainable impact analysis of freight pooling strategies on city crowdsourcing logistics platform. *Transportation Research Part D: Transport and Environment, 127*, 103851. https://doi.org/10.1016/j.trd.2024.103851

Kafle, N., Zou, B., & Lin, J. (2017). Design and modeling of a crowdsource-enabled system for urban parcel relay and delivery. *Transportation Research Part B: Methodological, 99*, 62–82.

Knott, D. (2015). *Hands-on mobile app testing: A guide for mobile testers and anyone involved in the mobile app business*. Addison-Wesley Professional.

Law, A. M., & McComas, M. G. (2001). How to build valid and credible simulation models. *Proceedings of the 2001 Winter Simulation Conference* –, (pp. 22–29, vol. 1). Arlington (USA), December 2001. doi:10.1109/WSC.2001.977242.

Michel, S., Bootz, J.-P., & Bessouat, J. (2023). Possible futures of crowd logistics for manufacturers: Results of a strategic foresight study. *Journal of Business & Industrial Marketing, 38*(10), 2019–2029. https://doi.org/10.1108/JBIM-12-2021-0548

Mkansi, M., & Nsakanda, A.L. (2023). Mobile application e-grocery retail adoption challenges and coping strategies: A South African small and medium enterprises' perspective. *Electronic Commerce Research, 25*, 419–464.

Mkansi, M., Nsakanda, A.L., Roux, J.P. (2024). Leveraging internet of things technologies to drive societal-based decarbonisation in the transport sector: Green mobile application. In X. S. Yang, R. S. Sherratt, N. Dey, A. Joshi (Eds), *Proceedings of eighth international congress on information and communication technology. ICICT 2023*. Singapore: Springer. Lecture Notes in Networks and Systems, vol. 696. https://doi.org/10.1007/978-981-99-3236-8_5.

Mohri, S.S., Ghaderi, H., Nassir, N., & Thompson, R.G. (2023). Crowdshipping for sustainable urban logistics: A systematic review of the literature. *Transportation Research Part E: Logistics and Transportation Review, 178*, 103289. https://doi.org/10.1016/j.tre.2023.103289

Natural Resources Canada. (2014). *Auto$mart – Learn the facts: Fuel consumption and CO_2*. https://natural-resources.canada.ca/sites/nrcan/files/oee/pdf/transportation/fuel-efficient-technologies/autosmart_factsheet_9_e.pdf

Paul, P., & Aithal, P. S. (2019, October). Mobile applications security: An overview and current trend. *Proceedings of National Conference on Research in Higher Education, Learning and Administration* (Vol. 1, No. 1, pp. 112–121). SSRN: https://ssrn.com/abstract=3484091Rai, H.B., Verlinde, S., Merckx, J., & Macharis, C. (2017). Crowd logistics: An opportunity for more sustainable urban freight transport? *European Transport Research Review, 9*(3), 39. https://doi.org/10.1007/s12544-017-0256-6

Rześny-Cieplińska, J., & Szmelter-Jarosz, A. (2019). Assessment of the crowd logistics solutions—The stakeholders' analysis approach. *Sustainability, 11*(19), 5361. https://www.mdpi.com/2071-1050/11/19/5361

Samad, T.A., Ganguly, K.K., & Das, D. (2023). Towards a framework for development of crowd logistics: Paving the way for sustainable logistics. *Computers & Industrial Engineering, 176*, 109008. https://doi.org/10.1016/j.cie.2023.109008

Savelsbergh, M., & Ulmer, M.W. (2024). Challenges and opportunities in crowdsourced delivery planning and operations—An update. *Annals of Operations Research, 343*(2), 639–661. https://doi.org/10.1007/s10479-024-06249-1

Shahin, R., Shahin, A., & Dulebenets, M.A. (2024). A survey of crowdsourcing in last-mile delivery in the optimization literature. In Štarchoň, P., Fedushko, S., & Gubíniová, K. (Eds.), *Data-Centric Business and Applications. Lecture Notes on Data Engineering and Communications Technologies* (Vol. 208, pp. 371–392). Cham: Springer. https://doi.org/10.1007/978-3-031-59131-0_17

Simoni, M.D., Marcucci, E., Gatta, V., & Claudel, C.G. (2020). Potential last-mile impacts of crowdshipping services: A simulation-based evaluation. *Transportation, 47*(4), 1933–1954.

Srivastava, A., Rapti, A., & Phuyal, R.K. (2022). Green supply chain management in last-mile logistics: A systematic review. *Sustainability, 14*(8), 4561.

Vallas, S., & Schor, J.B. (2020). What do platforms do? Understanding the gig economy. *Annual Review of Sociology, 46*, 273–294.

Wulf, J., & Blohm, I. (2020). Fostering value creation with digital platforms: A unified theory of the application programming interface design. *Journal of Management Information Systems, 7*(5), 694–709.

Zalia, A.-H., Zhang, Y., Aishatu, A., Akosua, A.S., & Fathia, B.V. (2021). Crowd logistics' impact on environmental sustainability in a developing economy: An analysis in Ghana. *American Journal of Industrial and Business Management, 11*, 416–436.

5

Exploring the Use of Crowdsourcing Platform Truck Transport in the Specialized Sectors in Kenya: The Case of Amitruck Application

John Michael Maxel Okoche and Wellington Chakuzira

Department of Operations Management, College of Economics and Management Sciences, University of South Africa, Unisa, South Africa

5.1 Introduction

Crowdsourcing platforms have established mechanisms that facilitate engagement of participants in the public domain (Diop et al., 2022). Interestingly, these platforms play a pivotal role in the realm of product development, supplying essential data and information pertinent to manufacturing, innovation, crowd sensing, and problem resolution (Vianna et al., 2020). Moreover, crowdsourcing platforms are instrumental in product development, transportation, funding, innovation, the generation of novel solutions, and the execution of tasks (Niu & Qin, 2021). Critically, in the contemporary business world, crowdsourcing platforms have been integral to the transformation of business entities and the execution of commercial activities (Xie et al., 2022).

Nevertheless, even with significant usefulness demonstrated by crowdsourcing platforms, they are not without limitations in certain sectors due to their inherent complexities. As argued by Prpić et al. (2015) the literature critically assessed the benefits of crowdsourcing platforms that revolve around enhancing individuals' livelihoods as exemplified by the following:

- Improving business profitability (Goudin, 2016; Hamari et al., 2015),
- Harnessing environmental sustainability (Mohri et al., 2023),
- Use of crowdsourcing platforms in agriculture for the generation of ideas (Posadas et al., 2021; Steinke, 2019),

Green Innovations in Supply Chain Management: Case Studies and Applications, First Edition. Edited by Marcia Mkansi, Godfrey Mugurusi, Aaron L. Nsakanda, Temidayo Akenroye, and Frank Tietze.

Published 2026 by John Wiley & Sons, Inc.

- support manufacturing in problem-solving (Beck et al., 2022),
- support construction industry by improvement in idea generation (Liu & Golparvar-Fard, 2015),
- Support the fund raising and waste management in municipalities (Mavakala et al., 2017),
- waste management initiatives aimed at fundraising (Testa et al., 2023; Testa et al., 2024),
- useful in the energy sector for enhancement of energy efficiency and cost reduction (Lane et al., 2013), and
- have proved useful for product development in manufacturing (Vianna et al., 2020).

Paradoxically, despite the advantages of crowdsourcing platforms in the manufacturing, waste management, energy, and agricultural sectors, most research has focused on passenger transportation services like Piggy Baggy, Uber, Taxify, and inDriver (Kafle et al., 2017; Castillo et al., 2018).

As such, this chapter examines the various facets of a crowdsourcing platform (Amitruck) and its contributions to the utilization of heavy trucks, particularly in relation to the complex management requirements inherent in a crowdsourcing framework. Amitruck, a crowdsourcing platform developed and being utilized in Kenya, offers a distinctive solution for transportation within the agriculture, waste management, manufacturing, construction, and energy sectors. The chapter expands upon this discussion by initially identifying the limitations associated with the diverse forms of crowdsourcing platforms in these industries regarding idea generation, problem-solving, and product development. Importantly, the chapter unpacks how Amitruck enhances the capabilities of crowdsourcing technology as it pertains to transportation in agriculture, waste management, manufacturing, construction, and energy sectors. Furthermore, the notion of green crowdsourcing is discussed in the chapter through utilization of Amitruck in reducing carbon emissions. Consequently, the main contributions of this study are twofold. First, the chapter provides a comprehensive study of the application of Amitruck crowdsourcing for transportation in agriculture, waste disposal, manufacturing, construction, and energy industries. Second, the chapter also discusses the role of Amitruck in promoting green innovation through the reduction of carbon emissions in the trucking business.

5.1.1 An Overview of Amitruck Application in the Kenyan Logistics Market

In Kenya, the trucking sector is witnessing significant expansion as the nation's economy grows, trade volumes increase, and logistics infrastructure evolves. Consequently, the trucking industry serves an indispensable function in the movement of goods, raw materials, and commodities throughout the nation, facilitating connections between production sites, distribution centers, and export terminals. Notably, with advancements in road networks, the adoption of digital logistics technologies, and the proliferation of e-commerce, there is an increasing demand for sustainable trucking services in Kenya. Thus, this chapter conducts an analysis of the Amitruck application in the Kenyan context. The chapter subsequently emphasizes Amitruck's beneficial influence across various sectors, including agriculture, waste management, manufacturing, construction, and energy, illustrating its capacity to enhance resource allocation and foster environmental sustainability by mitigating carbon emissions in comparison to conventional platforms. Furthermore, the advantages of the Amitruck application in fostering collaboration and partnerships with diverse stakeholders within the industry are elucidated in this chapter. Consequently, the study underscores the significance of collaboration, particularly with governmental bodies in Kenya, such as the National Environmental Management Agency, to guarantee compliance with safety, environmental, and regulatory standards.

"Amitruck" is a mobile application (app) that provides an innovative and multi-embedded solution for connecting trucks with different transporters using green crowdsourcing concepts for specialized services in agriculture, waste management, manufacturing, construction, and energy in Kenya. Amitruck is a digital green crowdsourcing platform and a marketplace for trucking, connecting transporters and clients for distribution of goods across Africa (Eke et al., 2023). The digital platform was founded in 2018 to connect customers and truck drivers (Yong & Festerling, 2023); it addresses a rapidly growing market, estimated at 23 billion USD in Kenya alone. The "Amitruck app" was developed with green crowdsourcing principles consisting of dynamic and flexible capabilities. Over the past years, the Amitruck app has dealt with cargo services, carpooling, and trip planning grounded on the concept of shared economy. In this way the application is more environmentally friendly as the number of vehicles on the road can be reduced, resulting in less carbon emissions. Compared to traditional applications, the environmental advantage of the Amitruck app is its flexibility, which helps in the reduction of carbon emissions.

The digital application has already achieved considerable growth within technology-based companies in Africa and is marked by the typically African characteristic diversity (Yong & Festerling, 2023). In 2020, the digital application employed 67 employees (Bunde, 2021), contributing to Kenya's digital sector by accounting for 4.5% of Kenya's digital market (Mulligan and Jackson, 2022). The digital application has connected 12,000 vehicles, 200 corporate clients and more than 75,000 deliveries (Amitruck: Your Reliable Logistics Partner in Kenya & East Africa, 2025). In 2022, the digital platform raised capital of 4 million dollars for expanding operations in Tanzania and Uganda (Barakat, 2022). The chapter now turns its focus to the potential pillars which may be enhanced by crowdsourcing platforms.

5.1.2 Application of Crowdsourcing Platforms

This chapter examines the contribution of Amitruck crowdsourcing platform and its application in sectors such as agriculture, waste management, manufacturing, construction, and energy industry. It is critically important to understand that crowdsourcing uses online platforms for performing tasks requiring different viewpoints and problem-solving, including idea generation, product development, problem-solving, content creation, task performance, and crowdfunding (Beck et al., 2022). Table 5.1 provides a summary on the application of crowdsourcing platforms in the different sectors: agriculture, waste management, manufacturing, construction, and energy.

From Table 5.1, it is evident that the agricultural sector predominantly employs crowdsourcing platforms for the purposes of idea generation and problem-solving (Dickinson et al., 2012; Kamilaris et al., 2017; Posadas et al., 2021; Steinke, 2019). Posadas et al. (2021) contend that crowdsourcing platforms have emerged as indispensable tools for optimizing precision agriculture through the collection of relevant information. Importantly, these platforms facilitate innovative ideas aimed at enhancing agricultural practices, while simultaneously empowering citizen scientists to derive insights regarding environmental issues by classifying preexisting datasets (Kamilaris et al., 2017; Steinke, 2019). In the context of waste management, the evidence presented in the literature matrix in Table 5.1 substantiates the assertion that crowdsourcing is vital for idea generation, problem-solving, and crowdfunding initiatives. Johnson and Grey (2016) underscore the significance of crowdsourcing platforms in promoting public engagement to leverage existing data pertaining to the dimensions and geographic distribution of current landfills within the United States, while concurrently offering users a novel framework to visualize and investigate landfill sites. To conceptualize crowdsourcing as an effective problem-solving mechanism, Dimitriadis et al. (2019) formulated a waste crowdsourcing framework and model that facilitates an in-depth evaluation and filtering of waste management strategies in developing nations. Consequently, crowdsourcing platforms can be utilized by urban planners for monitoring locations and data related to waste management, to enhance governance and sustainability practices in waste management, to evaluate the prerequisites for the establishment of municipal solid waste management systems, and as crowdfunding mechanisms that provide innovative and secure alternatives for waste disposal, thereby creating new avenues for improving the economic efficiency of waste management practices.

Furthermore, Table 5.1 affirms that crowdsourcing platforms assume a pivotal role in the manufacturing sector through their applications in idea generation, problem-solving, and product development. In this regard, various platforms may prove essential in the manufacturing domain for several reasons: (i) they offer suitable frameworks for collaboration; (ii) they facilitate the capture of explicit knowledge from extensive cohorts of colleagues and teams; (iii) they possess the capability

Table 5.1 Sectoral impact of crowdsourcing platforms

Literatures	Purpose	Idea generation	Problem-solving	Content creation	Product development	Task performance (transport)	Crowd funding
Posadas et al. (2021), Steinke (2019), Kamilaris et al. (2017), Minet et al. (2017), Lokers et al. (2016), Steinke (2019), Dickinson et al. (2012), Wolfert et al. (2017), Sykuta (2016), Fritz et al. (2012), Bey et al. (2016), Rossiter et al. (2015), Muller et al. (2015), Franzoni and Sauermann (2014), and Bruce (2016)	Agriculture	●	●	○	○	○	○
Johnson and Grey (2016), Ananta et al. (2023), Mavakala et al. (2017), Singh et al. (2017), Krishnan et al. (2023), Dimitriadis et al. (2019), Certoma et al. (2015), Kotsiuba et al. (2023), Chaves et al. (2019), Tuzhikov (2024), Testa et al. (2023, 2024), and Corsini and Frey (2024).	Waste Management	●	●	○	○	○	●

(Continued)

Table 5.1 (Continued)

Literatures	Purpose	Idea generation	Problem-solving	Content creation	Product development	Task performance (transport)	Crowd funding
Gao et al. (2015), Evans et al. (2015), Vianna et al. (2020), Doan et al. (2011), Gong et al. (2021b), Niu and Qin (2021), Niu et al. (2019), Shang et al. (2019), Cheng et al. (2018), Gong et al. (2021a), Yao and Lin (2016), and Pilloni (2018).	Manufacturing	⊖	●	⊖	●	⊖	○
Liu and Golparvar-Fard (2015), Han and Golparvar-Fard (2017), Wu et al. (2012), Bollacker et al. (2008), Karampinas and Triantafillou (2012), Cao, Zhang, et al. (2014), Elhamshary et al. (2019), Cardonha et al. (2013), Kondreddi et al. (2014), Liu and Dang (2014).	Construction	●	●	⊖	○	○	○
Flostrand et al. (2019), Wang et al. (2019), Wu et al. (2019), Pritoni et al. (2015), Wagy et al. (2017), and Lakhdari et al. (2018).	Petroleum industry	⊖	●	○	○	○	○

Assessment scale: Evidence of notable impact: ●. Fairly notable evidence of impact: ⊖. No evidence of notable impact: ○.

to unearth previously latent tacit knowledge within a less formalized virtual milieu; (iv) they create an optimal environment for engagement among diverse stakeholders; (v) they leverage advanced Industry 4.0 technologies to enhance product development, innovation, data provision for manufacturing, crowd sensing, problem-solving, and troubleshooting; and (vi) they furnish a secure cyber infrastructure for member production through the utilization of cloud-stored data. Significantly, Table 5.1 illustrates that the academic literature provides nuanced insights regarding the contributions of crowdsourcing to idea generation and problem-solving within the construction industry. Crowdsourcing has been demonstrated to be crucial for on-site construction processes, which are overseen via mobile platforms that also streamline the data collection process necessary for generating annotated material patches that are informed by Building Information Modeling (BIM) overlays. Additional noteworthy applications encompass the scaling of data collection efforts, which are prevalent in machine-based automated construction, as well as the enhancement of human-centric tasks such as data cleaning, the participatory sensing approach for the automatic construction of precise motion traces, and the push–pull dispatch model for fostering community engagement in idea generation.

Ultimately, the matrix presented in Table 5.1 elucidates the significant impact of crowdsourcing within the energy sector, particularly regarding enhancements in energy efficiency and reductions in associated costs. To elaborate on this, Lane et al. (2013) introduced a crowdsourcing initiative known as the Piggyback Crowdsensing system, which is designed to gather sensor data from smartphones to minimize the energy burden of user involvement. Notably, Bazilian et al. (2012) established an open crowdsourcing platform that serves as a comprehensive energy analysis tool, providing vital insights pertinent to policy formulation in developing nations. Additionally, the crowdsourcing platform may be leveraged to augment the strategic coordination of energy utilities, as well as to improve energy efficiency and retrofitting measures aimed at cost reduction. The chapter itself is organized as follows. The methodology is presented in Section 5.2. Then, in Section 5.3, we present the results. The discussion and implications are presented in Sections 5.4 and 5.5 respectively.

5.2 Research Methods

This research study was grounded in interpretative qualitative method, providing a deep insight into "the complex world of lived experience from the point of view of those who live it" (Schwandt, 1994). The chapter used an exploratory case study research design for unearthing specialized services offered by the Amitruck Crowdsourcing platform. Amitruck crowdsourcing platform provided single case for exploration of specialized services using crowdsourcing platform in Kenya. Combining exploratory and case study was critical in understanding the unique perspectives of this crowdsourcing platform. An exploratory research design is appropriate when studying emerging phenomena with limited understanding (Singh, 2007). We used case study research design for flexibility and contextualization of specialized services offered by Amitruck crowdsourcing platform (Yin, 2009). In addition, the case study is "an empirical inquiry that investigates a contemporary phenomenon within its real-life context" (Yin, 2009), for process comprehension (Benbasat et al., 1987; Guba & Lincoln, 1994), and holistic understanding (Creswell, 1998; Eisenhardt, 1989). The exploratory case study design was, therefore, critical for in-depth understanding (Yin, 2009), flexibility (Walsham, 1995), contextualization (Yin, 2018), discovery (Ponelis, 2015), understanding rich data (Niu et al., 2011), and interactive learning (Mahmood & Ferneley, 2006). In addition, Yin (2003) argues that case studies offer rich and distinct knowledge appropriate for green crowdsourcing applications in different unique and dynamic environments. In the case of Amitruck, the unique case provides insights on application in Kenya and East Africa region.

In addition, our study is premised in the interpretive phenomenology appropriate for understanding subjective worldviews. The research design was used for understanding the use of Amitruck platform for coordination of trucks in Kenya. In our study, critical theoretical and practical risks have emerged for advancement of theory and practice. Theoretically, green crowdsourcing platforms provide unique mechanisms for coordinating specialized transportation of agricultural products, waste management, manufacturing, construction, and energy. The specialized products require unique trucks with different individual and corporate customers. The traditional crowdsourcing platforms mainly provide services for individual transport needs and regular transport requirements. In practice, the Amitruck crowdsourcing platform in Kenya has to develop strategies for improvements in coordinating between truck owners and customers.

An application for crowdsourcing developed in Kenya, Africa, was selected for the study to allow it to be replicated and the findings compared (Eisenhardt, 1989) and to improve external validity and prevent observer bias (Barratt et al., 2011). The Amitruck application is explicitly for trucks in Kenya. In addition, the crowdsourcing platform enables flexibility, connectivity to match registered members' demands, and supply of truck requirements, thus improving environmental sustainability. The case study approach taken in the study of the Amitruck platform contributes to an understanding of specialized services in different sectors: agriculture, waste management, manufacturing, construction, and energy services in East Africa. In addition, there is collaboration between the corporate customers and truck drivers.

5.2.1 Sample and Data Collection

The purpose of our study was exploration of the specialized services offered by the Amitruck green crowdsourcing platform in Kenya. We purposively interviewed 11 participants, who were: customers from agriculture, manufacturing, construction, waste management, and energy transportation sectors together with the Amitruck crowdsourcing platform. The respondents who were interviewed were knowledgeable with rich and in-depth understanding of Amitruck green crowdsourcing application in Kenya. Despite the limitations of purposive sampling, such as bias and subjectivity, this was strengthened through triangulation, data saturation, and transparency (Table 5.2; Charmaz, 2014).

In addition, the study research objectives were clearly defined and the findings validated through triangulation. The interviews were recorded, and transcripts were made. In certain instances, notes were taken to capture the critical contextual risks in green crowdsourcing. The focus of the interviews was on gaining an understanding of participants' views on the specialized application of Amitruck crowdsourcing platform. We conducted our study by adhering to the highest level of ethical considerations through the phases of the study (Creswell & Creswell, 2017).

5.2.2 Data Analysis

The interviews with the different respondents were recorded through Microsoft Teams or using the laptop's audio instruments and were later transcribed, coded, and analyzed using MAXQDA (a computer-aided qualitative data analysis software). The transcripts were uploaded to the software and subsequently open coded, followed by inputting coding groups or themes (also known as axial coding). The perspectives of the respondents in view of the impact of the mobile application on the internal supply chain were captured. Open coding was undertaken using MAXQDA. In this context, words and phrases reflect the narrative of the participants who were involved in development and use of the mobile application, for evaluation of the application. After the open coding process, using the software hierarchical axial coding was used to identify and connect multiple codes to map out the impact of the mobile application (Babbie, 2020). The coded themes from hierarchical axial coding were combined as significant inputs for thematic development and the iterative process between themes that emerged from the data and the literature for crowdsourcing platforms. Yin (2018) guided the process for applicability or transferability with background description, the outline of the methodology, and the preservation of transcripts. The audio record and transcripts of the perspectives of participants were appropriate and relevant for convergence and support of the constructs discussed by Yin (2018).

Table 5.2 Respondent descriptions

Respondent	Nature	Description	Use of the platform
Respondent 1	Employee (Amitruck platform)	• Amitruck company employs more than 100 employees, connected to more than 12,000 vehicles, with more than 200 corporate clients, and undertakes more than 75,000 deliveries in different sectors	• Coordination of the Amitruck platform for different users: agriculture, waste disposal, manufacturing, construction, and petroleum • Development of Amitruck platform
Respondent 2	Transporter (agricultural products)	• The company supports more than 10,000 farmers in Kenya. They undertake transportation of fertilizers in Kenya	• Buying and selling of agricultural production and distribution • Extension of land through relations with local chiefs
Respondent 3	Exporter (agricultural products)	• The company is involved in rearing of cattle, goats, sheep • The company processes milk products for export to Indian Ocean Islands and East Africa	• The company uses the Amitruck platform for transportation of processed products to the export markets
Respondent 4	Household (waste management)	• A family in Nakuru supports waste disposal • Households basically participate in waste analysis and disposal	• Households use the Amitruck platform to support household waste disposal • The household participates in the process for waste management
Respondent 5	Truck driver (waste management)	• Truck driver is involved in waste transportation • The driver uses a specialized transport for transportation of waste to different sites	• The truck driver uses Amitruck platform for transportation of waste from households to the disposal areas, processing sites, and recycling sites

Table 5.2 (Continued)

Respondent	Nature	Description	Use of the platform
Respondent 6	Manufacturer	• The manufacturer offers 100 international and local brands with diverse transport needs like raw materials, finished products for distribution across the different market segments in Kenya	• The company is in partnership with Amitruck for transportation of finished products to supermarkets, and distributors in Kenya • The transport needs for goods across the different manufacturing plants
Respondent 7	Contractor	• This is a medium size local company employing 20 skilled staff and 400 unskilled laborers depending on the requirements	• The construction company uses Amitruck crowdsourcing platform for sourcing transport for the construction materials: manufactured and local materials
Respondent 8	Truck driver (construction)	• The truck driver transports local materials like sand, bricks, cement, and stones for the different contracts and individuals within Nairobi City	• The driver uses Amitruck application for providing services to potential individuals and contractors who need to transport construction materials
Respondent 9	Truck business owner	• The businessman has 30 tank trucks for transportation of petroleum products from the refinery in Mombasa to different dealer outlets in Nairobi	• The truck owner uses Amitruck for sourcing customers for these trucks in the different sectors, such as construction and manufacturing
Respondent 10	Manager of oil dealer	• The dealer has 12 petrol stations in the capital city Nairobi dealing in petroleum products, mainly petrol, diesel, paraffin, oils, and lubricants	• The dealer uses the Amitruck platform for sourcing trucks to facilitate the transportation of petroleum products from the refinery to its petrol stations

(Continued)

Table 5.2 (Continued)

Respondent	Nature	Description	Use of the platform
Respondent 11	Petroleum truck driver	• The truck driver transports petroleum products from the refinery in Mombasa to the dealers in Nairobi using the Amitruck application	• The truck driver uses the Amitruck platform for transportation of products from the refinery to the different petrol stations in Kenya

5.3 Results

This study used the MAXQDA software for analysis of document and key informant interviews on the specialized services provided by the Amitruck crowdsourcing application in Kenya. The findings revealed that the mobile application had a transformative impact on manufacturing by increasing production and reducing costs. Table 5.3 clearly highlights the contribution of the Amitruck crowdsourcing platform for agriculture, environment, manufacturing, construction, and energy industry. The responses from the respondents are also depicted in Table 5.2.

From Table 5.3, it is seen that the Amitruck platform has instituted mechanisms for the transportation of both production and distribution within the agricultural sector. It is evident that agricultural inputs and harvests necessitate meticulous handling and management to guarantee that produce arrives at the markets in optimal condition. As articulated by participants 2 and 3, Amitruck meticulously supports the transportation of agricultural products, adhering to the essential requirements for their handling. Moreover, participants 1, 2, and 3 corroborated that the Amitruck platform offers a distinctive and specialized service designed for the effective and efficient transportation of agricultural inputs and products. Additional corroboration from key informants and document analysis underscores that the platform facilitates appropriate sourcing, optimization methodologies, and specialized locking mechanisms to ensure the security of agricultural products.

Clearly, Amitruck facilitates sourcing of appropriate vehicles to meet the specific requirements for transportation of agricultural inputs and products. As such, the crowdsourcing agents systematically target transportation of agricultural products with customization of services for appropriate handling of agricultural produce for optimization of yield, and export management. The results also indicate that Amitruck has proved useful for transportation of agricultural inputs and products with modern techniques. Amitruck also uses optimization techniques to facilitate effective management of agricultural products. The advanced techniques provided

Table 5.3 Summary of results

Type of industry	Specialized services offered (second-order concepts)	Respondents' quotes (first-order quotes)
Agriculture	Production and harvest	"When I was first contacted by the marketing people about using Amitruck I couldn't see the possibility but later on, I tried and found it to be useful as the trucks are specialized to support the transportation of perishables like tomatoes. I found the application very convenient and reliable. I have never turned back and have continued using the application for transportation of inputs and harvest in my farm." **Respondent 2**
	Specialized trucks	"Our company undertakes the sourcing of specialized trucks to support in the transportation of agricultural inputs. We use our Amitruck mobile application for sourcing the right vehicles for transportation of agricultural products and materials to support the agricultural sector. This has been significant in provision of specialized agricultural services to the communities." **Respondent 2**
	Optimization techniques	"We undertake transportation of agricultural products and materials with great optimization techniques for the local and export corridor for clients with modern and advanced techniques. In addition, we use advanced techniques for transportation of agricultural products using techniques such as specialized locking mechanisms, seals, packing methodologies and weight distribution for efficient, reliable agricultural transport." **Respondent 1**
	Advanced security	"The trucks have always provided all the required professional support for transportation of agricultural products. The trucks have seals and locks, which promotes the security of your products during transportation. I have always transported my avocado to Mombasa port for onward export to Oman and United Arab Emirates. This application is more reliable for my transport needs." **Respondent 3**
	Use of extra capacity	"The Amitruck crowdsourcing platform is useful because it helps in combining the different goods, the application can help in the use of extra capacity. This is very helpful in environmental sustainability. We received funding because we are promoting reduction of carbon emissions. Previously we used to transport goods for one but combining deliveries is acceptable." **Respondent 1**

(Continued)

Table 5.3 (Continued)

Type of industry	Specialized services offered (second-order concepts)	Respondents' quotes (first-order quotes)
Waste management	Waste stream analysis	"This application has helped in improvement of waste management in our town by assessing and characterizing the waste generated from source. We identify the sources and quantities of waste generated. In addition, we sort and analyze the waste materials to determine their composition including the types and amounts of organic waste, recyclables, and hazardous waste. This has significantly helped us to reduce the amount of waste taken to landfill." **Respondent 4**
	Route optimization	"I used to privately provide my truck for waste management but the coming of Amitruck enabled us to discover better means of waste transportation. Our major approach was basically to transport waste from the different households to the landfill. However, Amitruck come up with route optimization technology to help us in planning the most efficient routes. Using this application automatically determines efficient collection routes, minimizing travel distances and fuel consumption." **Respondent 5**
	Vehicle technology	"Our company has registered vehicles with specialized waste management capabilities. Amitruck application helps in deployment of trucks with features like compaction mechanisms, automated loading systems, GPS tracking for enhancing collection efficiency. Trucks with compaction capabilities compact waste using hydraulic system. Specialized vehicles help in addressing the unique needs of waste transportation. The application facilitates in selection of the appropriate truck ideal for either recyclable, organic, hazardous, and no-recyclable waste. This has been fundamental in improvement of waste management." **Respondent 6**
	Partnerships	"Our company undertakes waste transportation in conjunction and collaboration with National Environmental Management Agency. In addition, the partnerships have helped in improving waste management, which requires stakeholder participation. We usually assist our clients with information on NEMA permits and specific industry knowledge in waste management logistics." **Respondent 1**

Manufacturing	Production to distribution	"Amitruck supports more than two hundred corporates on their transport requirements. We have supported companies like Unilever to deliver the raw materials from Mombasa port to the processing plants. We have also supported within the plant deliveries during processing and the final products to the different customers. This partnership has helped us to contribute to the transportation for manufacturing companies in Kenya." **Respondent 1**
	Processing	"Our manufacturing has been made easy by the support from Amitruck which helps our company to transport raw materials. The tracking capacity provided by Amitruck has tremendously improved our processes. We have found Amitruck as a reliable application for transportation of our raw materials and goods to distributor outlets, warehouses, and supermarkets." **Respondent 6**
	Specialized industry	"We have found the Amitruck application very useful for our requirements; during the manufacturing process, the application helps in sourcing specialized trucks like for the manufacturing of Unilever products we need trucks specialized in transportation of delicate chemicals for manufacture of soap and other detergents. We have found this very relevant and appropriate for our manufacturing." **Respondent 6**
	Extra capacity and return trips	"Amitruck has been useful in the transportation of manufactured goods while at the same time contributing to environmental sustainability because of use of extra capacity when available and return trips, which help in reduction of carbon emissions. Our innovation has been appreciated by different stakeholders in Kenya." **Respondent 1**
Construction	Relevant trucks	"We use our Amitruck mobile application for sourcing the right vehicles for transportation of construction materials. We specialize in sourcing the right type of vehicle for the needs of the clients. Our company tailors the transportation to construction needs of the clients, outfitting vehicles as needed." **Respondent 1**
	Transporting construction materials	"We have always found Amitruck to be cost effective and reliable in the transportation of our construction needs right from stones, sand, bricks, cement, and other industrial products at a more affordable price. The application helps you to see the different offers by the different truck drivers and is negotiable." **Respondent 7**
	Fragile goods	"We use Amitruck because of the reliability of the application in the provision of the transport requirements for delicate and fragile construction materials like glass, tiles, and most of the delicate plumbing materials. You can monitor the transportation of your materials from warehouse to the construction site. This is very important for us in the construction industry where losses can easily be incurred." **Respondent 7**

(*Continued*)

Table 5.3 (Continued)

Type of industry	Specialized services offered (second-order concepts)	Respondents' quotes (first-order quotes)
	Return trips	"I have used this application ever since the time of Covid and we found this application to be good because of the ability to help you get business on return as we take manufactured goods from Nairobi to the rural towns, which have local materials like sand and bricks. So, we don't come back empty but instead there will be requests for delivery of local materials to the city. This increases our profits, with reduced fuel consumed and increased income from the same distance travelled. We are also reducing polluting our environment." **Respondent 8**
Energy	Unique requirements	"Our company undertakes reliable transportation of goods and products in the volatile energy sector in Kenya and East Africa. We undertook this innovation by working closely with truckers with a good track record in transportation of energy products. This enabled us to meet customer expectations in the unique and challenging industry." **Respondent 1**
	Safety	"Amitruck provides a simple and more reliable solution because you negotiate for the price directly with client. This is more profitable because previously we used to get business through middlemen who used to cheat us. Removal of the middlemen is making us to realize more profits as middlemen have no role and our petroleum trucks do good business. This application is helpful for truck owners, and clients do business openly and transparently." **Respondent 9**
	Security	"We found Amitruck offer very useful for the transportation of energy requirements because of the due considerations which the application offered. First, the security considerations were taken care of by the application. Our evaluation of application discovered that safety measures were properly provided for by the application. We have continued using Amitruck for our offering." **Respondent 10**
	Procedures and controls	"Amitruck employs strict procedures for registration of trucks that transport the petroleum products. The drivers also take strict procedures to be admitted to transport energy products. We went through rigorous testing before admission and subsequent training. We went through training in oil and gas procedure because of risks in this industry." **Respondent 11**
	Flexibility	"Our trucks usually have several compartments for carrying different amounts of the different petroleum products. I have come to realize that the flexibility afforded by Amitruck enables us to transport the different requirements flexibly. Previously, you had to transport for a particular customer, but we can combine orders. This has proved to be very useful for environment protection." **Respondent 11**

by Amitruck are critical for transportation of agricultural products for the local and export corridors to meet the export standards. Finally, the crowdsourcing platform uses advanced techniques for transportation of agricultural products from farms, markets, and different export ports and facilitates transportation of agricultural products using specialized locking mechanisms and seals, which is critical for maintenance of quality.

Respondents 4, 5, and 6 in Table 5.3 confirmed that Amitruck provides appropriate capacity for waste transportation and effective management. More importantly, evidence from document review and respondent 5 established that the Amitruck innovation strengthened waste management in urban places in Kenya. As such, Amitruck undertakes waste management, waste stream analysis, route optimization techniques, and waste vehicle technology in Kenya. This has proved vital in waste management, contributing to clean and sustainable environment. Respondent 4 further confirmed that Amitruck has proved useful for waste stream analysis, systematically improving waste disposal as it undertakes waste stream analysis using technical people to guide the participants in waste generation, transportation, and disposal. The waste management agents support the identification, quantification, and characterization of waste generated from households, business enterprises, public and private institutions. This has been beneficial in providing appropriate opportunities for waste reduction, reuse, recycling, and proper disposal.

The results further confirm that Amitruck has improved waste management by enabling waste segregation, helping different stakeholders reduce waste, and promoting sustainable environment; use of route optimization software has revolutionized waste management by urban authorities partnering with Amitruck. The route optimization software has proved useful in creating more efficient collection routes for solid waste collection. The use of advanced route planning software to create efficient collection routes minimizes travel distances and fuel consumption. In addition, the crowdsourcing platform sources appropriate vehicle technology for collection, transportation, and disposal of recyclable, organic, hazardous, and non-recyclable wastes. Appropriate vehicles such as rear-loader compactors and front loader compactors have been critical in sustainable environment, cost reduction, efficiency, and social acceptability. The crowdsourcing agents systematically target transportation of waste for urban areas, commercial, and residential areas. This helps in professional disposal of waste. Appropriate vehicle technology for waste management considers factors such as environmental impact, cost, efficiency, and social acceptability.

Amitruck delivers essential logistics services for transportation within the manufacturing sector. As substantiated by respondents 1 and 6 in Table 5.2, Amitruck effectively addresses transport needs for the manufacturing industry, encompassing the entire spectrum from raw materials to finished goods. Moreover, it offers internal transportation solutions between warehouses, production facilities, and distribution centers. Furthermore, Amitruck facilitates the movement of manufacturing products, encompassing both raw inputs to the industry and finished products from manufacturing plants to market. The platform enhances the coordination of the manufacturing supply chain, spanning from initiation through to completion. This capability is pivotal for promoting advancements in manufacturing within less developed nations. The findings further corroborate that Amitruck adeptly fulfils the diverse and specific transport requirements of the manufacturing sector, accommodating the varied stakeholders involved from inception to conclusion. The collaborative methodology employed during transportation is essential for the effective management of the transport sector. The manufacturing process necessitates a comprehensive understanding of the distinct requirements, such as the handling of large cargo for specialized industries.

Respondents 1, 7, and 8 in Table 5.3 affirmed that Amitruck supports the transportation of inputs and materials within the construction industry in Kenya. The construction sector necessitates unique and critical transport solutions for construction materials. The company addresses the diverse and specific transportation needs inherent to the construction industry. Significantly, Amitruck facilitates the identification of suitable vehicles required for transporting construction materials in various forms. The crowdsourcing platform employs staff to coordinate the provision of transport for a wide range of construction needs. Additionally, Amitruck ensures secure transportation to satisfy the safety requirements of its clientele. Respondent 8 emphasized that the company implements advanced security and safety protocols, which are particularly suitable for the high-risk nature of the industry. This is crucial for addressing the construction needs associated with fragile materials. Ultimately, the crowdsourcing platform has enhanced the transportation of construction materials by the unique considerations pertinent to the industry.

Amitruck offers adequate capacity for the transportation of petroleum products within Kenya and the broader East African region. The energy sector is characterized by unique and evolving transport requirements that are dynamic in nature. The transportation of petroleum products necessitates significant investment in advanced safety measures, security protocols, and specialized hazard-focused services, alongside rigorous operating procedures. The energy industry is distinctive in its various forms, including liquid, gas, and concentrated components; thus, the transportation of energy necessitates different types of vehicles specifically designed to handle the diverse chemical compositions of energy products. Furthermore, energy is inherently volatile and presents various risks. As confirmed by respondents 1, 9, 10, and 11, Amitruck has been tailored to meet the transportation requirements of energy products within East Africa. Additionally, the truck owner (respondent 9) noted that the Amitruck crowdsourcing platform offers a more dependable environment for transactions between truck owners and clients. The platform facilitates direct contact with clients, thereby eliminating middlemen who previously appropriated a substantial portion of the profits. Amitruck supports the transportation of energy products by employing advanced security measures and specialized hazard services. Amitruck adheres to the highest established protocols for the transportation of energy products. The utilization of advanced security measures and specialized hazard services has proven to be critical for the crowdsourcing of transportation within the energy sector.

5.4 Discussion

Literature confirms that, crowdsourcing platforms are critical for idea generation, product development, problem-solving, content creation, task performance and crowdfunding (Beck et al., 2022). However, there are complexities in agriculture, waste management, manufacturing, construction, and energy industry; the application of crowdsourcing platforms has been mainly limited to idea generation, product development, and funding. Results from the current study confirms that Amitruck innovation extends the use of crowdsourcing applications for transportation in this unique and dynamic manner in the following sectors.

5.4.1 Agriculture

Results from the study confirm that Amitruck is critical for transportation of agricultural production and distribution through the use of advanced techniques for transportation of agricultural products. The results corroborate with literature from Kamilaris et al. (2017), Minet et al. (2017), and Lokers et al. (2016) who emphasized the use of crowdsourcing applications for idea generation, product development, and problem-solving but with limited emphasis regarding transportation. The study results further confirm that Amitruck application uses appropriate transport mechanisms for agricultural inputs and harvests, making the application critical for transportation of agricultural products right from production to distribution. However, earlier in the literature there was more emphasis on crowdsourcing platforms for idea generation in precision agriculture, gathering, processing and analyzing information for agricultural operations (Posadas et al., 2021).

Critically, the study established that Amitruck effectively addresses the unique transportation needs of agricultural products, optimizing sourcing of appropriate vehicles, implementing advanced handling techniques, and utilizing specialized locking mechanisms to ensure product quality and safety. This finding is in line with Dickinson et al. (2012), Steinke (2019), and Kamilaris et al. (2017) who argued that crowdsourcing platforms facilitate engagement of nonprofessionals in scientific research, central management of collected data for analysis, and citizen scientists to become mindful of environmental issues and help with classification of existing datasets. Key highlights to the findings is facilitation of efficient transport from farm to market and export and advanced security techniques that revolutionize agricultural transport while streamlining environmental protection through green crowdsourcing. In line with the key findings, past studies have also identified crowdsourcing platforms being extensively used for improvement of agriculture by collecting agricultural land use (Bey et al., 2016; Fritz et al., 2012), soil data (Rossiter et al., 2015), weather data (Muller et al., 2015), phenology and crop calendar information (Shakya et al., 2023), weed, pests and disease control (Bruce, 2016; Franzoni & Sauermann, 2014), yield and vegetation (Oteyo et al., 2021). Therefore, Amitruck clearly facilitates the contribution of innovative ideas aimed at enhancing agricultural practices, while simultaneously empowering citizen scientists to derive insights regarding environmental issues by classifying preexisting datasets.

5.4.2 Waste Management

This chapter has established that the Amitruck Mobile Application contributes to environmental protection. Similarly, Olairu (2021) argued that green crowdsourcing applications use analytics to reduce traffic congestion, which decreases pollution and mitigates negative impacts on the environment. The chapter demonstrates that Amitruck's flexible and dynamic nature delivers significant environmental benefits compared to traditional platforms like Uber and Bolt, particularly in reducing the number of vehicles on the road and emissions. Despite using similar unstructured green crowdsourcing principles, the use of this platform for specialized services advances the concept for environmental management in waste disposal. Previous studies highlight the importance of crowdsourcing platforms for waste management harnessing: participation (Johnson & Grey, 2016), evaluation of strategies (Dimitriadis et al., 2019), improved governance (Certoma et al., 2015), and funding (Tuzhikov, 2024). Johnson and Grey (2016) used crowdsourcing to facilitate public participation for harnessing existing data about the size and location of current landfills in the United States while simultaneously providing users an innovative way to see and explore landfills. Mavakala et al. (2017) identified the use of crowdsourcing platforms for identification, localization, characterization of solid waste in Kinshasa. Importantly, crowdsourcing is critical for improvement of waste governance and sustainability of waste management using crowd wisdom (Certoma et al., 2015). Kotsiuba et al. (2023) highlighted the use of crowdsourcing platforms for the development of appropriate crowdsourcing waste management strategies in urban areas. In addition, Chaves et al. (2019) discuss the use of crowdsourcing tools for urban planning and propose a five-dimension typology of quality in crowdsourcing. Finally, crowdsourcing has been extensively used for the generation of funds for solid waste management using crowdfunding (Testa et al., 2024).

In this chapter, environmental management using crowdsourcing applications can be strengthened using specialized teams, collaboration and partnerships, and customer support. This is critical in strengthening existing waste management strategies using the crowdsourcing applications. The findings contribute to a better understanding of the unique capabilities of Amitruck application in addressing the challenges in developing countries for improving the environment.

5.4.3 Manufacturing

The next finding highlights the use of Amitruck crowdsourcing application for transportation of manufacturing products. The chapter findings demonstrate that Amitruck is critical for manufacturing in collaboration with corporations. However, previous literature demonstrates the use of crowdsourcing platforms mainly for idea generation, problem-solving, and product development in manufacturing.

Previous research on the use of crowdsourcing in manufacturing highlights collaboration between manufacturers and other stakeholders (Evans et al., 2015). The crowdsourcing platforms mainly use idea generation and problem-solving activity (Beck et al., 2022). Vianna et al. (2020) highlight the use of crowdsourcing for product development, innovation, provision of data and information for manufacturing, crowd sensing, problem-solving, and troubleshooting. In addition, crowdsourcing facilitates problem-solving and process improvement (Gao et al., 2015), data generation (Doan et al., 2011), and data-driven manufacturing by delivering manufacturing as a service (Gong et al., 2021b). Furthermore, academic literature highlights how crowdsourcing platforms help manufacturers to build a physical cyber system (Yao & Lin, 2016). Despite the existence of studies that highlight the use of crowdsourcing application for transport (Chandra & Kumar, 2020; Dupljanin et al., 2019; Kumar et al., 2018; Mohri et al., 2023; Wu et al., 2024), the crowdsourcing application has been used in a limited way for manufacturers, mainly for delivery of groceries. In our study, the introduction of Amitruck extends the use of crowdsourcing to the entre supply chain in manufacturing by addressing all the unique requirements for transportation in manufacturing. In our chapter, the Amitruck mobile application streamlines the transportation of raw materials and finished goods throughout the manufacturing process, handling diverse product forms and managing specialized logistics needs, including large cargo transportation. The platform improves supply chain coordination and efficiency. In addition, the crowdsourcing application supports addressing the construction industry's unique transport needs by sourcing suitable vehicles for various materials and prioritizing advanced security measures to mitigate high-risk environments.

5.4.4 Construction Industry

Findings in our study confirm the use of Amitruck crowdsourcing application for transportation of construction materials in Kenya; however, previous studies highlight the use of crowdsourcing platforms for idea generation and problem-solving.

Academic literature generally elaborates on the benefit of crowdsourcing applications to the economy (Bassi et al., 2020), peoples' livelihoods (Prpić et al., 2015), profit (Goudin, 2016; Hamari et al., 2015), and the environment (Mohri et al., 2023) but does not demonstrate specialized use in construction. Our study provides insight on the approach for effective crowdsourcing in the construction industry meeting its critical needs. Amitruck crowdsourcing application considers the unique relevant needs and security needs of the construction industry. Previous studies emphasized the importance of crowdsourcing for improvement of construction activities: monitoring (Liu & Golparvar-Fard, 2015) and data collection (Han & Golparvar-Fard, 2017). Liu and Golparvar-Fard (2015) argued that crowdsourcing is critical for monitoring on-site construction using the mobile platforms. In addition, crowdsourcing applications are used in machine-based automatic construction (Suchanek et al., 2007; Wu et al., 2012) and human-based manual construction (Bollacker et al., 2008; Karampinas & Triantafillou, 2012). Furthermore, previous literature highlights that hybrid machine-crowdsourcing improves taxonomy accuracy and coverage. Finally, Cardonha et al. (2013) highlight crowdsourcing applications for accurately generating traces by reducing errors in the inertial motion data by using the points of interest in the indoor environment for error resetting. Despite the application, the studies don't provide insight into the extent to which crowdsourcing is used for transportation in developing countries. In our study the Amitruck crowdsourcing application addresses the unique construction industry requirements for the construction industry. Amitruck crowdsourcing application facilitates sourcing of appropriate vehicles for construction materials in the different forms. The crowdsourcing application uses staff to coordinate provision of transport for the different range of construction needs. In addition, the crowdsourcing application provides secure transport to meet the safety needs of the customers.

5.4.5 Energy Industry

Finally, our study established that Amitruck crowdsourcing application facilitates the transportation of specialized energy products which were primarily under traditional transport. Previous literature mainly examined the influence of crowdsourcing platforms for problem-solving using techniques such as crowdsensing (Lane et al., 2013), robust energy analysis (Bazilian et al., 2012), coordination of energy activities (Flostrand et al., 2019), robust efficient energy management (Liu et al., 2014), energy efficiency (Wu et al., 2019), and satisfaction of the energy customers (Lakhdari et al., 2018). Lane et al. (2013) proposed a crowdsourcing application known as Piggyback CrowdSensing system for collecting mobile sensor data from smartphones for lowering energy overhead user participation. In addition, Bazilian et al. (2012) developed an open, publicly accessible software and data as well as crowdsourcing technique with a robust energy analysis tool that can deliver crucial, policy-relevant insight, particularly in developing countries. Furthermore, Flostrand et al. (2019) advocated for the use of crowdsourcing for strengthening optimal participation of stakeholders in energy. Finally, Liu et al. (2014) harnessed the use of crowdsourcing for proposing a novel efficient network management framework for participatory crowdsourcing. Lakhdari et al. (2018) describe the use of an algorithm for composing energy services to satisfy a user's energy requirement. The study conducted preliminary experiments for demonstrating the performance effectiveness and approach for crowdsourcing for energy. However, the approach for transportation of energy products using crowdsourcing applications has not been explored. In our study we extend the discussion to the use of crowdsourcing applications for transportation of energy products in Kenya and East Africa region. The crowdsourcing application tackles the transportation of energy products (liquids, gases) with specialized handling, advanced security protocols, and stringent operating procedures to manage volatility and safety concerns. This has proved critical toward advancement of crowdsourcing to meet the unique needs and requirements in developing countries.

5.5 Implications

5.5.1 Practical Contribution

In practice, this chapter provides practical insights for use of crowdsourcing applications for provision of specialized services: agriculture, waste management, manufacturing, construction materials, and energy transportation. The practical contribution lies in presenting a comprehensive green crowdsourcing application operating with the principles of flexible crowdsourcing for the different specialized services in a developing country. Moreover, it offers practical insights for streamlining crowdsourcing concepts for effective coordination of specialized services.

First, the mobile application provides a practical mechanism for improved coordination and management of the environment in Africa. Different stakeholders can use the Amitruck application for environmental management and improvement. The green crowdsourcing application is critical for environmental management using route optimization and technology for reduction of carbon emissions. Our study emphasizes the use of route optimization techniques within Amitruck to reduce transportation costs, improve efficiency, and minimize environmental impact through lower fuel consumption. The application's use of analytics and technology is crucial for real-time monitoring and efficient resource allocation.

Secondly, the practical implication is that the chapter provides guidance for practitioners seeking to extend the operation of crowdsourcing applications across different sectors of the economy. Specifically, the Amitruck crowdsourcing application has been used in various sectors by utilizing specialized vehicles, handling and transportation methods, and advanced techniques. This is critical for developers of crowdsourcing applications who have continued to experience challenges toward the use of mobile applications in the different sectors. Amitruck's unique contribution lies in its focus on specialized services, which traditional structured crowdsourcing platforms often lack. The app addresses specific needs within various sectors by utilizing specialized vehicles, handling and transportation methods, and advanced techniques. The study highlights Amitruck's positive impact across various sectors (agriculture, waste management, manufacturing, construction, and energy), demonstrating its ability to optimize resource allocation and promote environmental sustainability by reducing carbon emissions compared to traditional platforms.

Finally, the application provides innovative mechanisms for collaboration and partnerships between the different actors in the provision of crowdsourcing. The new innovation provides simple techniques for collaboration and partnerships with the different stakeholders in the industry. The successes registered by Amitruck application have been predominantly harnessing collaboration and partnerships with the different stakeholders in the industry. Our study highlights the value of collaboration, particularly with governmental agencies (like the National Environmental Management Agency), to ensure adherence to safety, environmental, and regulatory standards.

5.5.2 Theoretical Contribution

The study provides insight into the contribution of Amitruck application for a successful model for green crowdsourcing in specialized sectors within a developing country. Its innovative approach to transportation and logistics, particularly its focus on route optimization and collaboration, contributes to both economic and environmental sustainability. This is critical toward improvement of the economies, people, and environment. In addition, this study highlights Amitruck's positive impact in agriculture, waste management, manufacturing, construction, and energy, demonstrating its ability to optimize resource allocation and promote environmental sustainability by reducing carbon emissions compared to traditional platforms. Finally, Amitruck application harnesses collaboration and partnerships with the different stakeholders in the industry.

References

Amitruck (2025). *Your reliable logistics partner in Kenya & East Africa*. Amitruck Company.

Ananta, M. T., Rohidah, S., Brata, K. C., & Abidin, Z. (2023, October 24–25). Mobile crowdsourcing app design: Managing waste through waste bank in rural area of Indonesia. *Proceedings of the 8th International Conference on Sustainable Information Engineering and Technology* (pp. 664–672). Badung, Bali, Indonesia.

Babbie, E.R. (2020). *The practice of social research*. Cengage Au.

Barakat, H. (2022, February 2). *Amitruck, Kenya-based startup, raises $4 million, expands into Uganda, and Tanzania*. Tech With Africa.

Barratt, M., Choi, T. Y., & Li, M. (2011). Qualitative case studies in operations management: Trends, research outcomes, and future research implications. *Journal of operations management*, 29(4), 329–342.

Bassi, H., Lee, C.J., Misener, L., & Johnson, A.M. (2020). Exploring the characteristics of crowdsourcing: An online observational study. *Journal of Information Science, 46*(3), 291–312.

Bazilian, M., Rice, A., Rotich, J., Howells, M., DeCarolis, J., Macmillan, S., Brooks, C., Bauer, F., & Liebreich, M. (2012). Open source software and crowdsourcing for energy analysis. *Energy Policy, 49*, 149–153.

Beck, S., Bergenholtz, C., Bogers, M., Brasseur, T.-M., Conradsen, M.L., Di Marco, D., Distel, A.P., Dobusch, L., Dörler, D., Effert, A., Fecher, B., Filiou, D., Frederiksen, L., Gillier, T., Grimpe, C., Gruber, M., Haeussler, C., Heigl, F., Hoisl, K., ... Xu, S.M. (2022). The open innovation in science research field: A collaborative conceptualisation approach. *Industry and Innovation, 29*(2), 136–185.

Benbasat, I., Goldstein, D.K., & Mead, M. (1987). The case research strategy in studies of information systems. *MIS Quarterly, 11*, 369–386.

Bey, A., Sánchez-Paus Díaz, A., Maniatis, D., Marchi, G., Mollicone, D., Ricci, S., Bastin, J.-F., Moore, R., Federici, S., Rezende, M., Patriarca, C., Turia, R., Gamoga, G., Abe, H., Kaidong, E., & Miceli, G. (2016). Collect earth: Land use and land cover assessment through augmented visual interpretation. *Remote Sensing, 8*(10), 807.

Bollacker, K., Evans, C., Paritosh, P., Sturge, T., & Taylor, J. (2008, June 10). Freebase: A collaboratively created graph database for structuring human knowledge. *Proceedings of the 2008 ACM SIGMOD International Conference on Management of Data* (pp. 1247–1250). Vancouver, BC, Canada.

Bunde, A.O. (2021). The digital platform economy revolution and entrepreneurship ecosystem in Kenya. *SSRN*. https://doi.org/10.2139/ssrn.4954030

Cardonha, C., Gallo, D., Avegliano, P., Herrmann, R., Koch, F., & Borger, S. (2013, May 13). A crowdsourcing platform for the construction of accessibility maps. *Proceedings of the 10th International Cross-Disciplinary Conference,* New York, NY, USA. ANo.: 26, Pages 1–4 https://doi.org/10.1145/2461121.2461129

Castillo, V. E., Bell, J. E., Rose, W. J., & Rodrigues, A. M. (2018). Crowdsourcing last mile delivery: Strategic implications and future research directions. *Journal of Business Logistics*, *39*(1), 7–25.

Certoma, C., Corsini, F., & Rizzi, F. (2015). Crowdsourcing urban sustainability. Data, people and technologies in participatory governance. *Futures*, *74*, 93–106.

Chandra, S., & Kumar, V. (2020). Crowdsourcing as a social interaction tool to stimulate sustainable transportation mode use. *The Open Transportation Journal*, *14*(1), 109–119.

Charmaz, K. (2014). *Constructing grounded theory*. SAGE Publications Ltd.

Chaves, R., Schneider, D., Correia, A., Motta, C.L., & Borges, M.R. (2019). Crowdsourcing as a tool for urban emergency management: Lessons from the literature and typology. *Sensors*, *19*(23), 5235.

Cheng, Y., Tao, F., Xu, L., & Zhao, D. (2018). Advanced manufacturing systems: Supply–demand matching of manufacturing resource based on complex networks and internet of things. *Enterprise Information Systems*, *12*(7), 780–797.

Corsini, F., & Frey, M. (2024). *Crowdfunding for environmental sustainability and the circular economy: Empowered strategies for sustainable growth*. Springer Nature.

Creswell, J.W. (1998). *Qualitative inquiry and research design: Choosing among five traditions*. Thousand Oaks, CA: Sage London.

Creswell, J.W., & Creswell, J.D. (2017). *Research design: Qualitative, quantitative, and mixed methods approaches*. Sage publications.

Dickinson, J. L., Crain, R. L., Reeve, H. K., & Schuldt, J. P. (2013). Can evolutionary design of social networks make it easier to be "green"? *Trends in Ecology & Evolution*, *28*(9), 561–569.

Dimitriadis, I., Psomiadis, V. G., & Vakali, A. (2019, October 14–17). A crowdsourcing approach to advance collective awareness and social good practices. *IEEE/WIC/ACM International Conference on Web Intelligence – Companion* (pp. 200–207). Thessaloniki, Greece.

Diop, E.B., Chenal, J., Tekouabou, S.C.K., & Azmi, R. (2022). Crowdsourcing public engagement for urban planning in the global south: Methods, challenges and suggestions for future research. *Sustainability*, *14*(18), 11461.

Doan, A., Ramakrishnan, R., & Halevy, A.Y. (2011). Crowdsourcing systems on the World-Wide Web. *Communications of the ACM*, *54*(4), 86–96.

Dupljanin, D., Mirkovic, M., Dumnic, S., Culibrk, D., Milisavljevic, S., & Sarac, D. (2019). Urban crowdsourced last mile delivery: Mode of transport effects on fleet performance. *International Journal of Simulation Modelling*, *18*(3), 441–452.

Eisenhardt, K.M. (1989). Building theories from case study research. *The Academy of Management Review*, *14*(4), 532.

Eke, D.O., Wakunuma, K., & Akintoye, S. (Eds.) (2023). *Responsible AI in Africa: Challenges and opportunities*. Springer International Publishing.

Elhamshary, M., Alzantot, M., & Youssef, M. (2019). JustWalk: A crowdsourcing approach for the automatic construction of indoor floorplans. *IEEE Transactions on Mobile Computing*, *18*(10), 2358–2371.

Evans, R.D., Gao, J.X., Mahdikhah, S., Messaadia, M., & Baudry, D. (2015). A review of crowdsourcing literature related to the manufacturing industry. *Journal of Advanced Management Science, 4*(3), 224–321.

Flostrand, A., Eriksson, T., & Brown, T.E. (2019). Better together—Harnessing motivations for energy utility crowdsourcing activities. *Energy Research & Social Science, 48*, 57–65.

Franzoni, C., Poetz, M., & Sauermann, H. (2022) Crowds, citizens, and science: A multi-dimensional framework and agenda for future research. *Industry and Innovation, 29*, 251–284.

Fritz, S., McCallum, I., Schill, C., Perger, C., See, L., Schepaschenko, D., van der Velde, M., Kraxner, F., & Obersteiner, M. (2012). Geo-Wiki: An online platform for improving global land cover. *Environmental Modelling & Software, 31*, 110–123.

Gao, R., Wang, L., Teti, R., Dornfeld, D., Kumara, S., Mori, M., & Helu, M. (2015). Cloud-enabled prognosis for manufacturing. *CIRP Annals, 64*(2), 749–772.

Gong, X., Jiao, R., Jariwala, A., & Morkos, B. (2021a). Crowdsourced manufacturing cyber platform and intelligent cognitive assistants for delivery of manufacturing as a service: Fundamental issues and outlook. *The International Journal of Advanced Manufacturing Technology, 117*(5–6), 1997–2007.

Gong, X., Jiao, R., Jariwala, A., & Morkos, B. (2021b, December 13–16). Crowdsourced Manufacturing for delivery of manufacturing as a service. *2021 IEEE International Conference on Industrial Engineering and Engineering Management (IEEM)* (pp. 1617–1621). Singapore.

Goudin, P. (2016). *The cost of non-Europe in the sharing economy: Economic, social and legal challenges and opportunities.* EU publications.

Guba, E.G., & Lincoln, Y.S. (1994). Competing paradigms in qualitative research. *Handbook of Qualitative Research, 2*(163–194), 105.

Hamari, J., Huotari, K., & Tolvanen, J. (2015). *Gamification and economics.* The MIT Press.

Han, K., & Golparvar-Fard, M. (2017). Crowdsourcing BIM-guided collection of construction material library from site photologs. *Visualization in Engineering, 5*(1), 14.

Johnson, N.E., & Grey, F. (2016). Landfill hunter: Learning about waste through public participation. *Human Computation, 3*(1), 243–252.

Kafle, N., Zou, B., & Lin, J. (2017). Design and modeling of a crowdsource-enabled system for urban parcel relay and delivery. *Transportation Research Part B: Methodological, 99*, 62–82.

Kamilaris, A., Kartakoullis, A., & Prenafeta-Boldú, F. X. (2017, December 1). A review on the practice of big data analysis in agriculture. *Computers and Electronics in Agriculture, 143*, 23–37.

Karampinas, D., & Triantafillou, P. (2012). Crowdsourcing taxonomies. In Simperl, E., Cimiano, P., Polleres, A., Corcho, O., & Presutti, V. (Eds.), *The semantic web: Research and applications* (Vol. 7295, pp. 545–559). Berlin Heidelberg: Springer.

Kondreddi, S. K., Triantafillou, P., & Weikum, G. (2014, March 31–April 04). Combining information extraction and human computing for crowdsourced knowledge acquisition. *2014 IEEE 30th International Conference on Data Engineering* (pp. 988–999). Chicago, IL, USA.

Kotsiuba, I., Herasymchuk, O., Shamrai, V., Lukianova, V., Anpilova, Y., Rybak, O., & Lefter, I. (2023). A strategic analysis of the prerequisites for the implementation of waste management at the regional level. *Ecological Engineering & Environmental Technology*, *24*(1), 55–66.

Krishnan, K., Sahoo, K.C., Kalyanasundaram, M., Singh, S., Srinivas, A., Pathak, A., Stålsby Lundborg, C., Atkins, S., Rousta, K., & Diwan, V. (2023). Feasibility assessment of crowdsourcing slogans for promoting household waste segregation in India: A cross-sectional study. *Frontiers in Public Health, 11*, 1118331.

Kumar, H., Singh, M. K., & Gupta, M. P. (2018, April 4). Smart mobility: Crowdsourcing solutions for smart transport system in smart cities context. *Proceedings of the 11th International Conference on Theory and Practice of Electronic Governance* (pp. 482–488). Galway, Ireland.

Lakhdari, A., Bouguettaya, A., & Neiat, A.G. (2018). Crowdsourcing energy as a service. In Pahl, C., Vukovic, M., Yin, J., & Yu, Q. (Eds.), *Service-oriented computing* (Vol. 11236, pp. 342–351). Springer International Publishing.

Lane, N. D., Chon, Y., Zhou, L., Zhang, Y., Li, F., Kim, D., Ding, G., Zhao, F., & Cha, H. (2013, November 11). Piggyback CrowdSensing (PCS): Energy efficient crowdsourcing of mobile sensor data by exploiting smartphone app opportunities. *Proceedings of the 11th ACM Conference on Embedded Networked Sensor Systems* (pp. 1–14), Roma, Italy.

Liu, C.H., Fan, J., Hui, P., Wu, J., & Leung, K.K. (2014). Toward QoI and energy efficiency in participatory crowdsourcing. *IEEE Transactions on Vehicular Technology*, *64*(10), 4684–4700.

Liu, K., & Golparvar-Fard, M. (2015). Crowdsourcing construction activity analysis from jobsite video streams. *Journal of Construction Engineering and Management*, *141*(11), 04015035.

Liu, Y., & Dang, D.P. (2014). Research on the construction of crowdsourcing platform. *Applied Mechanics and Materials*, *602*, 3198–3201.

Lokers, R., Knapen, R., Janssen, S., van Randen, Y., & Jansen, J. (2016). Analysis of big data technologies for use in agro-environmental science. *Environmental Modelling & Software*, *84*, 494–504.

Mahmood, A.K., & Ferneley, E. (2006). Emboddied agents in e-learning environments: An exploratory case study. *Journal of Interactive Learning Research*, *17*(2), 143–162.

Mavakala, B., Mulaji, C., Mpiana, P., Elongo, V., Otamonga, J.-P., Biey, E., Wildi, W., Pote-Wembonyama, J., & Giuliani, G. (2017, October 2–6). Citizen sensing of solid waste disposals: Crowdsourcing as tool supporting waste management in a

developing country. *16th International Waste Management and Landfill Symposium,* Cagliari, Italy.
Minet, J., Curnel, Y., Gobin, A., Goffart, J. P., Mélard, F., Tychon, B., Wellens, J., & Defourny, P. (2017). Crowdsourcing for agricultural applications: A review of uses and opportunities for a farmsourcing approach. *Computers and Electronics in Agriculture, 142*, 126–138.
Mohri, S.S., Ghaderi, H., Nassir, N., & Thompson, R.G. (2023). Crowdshipping for sustainable urban logistics: A systematic review of the literature. *Transportation Research Part E: Logistics and Transportation Review, 178*, 103289.
Muller, C. L., Chapman, L., Johnston, S., Kidd, C., Illingworth, S., Foody, G., Overeem, A., & Leigh, R. R. (2015). Crowdsourcing for climate and atmospheric sciences: Current status and future potential. *International Journal of Climatology, 35*(11), 3185–3203.
Mulligan, G., & Jackson, T. (2022). *Kenya Startup Ecosystem Report* (pp. 1–49). Nairobi, Kenya: Disrupt Africa.
Niu, N., Lopez, A.Y., & Cheng, J.-R.C. (2011). Using soft systems methodology to improve requirements practices: An exploratory case study. *IET Software, 5*(6), 487.
Niu, X., & Qin, S. (2021). Integrating crowd-/service-sourcing into digital twin for advanced manufacturing service innovation. *Advanced Engineering Informatics, 50*, 101422.
Niu, X.-J., Qin, S.-F., Vines, J., Wong, R., & Lu, H. (2019). Key crowdsourcing technologies for product design and development. *International Journal of Automation and Computing, 16*(1), 1–15.
Olariu, S. (2021). Vehicular crowdsourcing for congestion support in smart cities. *Smart Cities, 4*(2), 662–685.
Oteyo, I. N., Marra, M., Kimani, S., Meuter, W. D., & Boix, E. G. (2021). A survey on mobile applications for smart agriculture: Making use of mobile software in modern farming. *SN Computer Science, 2*(4), 293.
Pilloni, V. (2018). How data will transform industrial processes: Crowdsensing, crowdsourcing and big data as pillars of industry 4.0. *Future Internet, 10*(3), 1–11.
Ponelis, S.R. (2015). Using interpretive qualitative case studies for exploratory research in doctoral studies: A case of information systems research in small and medium enterprises. *International Journal of Doctoral Studies, 10*, 535.
Posadas, B. B., Hanumappa, M., Niewolny, K., & Gilbert, J. E. (2021). Design and evaluation of a crowdsourcing precision agriculture mobile application for Lambsquarters, mission LQ. *Agronomy, 11*(10), 1951.
Pritoni, M., Meier, A.K., Aragon, C., Perry, D., & Peffer, T. (2015). Energy efficiency and the misuse of programmable thermostats: The effectiveness of crowdsourcing for understanding household behavior. *Energy Research & Social Science, 8*, 190–197.
Prpić, J., Taeihagh, A., & Melton, J. (2015). The fundamentals of policy crowdsourcing: The fundamentals of policy crowdsourcing. *Policy & Internet, 7*(3), 340–361.

Rossiter, D. G., Liu, J., Carlisle, S., & Zhu, A. X. (2015). Can citizen science assist digital soil mapping?*Geoderma*, *259*, 71–80.

Schwandt, T.A. (1994). Constructivist, interpretivist approaches to human inquiry. In Denzin, N.K. & Lincoln, Y.S. (Eds.), *Handbook of qualitative research* (pp. 118–137). Sage Publications, Inc.

Shakya, A., Tripathi, G., & Ningombam, D. D. (2023). Crowdsourcing applications in agriculture. In S. Chatterjee, T. P. Singh, S. Lim, & A. Mukhopadhyay (Eds.), *Social-media and crowdsourcing* (pp. 49–70). Auerbach Publications.

Shang, X., Shen, Z., Xiong, G., Wang, F.-Y., Liu, S., Nyberg, T.R., Wu, H., & Guo, C. (2019). Moving from mass customization to social manufacturing: A footwear industry case study. *International Journal of Computer Integrated Manufacturing*, *32*(2), 194–205.

Singh, K. (2007). *Quantitative social research methods.* Sage.

Singh, S., Mehta, K. S., Bhattacharya, N., Prasad, J., Lakshmi, S. K., Subramaniam, K. V., & Sitaram, D. (2017, October 19). Identifying uncollected garbage in urban areas using crowdsourcing and machine learning. *2017 IEEE Region 10 Symposium (TENSYMP)* (pp. 1–5). Cochin, India.

Steinke, J., van Etten, J., Müller, A., Ortiz-Crespo, B., van de Gevel, J., Silvestri, S., & Priebe, J. (2021). Tapping the full potential of the digital revolution for agricultural extension: An emerging innovation agenda. *International Journal of Agricultural Sustainability*, *19*(5–6), 549–565.

Suchanek, F. M., Kasneci, G., & Weikum, G. (2007, May 8). Yago: A core of semantic knowledge. *Proceedings of the 16th International Conference on World Wide Web* (pp. 697–706). Alberta, Canada.

Sykuta, M. E. (2016). Big data in agriculture: Property rights, privacy and competition in ag data services. *International Food and Agribusiness Management Review*, *19*, 57–74.

Testa, S., Troise, C., Cincotti, S., & Camilleri, M.A. (2023). Exploring the role of e-waste management solutions and message framing in influencing consumer behaviours: The crowdfunding context. *Business Strategy and the Environment*, 26. https://doi.org/10.1002/bse.3526

Testa, S., Troise, C., Cincotti, S., & Camilleri, M.A. (2024). The role of electronic waste management solutions and message framing in influencing consumer behaviours: Exploring the crowdfunding context. *Business Strategy and the Environment*, *33*(2), 917–929.

Tuzhikov, A. (2024). Innovative technologies in waste management for improving urban environment comfort. *BIO Web of Conferences*, *130*, 03010.

Vianna, F.R.P.M., Graeml, A.R., & Peinado, J. (2020). The role of crowdsourcing in industry 4.0: A systematic literature review. *International Journal of Computer Integrated Manufacturing*, *33*(4), 411–427.

Wagy, M.D., Bongard, J.C., Bagrow, J.P., & Hines, P.D. (2017). Crowdsourcing predictors of residential electric energy usage. *IEEE Systems Journal, 12*(4), 3151–3160.

Walsham, G. (1995). Interpretive case studies in IS research: Nature and method. *European Journal of Information Systems, 4*(2), 74–81.

Wang, S., Taha, A.F., Wang, J., Kvaternik, K., & Hahn, A. (2019). Energy crowdsourcing and peer-to-peer energy trading in blockchain-enabled smart grids. *IEEE Transactions on Systems, Man, and Cybernetics: Systems, 49*(8), 1612–1623.

Wolfert, S., Ge, L., Verdouw, C., & Bogaardt, M. J. (2017). Big data in smart farming—A review. *Agricultural Systems, 153*, 69–80.

Wu, W., Li, H., Wang, H., & Zhu, K. Q. (2012, May 20). Probase: A probabilistic taxonomy for text understanding. *Proceedings of the 2012 ACM SIGMOD International Conference on Management of Data* (pp. 481–492). New York, USA.

Wu, X., Wang, J.Y., Lin, Z., & Hu, D. (2024). Crowdshipping with integrated bidding, pricing, matching and routing optimisation. *SSRN*, 48. https://doi.org/10.2139/ssrn.5059661

Wu, Z., Li, Q., Wu, W., & Zhao, M. (2019). Crowdsourcing model for energy efficiency retrofit and mixed-integer equilibrium analysis. *IEEE Transactions on Industrial Informatics, 16*(7), 4512–4524.

Xie, X., Han, Y., Anderson, A., & Ribeiro-Navarrete, S. (2022). Digital platforms and SMEs' business model innovation: Exploring the mediating mechanisms of capability reconfiguration. *International Journal of Information Management, 65*, 102513.

Yao, X., & Lin, Y. (2016). Emerging manufacturing paradigm shifts for the incoming industrial revolution. *The International Journal of Advanced Manufacturing Technology, 85*(5–8), 1665–1676.

Yin, R.K. (2003). Designing case studies. *Qualitative Research Methods, 5*(14), 359–386.

Yin, R.K. (2009). *Case study research: Design and methods*. Sage.

Yin, R.K. (2018). *Case study research and applications*. Thousand Oaks, CA: Sage.

Yong, E., & Festerling, T. (2023). Success factors for startups in Africa. In Schmidt, T., Pfaffenberger, K., & Liebing, S. (Eds.), *Practical handbook business in Africa* (pp. 179–189). Springer Fachmedien.

6

A Multilevel Analysis of Carbon Emission Tracking Technology Adoption in South Africa

John Michael Maxel Okoche[1] and Marcia Mkansi[2]

[1] *Department of Operations Management, College of Economics and Management Sciences, University of South Africa, Unisa, South Africa*
[2] *Department of Operations Management, University of South Africa, Pretoria, South Africa*

6.1 Introduction

Carbon emission (CE) contributes to global warming, and controlling emission has become a challenge since the last few decades (Lee et al., 2013). Holappa (2020) examined diverse measures for reducing CE from the iron and steel industry, enhancing energy efficiency, alternative clean energy fuels, utilization of scrap, and collection and deposition of CO_2. In another study by Zhang et al. (2012), the CO_2 mitigation applications with relative impacts were discussed. Later, Qin et al. (2025) in their study developed a model to facilitate the understanding of differences in energy systems and trends during CEs. Hai et al. (2023) approached the concept of reducing CEs by developing an algebraic framework for industrial site planning. Further, Cao et al. (2019) assessed the impacts of the production and consumption of industrial products on CEs. However, relatively little attention has been paid to the impact that individual-organizational behavioral innovation (micro–macro dynamics) has on CEs, specifically due to their choice of transport mode.

In addition to industry efforts, governments across the globe introduced a tax levied on the CEs for producing goods or services, known as carbon tax. However, previous literature on the impact of carbon tax on carbon emissions shows a very little impact on carbon reductions. For example, Calderón et al. (2016) undertook the assessment of the achievement of CO_2 reductions and established that despite

Green Innovations in Supply Chain Management: Case Studies and Applications, First Edition. Edited by Marcia Mkansi, Godfrey Mugurusi, Aaron L. Nsakanda, Temidayo Akenroye, and Frank Tietze.

Published 2026 by John Wiley & Sons, Inc.

introducing the carbon tax, CE is still on the increase until 2050. Feijoo et al. (2019) established that cumulative CEs instantaneously increase global temperatures, therefore advocating the adoption of technologies for reducing CEs. Floros and Vlachou (2005) used a theoretical model for analyzing the impact of carbon credit on carbon emissions in Greece and established that the policy was effective in mitigating global warming. Mardones and Flores (2018) assessed the effectiveness of CO_2 tax on industrial emissions in Chile. The study concluded that taxes that are too low or too high are effective in raising revenue but not in reducing emissions. The studies mainly examined the policy perspective of CEs, which are at macro-level, both nationally and internationally.

In this chapter, our focus is on incentive-based wearable technologies that track individual CEs and influence both individual and organizational behaviors to move toward a decarbonized society.

Previous studies predominantly focus on the macro-level analysis of CE reduction strategies: market power and oligopoly behavior (El Khatib & Galiana, 2019; Tanaka & Chen, 2012), risks and uncertainties (Ji et al., 2015; Li et al., 2011), implementing block carbon tax (An et al., 2022), long-term capacity investment generation (Chen & Tseng, 2011; Rocha et al., 2015), impact of carbon taxes on non-carbon emissions (Mortha et al., 2021), country level analysis: EU, the United States, and China (Feijoo et al., 2020; Lise et al., 2010; Liu et al., 2017), efficiency of carbon market transactions (Zhu et al., 2020), and multinational integration of the carbon credit (Anjos et al., 2022). This chapter highlights the absence of sufficient research examining individual and corporate-level behaviors and their impact on overall emission reduction efforts. It aims to fill this gap by exploring the micro-level dynamics of adopting the CE tracking technology. This is critical in strengthening carbon markets by bringing in the new appropriate measures for both micro and macro contributors of CE. Anjos et al. (2022) studied the integration of multinational carbon credit but this is still at a higher-level national level but the individual employee actions and corporate initiative are critical in determining the macro- and meta-level integration. Worldwide, companies are increasingly making claims about their current climate efforts and their future mitigation commitments (Trouwloon et al., 2023). According to Lokuge and Anders (2022), carbon credit claims are underpinned by carbon credits issued in voluntary carbon markets to offset emissions.

The main contributions of this study are twofold. First, it provides a detailed micro-level analysis of technology absorption and adoption. Second, it incorporates individual and organizational factors and demonstrates how these factors interact within the broader policy and regulatory contexts to achieve effective CE reduction. The paper itself is organized as follows. The literature is presented in Section 6.2. Then, in Section 6.3, we present the methods used. The results are then covered in Section 6.4, after which the discussion and implications are presented in Sections 6.5 and 6.6, respectively.

6.2 Literature Review

In recent years, the use of applications for environmental mitigation has generated debate on the implications of Information Communication and Technology (ICT) applications on CEs. The first cohort of scholars argues that ICT application reduces CEs: ICT lowers carbon intensity (Sun & Kim, 2021), use of ICT has symmetric effects on CO_2 emissions (Usman et al., 2021), digitalization empowers green development by adjusting energy structure (Wang et al., 2021), autonomous robots reduce emissions (Figliozzi & Jennings, 2020), and use of AI reduces carbon intensity (Liu et al., 2022). Second, ICT applications increase CEs: internet penetration increases emissions (Chen et al., 2019), and digitization does not decouple economic growth from energy consumption (Lange et al., 2020). This study validates the context of the application benefits in the contribution toward the reduction of CEs. The use of ICT, on the other hand, increases energy demand through the growth of ICT, while decreasing energy demand through improved energy efficiency (Lange et al., 2020). Finally, ICT applications play a dual role in increasing and decreasing CEs through increased energy efficiency and economic activity (Lange et al., 2020), and limited impact (Moyer & Hughes, 2012). These studies are generic, describing the aspects of how the carbon applications impact CEs. In addition, the studies have an insufficient understanding of individual and corporate absorption dynamics. There is limited understanding regarding the specific factors influencing the absorption of CE tracking technologies at the individual and corporate levels. The study seeks to identify, analyze, and discuss these crucial factors. In this chapter, we explore, identify, analyze, and discuss the crucial factors. While research such as Sun and Kim (2021) found a positive impact of ICT on lowering carbon intensity under the consideration of spatial dependence, particularly in developing regions, there are gaps on the specific applications, which is directly determined by individual behavior and corporate clients. In addition, Usman et al. (2021) examined the impact of ICT on the CO_2 emissions in the nine selected economies, tested the symmetric impact of ICT on CO_2 emissions, and established that symmetric impacts were strengthened and observed in more than a half of the countries.

In this study, our goal and purpose are to extend this discussion by first identifying the features of CE application and highlighting the benefits of the application for both individuals and corporate clients. Anjos et al. (2022) despite recognizing the benefits of the carbon credit concept, laments about the imperfections and non-reliability of the information used for calculating carbon credit. This gap demands the development of accurate and reliable applications, which are acceptable for the different stakeholders in carbon credit emissions. Song and Wu (2023) emphasized the need for the development of an appropriate link between companies for effective carbon credit reduction. The goal is to highlight the critical features of the CE application and benefits for individual and corporate stakeholders and to advance new ideas for building a more robust mechanism for carbon trading for the micro-, macro-, and meta-actors in carbon credit trading. This is critical for strengthening carbon credit trading as it provides an appropriate, effective, and reliable mechanism for bearing responsibilities between individuals and corporates for the achievement of both economic and environmental gains. TAM is used as a lens to study micro–macro dynamics adoption and how such innovation would impact carbon emission reduction.

6.2.1 Technology Acceptance Model

In this chapter, we adopted TAM to advance explanations for benefits and challenges toward the acceptance and use of the IoT application for CEs in South Africa. While the adoption, use, and application of IoT are beneficial, the dynamics of use can be technically challenging as they are also influenced by several interacting factors. The TAM may account for behavioral patterns exhibited by various actors during the use the new technology application. According to Granić and Marangunić (2019), it has been widely established that TAM is among the most well-known models for predicting technology adoption and usage behavior. In particular, TAM has witnessed an increased level of use by scholars for the purpose of predicting learners' acceptance of the e-learning platforms by government school teachers (Rafique et al., 2024), adoption of AI in e-commerce (Wang et al., 2023), acceptance of Metaverse (Wu & Yu, 2024), acceptance of health information systems (Tetik et al., 2024), acceptance of halal tourism (Berakon et al., 2023), and students' actual use of AI system metaverse-based learning platforms. In addition, other scholars have used it for establishing the determinants of college students' actual use of AI systems (Li, 2023) and factors for the acceptance of e-wallets among the genZ from the lens of extended technology (Rosli et al., 2023). This is crucial for understanding the micro and macro benefits of the IoT technology to strengthen or weaken the use of the IoT-based CE tracker application. Chau (1996) highlights that TAM is critical for assessing perceived usefulness and perceived ease. Al-Adwan et al. (2023) developed a TAM model to identify the cognitive and psychological factors that influence users' acceptance of new technologies. In addition, Chau (1996) suggests that perceived usefulness can be of two distinct types: near-term usefulness and long-term usefulness. According to Zhang et al. (2022), TAM offers broad insights into users' willingness to adopt the technology.

Despite the applicability, TAM is criticized for ignoring intrinsic motivations and negative perceptions. Taherdoost (2018) argues that TAM embraces only extrinsic motivations while ignoring intrinsic motivations. Moreover, Dou et al. (2017) highlight that TAM-related studies have ignored users' negative perceptions in favor of focusing only on their positive perceptions. In our study, we examined the benefits and challenges of adopting the IoT technology. TAM has been critical in determining the behavior intention for technology adoption and actual behavior for use. In addition, Al-Adwan et al. (2023) state that individual behavior intention is "jointly determined" by individuals' attitude toward technology use (A) and their perception of usefulness (PU). In accordance to Lee et al. (2019), we integrated the extra potential factors that drive the adoption of IoT-based CE tracker technology: perceived risk and enjoyment. TAM was used to ascertain consistent prediction, explanation, and improvement of the user acceptance of the CE tracker technology.

6.2.1.1 Perceived Usefulness

PU describes the extent to which individuals perceive that the usage of a specific technology enhances their performance (Davis et al., 1989). An earlier research has revealed that PU positively affects technology adoption intention (Adwan et al., 2018; Akour et al., 2022; Al-Adwan et al., 2023; Al-Rahmi et al., 2022; Martinho et al., 2018).

In this study, PU is described as the extent to which individual employees and corporate organizations perceive that the use of the CE tracker application will be useful for themselves. A study by Al-Rahmi et al. (2022) revealed that the perceived mobile value, academic relevance, and self-management of Mobile Learning (M-learning) were the primary drivers of students' acceptance of M-learning platforms. Akour et al. (2022) identified PU to be an essential predictor of the factor of users' intention to use the new technology. In the study by Chen et al. (2021), the results showed that clothing brand green innovation can effectively promote green purchase intention and behavior as consumers produce purchase intention and behavior through the perception of novelty, usefulness, and greenness, and that highly innovative consumers are more likely to perceive novelty and are more willing to buy.

Specifically, studies in relation to green innovations have highlighted the impact of perceived usefulness for adoption (Adwan et al., 2018; Bhatti & Juhari, 2023; Chen & Lu, 2016; Gao et al., 2022). The findings by Gao et al. (2022) confirmed that farmer green values have a positive association with green innovation intention and adoption behavior. Bhatti and Juhari (2023) suggested that both sustainable attitudes and behaviors exert a significant impact on the adoption of green innovation. Chen and Lu (2016) green perceived usefulness has more powerful positive effects, and the perceived ease of use has a weak positive effect. In their study, Adwan et al. (2018) reveal that effort expectancy, performance expectancy, trust expectancy, self-management of learning, system functionality, and social influence are the significant determinants of mobile learning adoption. Martinho et al. (2018) tested a model for the assessment of perceived usefulness and established that it was essential for indirect and direct technology absorption. Sadriwala and Sadriwala (2022) in their study established that PU of artificial intelligence influenced marketing innovation.

6.2.1.2 Perceived Ease of Use

The perceived ease of use refers to the extent to which an individual perceives that using a given technology is easy and effortless (Davis et al., 1989). Previous studies on technology adoption indicated that users' perceived ease of use substantially impacted their PU (Mailizar et al., 2021).

Akdim et al. (2022) established that the two utilitarian factors, perceived ease of use and usefulness, were responsible for the use of social mobile apps. Tam et al. (2020) found out that the most important drivers of the continuance intention of mobile apps are satisfaction, habit, performance expectancy, and effort expectancy. Wang et al. (2024) established that extraversion and neuroticism positively affect perceived usefulness, conscientiousness negatively influences usefulness, and agreeableness and neuroticism have positive effects on perceived enjoyment. The Ma et al. (2017) study findings validated that the perceived ease of use, PU, attitudes, and purchase intention were associated with consumers' use of sustainability labels. Natalwala and Habibullah (2024) in their study on examining the adaption of the expected use in the banking conclude that performance expectancy has a significant impact on the perceived ease of use and PU. Sadriwala and Sadriwala (2022) established that PU and perceived ease of use of artificial intelligence influenced marketing innovation. Wu et al. (2019) green PU, perceived ease of use, environmental concern, and behavioral intention were critical for the adoption of the green innovation technology. Shahzad et al. (2022) adoption of the green innovation technology to accelerate sustainable development in the manufacturing industry depends on the perceived ease of use. On the one hand, some studies found out that the ease of use does not strongly influence the adoption of green innovation. Polas et al.'s (2023) intention to use green energy technology does not play a mediating role in the association between the perceived ease of use and adoption of green innovation. Chen and Lu (2016) study findings indicate that green perceived ease of use has a weak positive effect.

6.2.1.3 Self-Efficacy

Bandura (1986) defines self-efficacy of individuals' judgments in terms of "their capabilities to organize and execute a course of action required to attain designated types of performances." In addition, Mikusa (2015) defines technology self-efficacy as "personal belief in one's ability to successfully use technology to increase learning outcomes." It is seen as an individuals' belief in their own competence and typically reflects the level of capability that an individual believes he/she possesses (Christensen & Knezek, 2016). According to Bubou and Job (2020), self-efficacy is intricately connected to self-esteem and self-worth, denoting a person's level of confidence in a certain circumstance or event.

Venkatesh and Davis (1996) argue that individuals' self-efficacy improves the ability of the adoption as they find the technology easy to use and beneficial. In addition, Venkatesh et al. (2012) highlight that individuals with a higher level of confidence on the utilization of information technology find it easy to use and beneficial than those with less confidence. Different scholars have diversely examined the concept of self-efficacy using different contexts, domains, and perspectives: computer self-efficacy (Nardi & Ranieri, 2019), internet self-efficacy (Joo et al., 2000), technology self-efficacy (Durak, 2018), and online learning self-efficacy (Hong et al., 2019). Durak (2018) observed that technology self-efficacy is vital factor for the successful use of an online learning system as it facilitates effective access to lecture materials, interaction with instructors and classmates, use of virtual instructional tools, discussion, and problem-solving.

6.2.1.4 Personal Innovativeness

Fan et al. (2020) define personal innovativeness (PI) as the desire by an individual to try out new technology innovation. The notion of personal innovation in technology refers to measure the degree to which an individual has a tendency for experimenting with new technologies (Kogabayev & Maziliauskas, 2017).

The next cohort of studies established that personal innovation in technology is vital for strengthening personal efficacy: efficacy in e-learning in Nigeria (Bubou & Job, 2020), adoption of e-learning technology (Chahal & Rani, 2022), and Chatbot adoption intentions (Trapero et al., 2020). The studies advocate for cultivating personal innovation as a means of strengthening personal efficacy. Bubou and Job (2020) in their study on the impact of personal innovation for the adoption of e-learning readiness of the Yenagoa study center in the National Open University of Nigeria concluded that individual innovativeness improves self-efficacy. Chahal and Rani (2022) in their study established that there were direct and indirect effects of PI, social factors, and self-efficacy in the participation of acceptance of e-learning platform in India. Trapero et al. (2020) established that the quality of interactions between the customers and airline companies using chatbots was dependent on the personal innovation of technology. Furthermore, it is recommended to strengthen personal innovation for the improvement of the new competitive atmosphere in the airline industry, whose operation takes place in a completely globalized environment.

6.2.1.5 Perceived Risks

In adoption of emerging technology-perceived risks, privacy and data security are critical concerns (Lee et al., 2019); applications collect user information such as transactions, physical state, and facial images (Zhao et al., 2022). Academic literature highlights the major challenge hindering the growth and adoption in view to security and privacy (Wang et al., 2022). In addition, Zhao et al.'s (2022) emergent technology provides a mechanism for the collection of private information comes with privacy and security risks. Given that social engineering attacks currently make up the bulk of cyberattacks experienced online (Salahdine & Kaabouch, 2019), it is anticipated that such attacks from emerging applications are more powerful and very common (Di Pietro & Cresci, 2021). Wang et al. (2018), in their study, observed that security risks have policy implications for promoting the adoption of electric vehicles. Di Pietro and Cresci (2021) analyzed the foundations of the adoption of the metaverse technology and noted that security issues come with the new technology and other far-reaching consequences.

Salahdine and Kaabouch (2019) state that despite the advancement of digital technology, social networks and online systems increase vulnerabilities perpetuated by malicious users. Zhang et al. (2022) observed that severe privacy invasions and security breaches inherited from underlying technologies impede its wide deployment. In addition, they discussed the security and privacy threats, presented the critical challenges of systems, and reviewed the state-of-the-art countermeasures. In addition, Zhang et al. (2022) in their study of metaverse technology established the privacy and security risks as one of the challenges in technology adoption. Finally, Zhou et al. (2022) observed that one of the critical aspects for the adoption of technology is perceived system quality: security and risks as predictors for successful adoption of technology.

6.3 Methods

6.3.1 The Case Study

The CO_2 emission tracker is a mobile application that measures, monitors, and records individual greenhouse gas emissions by transport mode. The application can be used by individuals and organizations to understand their carbon footprint and identify areas for improvement. The CO_2 tracker records to automate the accurate computation of CE data collection, providing critical insights for individual and corporate strategies. First, individual CEs are captured, summarized, and analyzed for personalized use. Individuals can review saved CE and design appropriate strategies for the reduction of CEs. On the basis of the analysis, individual commitments and strategies are appropriately developed. Environmentally sensitive corporate organizations can build good social contracts with employees in reducing CEs. The personalized individual CE information is important for individual strategies and approaches for reducing CEs. The aggregate employee CEs can facilitate the development of CEs for organization. The data generated from corporate level is useful for carbon trading as the tracker provides accurate and reliable information CEs. Finally, the organizations can provide incentives to employees on the basis of reduction of carbon print.

Figure 6.1 shows the home screen design centered on the information that was deemed to be the most important at first glance, to give the user a concise

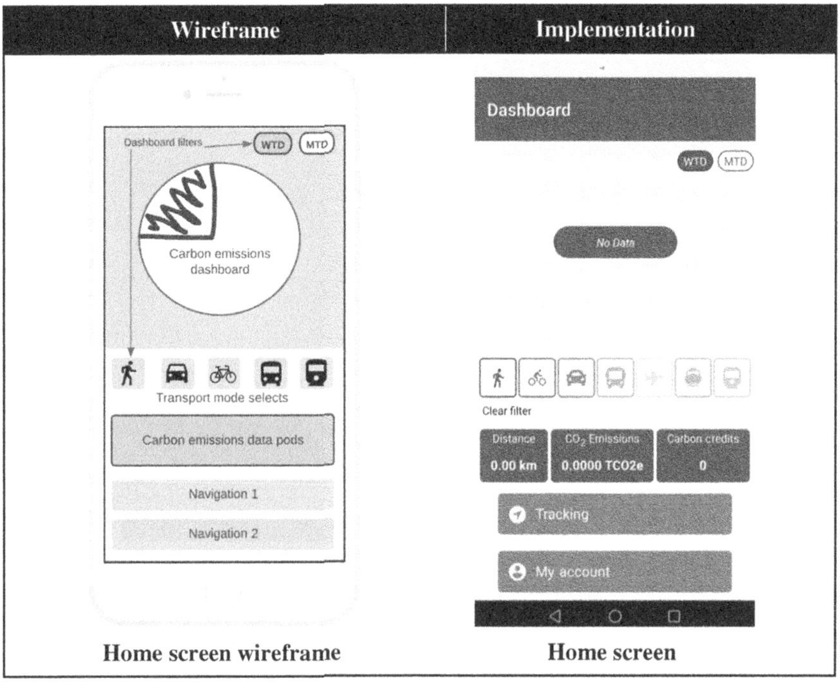

Figure 6.1 Home screen UI design.

overview of the CO_2 tracker, while not being too cluttered. The home screen gives the user a high-level overview of their current CE progress, with a few high-level filters to get the most important information with minimal input. Two navigational buttons were added to direct the user to the two most important features of the prototype. Hence, Figure 6.2 shows tracking the user interface design of the CO_2 tracker.

6.3.2 Research Design

This chapter used an exploratory case study research design with a qualitative approach, using the University of South Africa as single case for the assessment of technology absorption of the CO_2 emission tracker. Yin (2009) argues that a case study research design is critical in capturing the context and lived realities of the participants in relation to mobile applications. Finally, the study design was flexible in the qualitative exploration of the perspectives of the participants in reasons for the absorption of the new technology (Creswell & Creswell, 2017). The case study research design was critical in gaining insights of the absorption of the new green innovation technology as provided us with an opportunity of assessing absorption considerations by individual employees.

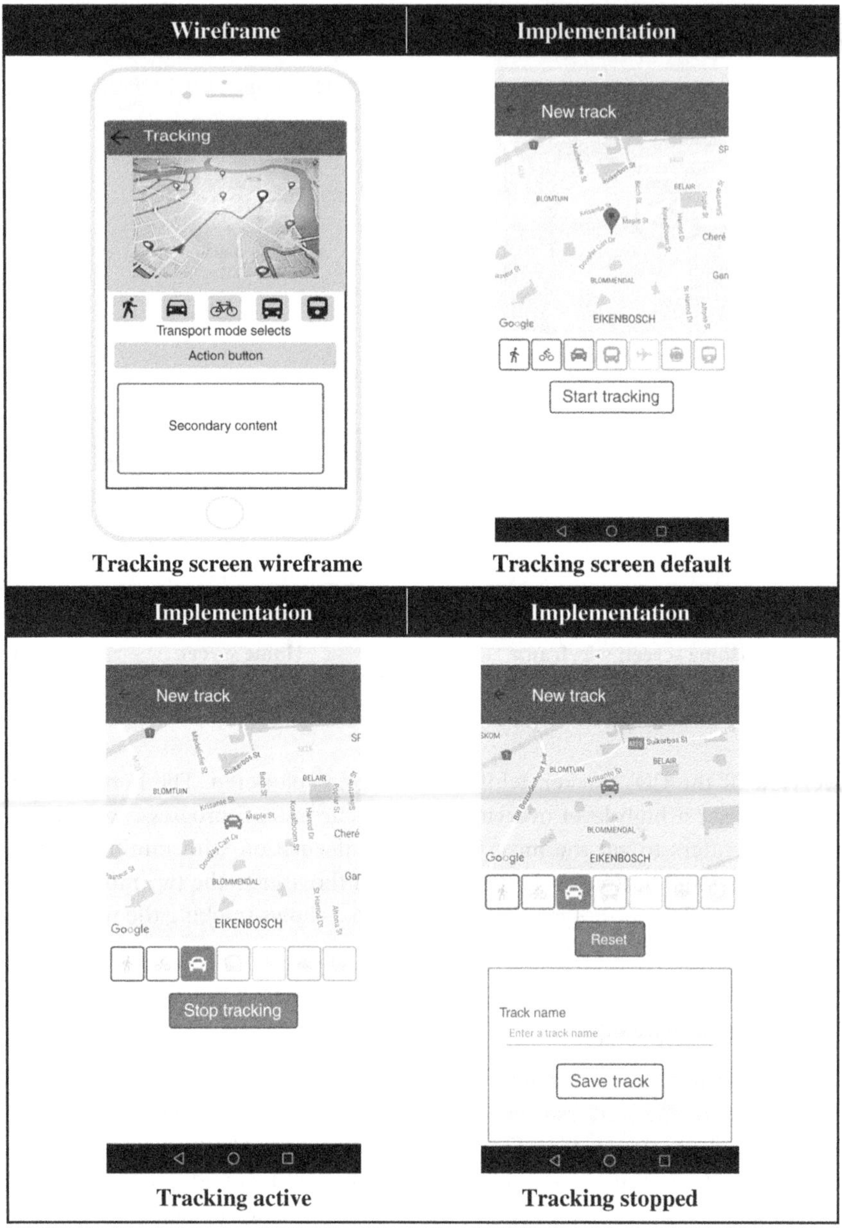

Figure 6.2 Tracking UI design.

The CE tracker technology is designed for tracking individual CEs. The new application provides a framework for monitoring and calculating individual CE contribution on the basis of transportation used. Anjos et al. (2022) studied the integration of multinational carbon credit, but this is still at a higher-level national level, but the individual employee actions have not yet been undertaken. The CE tracker consolidates the global, national, and corporate initiatives. Monitoring individual employee initiatives toward CE reduction is strengthening and consolidating corporate, national, regional, and global carbon trading. Worldwide, companies are increasingly making claims about their current climate efforts and their future mitigation commitments (Trouwloon et al., 2023). Understanding the individual employee absorption factors is critical in strengthening the new green innovation technology. In addition, the individual corporate employee efforts can easily be integrated with corporate and national efforts for the reduction of carbon print. The exploration of individual employee perspectives provided insights on individual perceptions about the new technology with critical factors for the absorption of this innovation.

In this chapter, we bring insights for drivers for the absorption of CE tracker at individual and corporate levels.

6.3.3 Sample and Data Collection

The major purpose of this chapter is understanding absorption factors for CE tracker in Africa. We conducted eight key informant interviews with employees of the University of South Africa with in-depth and rich information for individual technology adoption. The participants have always been involved in new technology adoption, therefore, critical on insights in relation to the absorption of the new technology at individual and corporate levels. Participants' perspectives were critical in assessing the key factors for the absorption of new green innovation CO_2 tracker. Recognizing that interviewees participated in the pilot and utilization of the mobile application in the operations of the corporate organization with employees. This was critical in framing and structuring the drivers for the absorption and utilization of the CE tracker by individuals in South Africa.

We purposively selected interviewees with a critical understanding of the individual employee considerations for the absorption of the new CE tracker. The interviews were recorded and transcribed to facilitate the process of interpretation.

The interview protocol focused on understanding the critical factors for the absorption of the new technology in South Africa. Follow-up questions were used to clarify responses. Creswell and Creswell (2017) in their study used anonymity for harnessing ethical considerations during the study. The views of the respondents were critical in understanding the technological considerations for the absorption of the CO_2 emission tracker by individual employees in corporate organizations in South Africa.

6.3.4 Analysis

The data analysis for this chapter adhered to a circular-spiral methodology, iteratively oscillating between empirical narratives, emergent themes as interpreted by the researcher, and theoretical constructs to synthesize these themes (Schwartz-Shea & Yanow, 2013). In this study, the emergent themes from the interviews of the respondents in relation to CO_2 tracker application absorption were developed. Subsequently, these themes were categorized into major domains, highlighting the primary drivers behind the absorption of CO_2 tracker technology. Moreover, the researcher documented the emerging themes, which included the influence of upbringing and culture on business associations and networks, the significance of close familial relationships in resource acquisition and business development, among others. Initially, themes pertaining to the initial decisions for adoption and effective utilization of CO_2 emission tracker technology. Distinct parallels began to surface regarding how interviewees establish the key critical drivers for the absorption of the new CE tracker application. Furthermore, preliminary analyses appeared to indicate that intrinsic and extrinsic factors determine the absorption of the new technology. TAM mainly outlined the extrinsic factors: PU and perceived ease of use. In addition, existing literature extended critical emphasis on efficacy and personal innovation. However, the critical issues like security and organization considerations were not adequately examined.

The chapter analysis facilitates the formulation of the critical factors for the absorption of the CE tracker technology. Insights on the absorption are critical for improvements in the design, development, and individual use of the CO_2 tracker. The University of South Africa employees were sampled to understanding the perceptions of employees on the absorption conditions for the application. The application was provided for the individuals to undertake the assessment. Interaction with the respondents helped in determination of the six categories of absorption of the technology: PU, perceived ease of use, self-efficacy, PI, privacy and security risks, and organization considerations.

Finally, the concepts for underlying green innovation absorption were categorized into data patterns; the chapter concentrated on the technology absorption dynamics that manifest within the various categories of the absorption of the CE tracker innovation. Ultimately, the data underwent a systematic organization, transitioning from raw data to initial codes, second-order and axial codes (interview excerpts) to initial empirical categories, and subsequently to second-order themes that are theoretically informed and aggregate dimensions. The transcripts were meticulously coded manually to maintain fidelity to the contextual nuances of the interviewed cases, and the interview excerpts were systematically categorized by expanding upon the descriptive matrices established earlier.

6.4 Results

Table 6.1 Data structure: factors of technology absorption of CO_2 emission tracker

First-order categories	Second-order themes	Aggregate dimensions
Influence of perceived usefulness on the absorption of technology for individuals and corporate organizations		
1) Performance improvement	"OK. Thank you for the opportunity of this interaction, this application is critical for us who believe in sustainability of our country and the world. My experience for procurement company applications usually on the usefulness of the application for individuals and corporations. Corporate managers and individuals look at the ability of the new technology to improve performance: effectiveness, efficiency and productivity. The new green innovation has a potential of improving efficiency of carbon credit calculation. I find this a very important factor for acquisition of a new technology. This is very interesting as it provides insight on the daily contribution per person making it easy to understand the contribution of each person to sustainability or destruction of the environment" **Respondent 1**	Perceived usefulness
2) Goal achievement	"Goal achievement is another factor absorption of a new technology. The decisions, that I have participated for selection a new technology innovation always considered achievement of the organisation goal. The new technology is considered useful if it can help in achievement of organisation or individual goal. This must consider alignment with goals, relevant to tasks and support organisation decision making. The new green carbon emission technology supports users in achievement of sustainability goals reduction of carbon emissions. This new technology will facilitate reliable and accurate carbon trading." **Respondent 4**	
3) Compatibility and flexibility	"Yes. I have participated in several technology acquisition decisions but compatibility of the application with existing system is paramount. I have been attended several management meetings where decisions are taken in relation acquisition of new technology. The compatibility and flexibility is one of the major considerations to be taken. The green carbon innovation has captures critical aspect of usefulness of being compatible with the existing systems. The innovation is compatible because it seamlessly integrates with mobile phones which are being used every day. In addition, the application has to be flexible and adaptable in different contexts and situations." **Respondent 4**	

(*Continued*)

Table 6.1 (Continued)

First-order categories	Second-order themes	Aggregate dimensions
4) Support resources	"Support resources help the user to appreciate usefulness of the new technology. The applications which have support and resources usually are useful for the individuals as their help in manoeuvring. My decision to install the Atlas Ti was backed by the availability of support during the use of the new application. The technology vendor provides adequate support and resources. Likewise, the new green innovation has to provide support resources with quality documentation. The technology has clear and comprehensive documentation. In addition, innovator has to offer training and education to help users get the most out of the technology. So, after the acquisition, there must be continued support for individual and corporate users." **Respondent 3**	
5) Social influence	"I have found myself acquiring new technology simply because of influence from my friends and colleagues. There are technologies which are acceptable in particular social circles. I come from Finance Field and QuickBooks is widely adopted and accepted among finance. Users are influenced by recommendations or endorsements from colleagues, friends, or family members. Usefulness is influenced by users. Likewise, absorption of the new carbon emission software will depend on the acceptability by environment society in South Africa." **Respondent 2**	
Influence of perceived ease of use on absorption of technology for individuals and corporate organisations		
1) System quality	"Yeah, I have been in the field of Information Communication and Technology (ICT) for now a period of more than 10 Years. I have observed that technology acceptance depends on the perception of management and users on the ease to use. Regardless of the innovation but the most critical aspect for adoption of the technology is system quality, my experience as IT expert has helped me to realise that the quality of information provided by the new technology motivates potential users. The quality of information provided by the technology, such as its accuracy, relevance, and timeliness, also influences perceived ease of use. If users feel they can trust the information provided, they're more likely to find the technology easy to use. We normally sit down to evaluate the compatibility of the new application with our system. Usually, applications with simple systems are preferred as opposed to the one which are complicated." **Respondent 1**	Perceived ease of use

2) Information quality	"I believe information quality is a very important factor when choosing the new application. Regardless of the innovation but the most critical aspect for adoption of the technology is information quality, my experience as IT expert has helped me to realise that the quality of information provided by the new technology motivates potential users. The quality of information provided by the technology, such as its accuracy, relevance, and timeliness, also influences perceived ease of use. If users feel they can trust the information provided, they're more likely to find the technology easy to use. We normally sit down to evaluate the compatibility of the new application with our system. Usually, applications with simple systems are preferred as opposed to the one which are complicated." **Respondent 1**	
3) User experience	"Yes, I believe that user experience in adopting new technologies as a critical factor for adoption of the new technology. I have always noted that individuals who have ever adopted new applications easily adopt new applications. The user experience motivates users to explore new technology in the same order." **Respondent 2**	
4) Cognitive absorption	"Honestly speaking there are now so many applications left and right. So determining the most appropriate application to be adopted depends on perceived user experience. In case the application is familiar with already existing applications it is usually easy to adopt the application. User experience is very important in adoption of a new technology. The new green innovation application will not be unique. I will normally go for a new application which I already have experience as opposed to completely a new one. The green innovation platform overall user experience, has to a familiar user interface, user support, and training, also plays a role in perceived ease of use. A positive user experience can make it easier for users to adopt and use the technology." **Respondent 4**	
Influence of self-efficacy on absorption of green carbon innovation technology for individuals and corporate organizations		
1) Task-specific self-efficacy	"My experience is relation to acquisition of applications are dependent on self believe in ability to use the new application. First the task specific believe to perform a particular task. This usually comes from individuals who have had similar success in using new technology." **Respondent 5**	Self-efficacy

(Continued)

Table 6.1 (Continued)

First-order categories	Second-order themes	Aggregate dimensions
2) Domain-specific self-efficacy	The next perspective self-believes by individuals to perform tasks in a particular domain like computing. Domain specific self-belief is one's ability to perform tasks specific domains like research. I have always discovered that individuals who have spent time doing tasks like research easily acquire another application in research. The people who have a belief in using technologies in a particular field can easily acquire the technologies" **Respondent 5**	
3) General self-efficacy	I have also observed that people with general belief in their abilities usually easily adopt the new technology. The general belief that they will be able use any technology: handle various tasks and situations. These individuals and organsiations easily acquire new technology." **Respondent 1**	
Influence of personal innovation on absorption of green carbon innovation technology for individuals and corporate organizations		
1) Curiosity	"I have always supervised research students and discovered that personal innovation is critical for adoption and absorption of new applications by the students. The individual ability to explore, experiment and adapt to new technology depends on the personal motivation. First, the personal motivation depends on curiosity to learn the new application. students and researchers who are curious in learning the applications have been able to use new applications. This is mainly driven by a desire to explore and learn the new technology. Similarly, individuals who are curious will find it easy to learn the new green carbon emission technology." **Respondent 6**	Personal innovation
2) Open mindedness	"The next driver for personal innovation is individuals and organisations who are open-minded. I have also discovered that open-mindedness is a virtue which has been critical for adoption of research applications by the students. People who are willing to consider new ideas and perspectives find it easy to learn new applications." **Respondent 6**	

3) Risk tolerance	"I am using my understanding for the adoption for use of the new PhD students for the research applications to relate and predict on possible personal innovation characteristics for absorption of new green carbon credit technology. I have discovered that people who are risk tolerant can easily learn and adopt to the new technology as opposed to the ones averse to risks. Students who have always willing to take calculated risks and experiment with new technology find themselves in absorption. Risk tolerance is a vital characteristic for personal motivation and technology absorption." **Respondent 6**	
4) Creativity	"Finally, creative individuals find themselves in a position to quickly adopt to new technology. Ability to think creatively and find new solutions helps in technology absorption. I have always found that creative people easily learn and adopt the new technologies. This inspires personal motivation to learn and adopt the new technology." **Respondent 6**	
Influence of organizational factors on absorption of green carbon innovation technology for individuals and corporate organizations		
1) Strategic	"I am in top management and usually applications which promote our company strategy are highly encouraged. Alignment with business strategy is potential risk which could discourage management from acquisition of the new green innovation. Potential mismatch between the technology and the organization's overall business strategy. Corporations encourage introduction of applications which don't endanger their competitive advantage. The green innovation is highly desirable because it in line with one of our corporate strategies of harnessing environmental protection." **Respondent 8**	Organization considerations
2) Financial	"On a strategic level financial risks play a critical role on management decision of acquiring a new technology. If we discover Chief Executives always consider impact of acquisition of the new technology on the financial standing. The new innovation will be weighed on financial benefits like costs, current expenses and opportunity cost. The new carbon emission technology promises to help companies to claim for carbon credit it will be highly appealing. Management will examine the uncertainty about the technology's potential return on investment". **Respondent 8**	

(Continued)

Table 6.1 (Continued)

First-order categories	Second-order themes	Aggregate dimensions
3) Technical	"The next perspective for us in Information technology are technical risks of the new application which can undermine use of the application by potential users. The most common technical risks are usually complexity, and compatibility of the new technology can hinder an organisation from acquiring the new technology. Complexity in the new application could have factors like difficulty in understanding, implementing, or using the technology. In addition, some technologies are not compatible with existing systems, hardware, or software." **Respondent 8**	
4) Operational	"My experience also operational risks have always featured prominent during the deployment of the new technology. The deployment of the new green innovation must be mindful of potential disruption university operations. the new application has bear in mind the potential disruption to normal business operations during implementation or transition. In addition, we are always mindful of training and support requirement, change process and re-engineering if there are any requirements. This is the reason we have centralized installation of applications." **Respondent 8**	
5) Environmental	"I am also seeing a new trend for environmental sustainability. Corporations and individuals favour technologies contribute to environmental protect. The new application is good for environmental protection and sustainability. South African companies are now encouraged to reduce carbon emissions. This application aligns well with our environmental objectives which is sustainability. At strategic level applications which are more environmentally friendly are easily adopted." **Respondent 1**	

Influence of cyber risks on absorption of green carbon innovation technology for individuals and corporate organizations		
1) Data privacy and security	"Thank you for this new application. It's important to assess the data privacy and security of the new application. Emergence of virtual systems come with personal data security and privacy due to connectivity. Digital officers while evaluating technology must ensure personal and organisation data will not be breached by the new applications. The new technologies come with data related risks as these applications collect personal and organisation information. The users are mindful about data breaches, data loss, data tampering and data exfiltration. Data breaches or unauthorized access to sensitive data is usually common. Accidental or intentional deletion or corruption of data, unauthorized modification of data together with unauthorized transfer of data outside the organization. In all our evaluations of the new technology data security and privacy has to be protected otherwise the new application will not be procured" **Respondent 3**	Perceived risks
2) Systems security	"I have been in ICT for more than a decade, and corporations are worrying of security risks. The corporate first objective ensuring that the new applications don't compromise system security. The most common security risks which come with these applications are malware, denial service attacks, distributed service attacks and system vulnerabilities. The new green innovation must ensure that the new application doesn't expose the corporation to security risks. Malware infections such as Viruses, Trojan horses, spyware, adware, and ransomware, can impact on company." **Respondent 3**	
3) Network-related security	"My personal experience in working with ICT for more than a decade, organisations are weary about network related risks in acquisition of new technology. The most common risk is unauthorized access, network intrusions, man-in the middle and Wi-Fi vulnerabilities. Unauthorized access through hacking, and phishing, prevalent as unauthorized parties can easily access your network and harm the organisation. These perceived risks can hinder acquisition of the green carbon innovation." **Respondent 3**	

(*Continued*)

Table 6.1 (Continued)

First-order categories	Second-order themes	Aggregate dimensions
4) Cloud security	"I think the next risk from the technical point of view is cloud-related security risks. The common vulnerabilities which come with new technologies are cloud data breaches, cloud service vulnerabilities, account hijacking and data loss. The technical evaluation of the new technology asses' vulnerabilities from cloud storage. Some technologies come with weaknesses, unauthorized access and data losses." expose the corporation to security risks. Malware infections such as Viruses, Trojan horses, spyware, adware, and ransomware, can impact on company." **Respondent 7**	
5) Identity-related risks	"In our daily interactions with applications identity related risks negatively impact on the organisations. The acquisition of new applications has to ensure security to avoid identity theft, authentication bypass, authorization vulnerability and credential stuffing. Technically, the new technology is acceptable therefore, high absorption identity security considerations." **Respondent 7**	
6) IoT-related security	"Most technologies use IoT devices for the different system performance. We have realized that these come with IoT related risks. In order to ensure that the green carbon innovation platforms IoT devices are protected from device vulnerabilities, data breaches, device hijacking and IoT network vulnerabilities." **Respondent 3**	
7) Supply chain security	"The technology domain has to assessed against vulnerabilities in the supply chain. In our experience of procurement of applications for the university, the common supply chain risks are third part vulnerabilities, supply chain data breaches, counterfeit components and supply chain disruptions. There are instances where third party software and hardware come with weaknesses. In addition, supply chain unauthorized access to supply chain data and disruptions to supply chain operations." **Respondent 7**	

6.5 Discussion

6.5.1 The Influence of Perceived Usefulness on the Absorption of the New Carbon Emission Technology

The concept of PU has also been critical in the adoption of the new CE technology for individual and corporate users. The PU of the new technology as described by both intrinsic and extrinsic factors determines the absorption of the new carbon credit innovation technology. The PU of the new technology as described in this chapter includes performance improvement, goal achievement, compatibility, support resources, and social influence, impacting the potential user to acquire the new technology. Individuals and corporate organization on deciding to acquire the new technology are usually inspired by the usefulness of the technology.

Previous scholars used the TAM for the acceptance of e-learning platforms by government school teachers (Rafique et al., 2024), the adoption of AI in e-commerce (Wang et al., 2023), and the acceptance of Metaverse (Wu & Yu, 2024). Davis et al. (1989) defined PU as the extent to which individuals perceive that the usage of a specific technology enhances their performance. Earlier research has revealed that PU positively affects technology adoption intentions (Adwan et al., 2018; Akour et al., 2022; Al-Adwan et al., 2023; Al-Rahmi et al., 2022; Martinho et al., 2018). Individuals and corporate management PU of new green carbon innovation is critical on the decision for acquisition and utilization of the new technology. This is critical in reaping organizational, economic, social, and environmental benefits. Furthermore, appropriate technology absorption is critical in driving growth, innovation and sustainability.

6.5.2 The Influence of the Perceived Ease of Use on the New Carbon Emission Technologies

The perceived ease of use factors for the absorption of the new green innovation examined in this study emerges from the design characteristics of the new technology. This chapter describes the critical design attributes and potential user attributes determinants for the ease of use of the new design application. These characteristics for perceived ease of use include system quality, information quality, user experience, and cognitive absorption. The intrinsic factors are inherent to the potential clients of the new technology, while the extrinsic factors are the design characteristics of the technology. The detailed attributes are elaborated in Table 6.1.

The perceived ease of use refers to the extent to which an individual perceives that using a given technology is easy and effortless (Davis et al., 1989). Previous studies on technology adoption indicated that users' perceived ease of use substantially impacted their PU (Akdim et al., 2022; Mailizar et al., 2021; Wang et al., 2024) and adoption of green technologies (Chen & Lu, 2016; Ma et al., 2017;

Natalwala & Habibullah, 2024; Polas et al., 2023; Sadriwala & Sadriwala, 2022; Shahzad et al., 2022). Ma et al. (2017) study findings validated that the perceived ease of use, PU, attitudes, and purchase intention were associated with the consumers' use of sustainability labels. Perceived ease of use, therefore, plays a critical role in the design, acquisition, utilization, and adoption of the new green CE technology. In order to harness environmental sustainability, it is imperative to cascade individual ease of use perceptions in the design of the new technology. Intrinsic and extrinsic perspectives on ease of use include self-efficacy, system quality, information quality, user experience, and cognitive absorption.

6.5.3 Influence of Self-Efficacy in the Absorption of the New Carbon Emission Technology

The chapter established that self-efficacy is critical in the absorption of the new CE technology in South Africa. The critical attributes of self-efficacy: task specific, domain specific, and general. In the acquisition of the new green innovation, the self-efficacy of individuals and corporate employees needs to be considered that will be provided with the new application. Specifically, individual's belief in their ability to succeed in specific situations. In other words, self-efficacy is about confidence in one's own ability.

Mikusa (2015) defines technology self-efficacy as "personal belief in one's ability to successfully use technology to increase learning outcomes." It is seen as an individuals' belief in their own competence and typically reflects the level of capability that an individual believes he/she possesses (Christensen & Knezek, 2016). Previous studies argued that self-efficacy as a critical factor includes computer self-efficacy (Nardi & Ranieri, 2019), internet self-efficacy (Joo et al., 2000), technology self-efficacy (Durak, 2018), and online learning self-efficacy (Hong et al., 2019). In addition, Bubou and Job (2020) state that self-efficacy is intricately connected to self-esteem and self-worth, which drive the absorption of the new technology. Compared to those with less confidence, individuals with a higher level of confidence on the utilization of information technology find it beneficial and easy to use (Venkatesh et al., 2012). This plays a critical role in technology absorption. Self-belief in the use of the green innovation plays a critical role in technology innovation.

6.5.4 Personal Innovation and Absorption of New Carbon Emissions in South Africa

In this chapter, personal innovation is a critical factor in the absorption of new technology. Personal innovation refers to an individual's ability to explore, experiment, and adapt new technologies to improve their performance or achieve their goals. In this study, the critical factors for personal innovation are curiosity, open-mindedness, risk tolerance, and creativity. In this chapter, personal innovation plays a critical role in the absorption of the new technology. The study findings highlight that the absorption of the CO_2 emission tracker depends on the PI of the individuals and corporate organizations. Specifically, curiosity, open mindedness, risk tolerance, and creativity are critical in the absorption of the new CO_2 emission tracker. By understanding the cyber risks, individuals, organizations, and developers of the new technologies can develop strategies for mitigating the cyber risks, ensuring a secure absorption of new technologies.

Fan et al. (2020) define PI as the desire by an individual to try out new technology innovation. Kogabayev and Maziliauskas (2017) define PI as the degree to which an individual has a tendency for experimenting with new technologies. Previous studies equally demonstrated the extent to which personal innovation on the technology adoption: mobile learning (Joo et al., 2014), virtual learning (Raaij & Schepers, 2008), and healthcare technologies (Fan et al., 2020; Wu et al., 2011). The integration of PI was critical in the success of mobile learning absorption technology in Cyber University China (Joo et al., 2014). It is imperative, therefore, that the process of acquisition, development, and use of the CO_2 tracker innovation incorporates the innovative capacity of intended users. The corporate companies targeted for the use of the tracker need to have to improve new technical abilities. In this chapter, the intrinsic drivers for technology inherent in PI are critical for the new CO_2 emission tracker technology. Similarly, Wu et al. (2011) observed that personal innovation was responsible for the adoption and the use of health-care technology in China. Consequently, this chapter posits that personal curiosity of the potential users of the technology is critical in the absorption of the new green innovation.

6.5.5 Perceive Privacy and Security Risks Influence on the Absorption of the New Carbon Emission Technology

The perceived privacy and security risks that come with the new technology play a critical role in the adoption of the innovation. In this chapter, the technical considerations for the acquisition of the new technology are usually hinged on the vulnerabilities, which could come with the acquisition of the new technology. By understanding the cyber risks, individuals, organizations, and developers of the new technologies can develop strategies for mitigating the cyber risks, ensuring the secure adoption of new technologies.

In this chapter, the developers of new green innovation have to be mindful of potential cyber risks. Second, corporate organizations absorbing the new technology likewise have to consider these cyber risks during the absorption of the new technology. The privacy and security concerns as elaborated in the chapter include data-related risks, system security, network-related risks, identity-related risks, cloud-related, IoT-related risks, and supply chain-related risks. The intrinsic factors are inherent on the potential clients of the new technology, while extrinsic factors are the design characteristics of the technology. The detailed attributes are elaborated in Table 6.1.

Wang et al. (2018) argued that the perception of risk is a major barrier to the user adoption of innovative technologies. Emerging technologies usually collect user information such as transactions, physical state, and facial images (Zhao et al., 2022), thus compromising individual and organization security and privacy (Wang et al., 2022). Consequently, the privacy and security risks are prevalent in devices and systems for processing information. Risks are prevalent in data, systems, cloud computing, IoT devices, and networks. The developers of the new technologies must ensure that vulnerabilities are eliminated in the new applications. In addition, individuals and organizations must assess the vulnerabilities of the new technology on the risks across the different domains. The inherent characteristics of the technology can compromise the value of the technology due to privacy and security concerns.

6.5.6 Organizational Considerations and the New Carbon Emission Technology Application

The study established that corporate clients must factor key organizational strategic considerations when acquiring the new technology. The developers must understand the dynamics driving an organization's success before developing applications to meet the organization's requirements. Second, the organizations absorbing the new technology must put in place critical considerations while acquiring the new technology. Failure to undertake a proper assessment of the new technology can hinder the achievement of the organizational objectives.

The organizational considerations discussed in this chapter are strategic alignment and financial, technical, operational, and environmental considerations. The elements identified in this chapter are critical for the success of an organization using the new technology. The detailed attributes are elaborated in Table 6.1. By acknowledging and addressing these organizational factors, businesses can create a conducive environment for successful technology absorption, driving growth, innovation, and competitiveness. The development of a new technology must take due considerations of corporate strategies, as well as financial, technical, operational, and environmental considerations. In addition, corporate entities have to undertake the assessment of the new CE technology before adoption.

6.6 Implications

6.6.1 Theoretical Implications

The study's theoretical contributions move beyond the limitations of existing models, offering a richer and more nuanced understanding of technology adoption within the specific context of green innovation initiatives focused on carbon reduction. Its focus on the micro-level and its incorporation of contextual factors make it a significant contribution to the field.

First, the chapter study extends the TAM by incorporating additional factors not typically considered. While TAM primarily focuses on PU and perceived ease of use, this research adds intrinsic motivational factors, contextual factors, and self-efficacy. The study highlights the role of personal innovation (curiosity, open-mindedness, risk tolerance, and creativity) in influencing technology adoption, going beyond the extrinsic motivators emphasized in traditional TAM. In addition, the chapter explicitly considers the influence of social norms, organizational context (strategic alignment, financial resources, technical capabilities, operational feasibility, and environmental considerations), and security/privacy concerns. These are crucial contextual factors often overlooked in broader technology adoption models. The chapter incorporates self-efficacy (task-specific, domain-specific, and general) into the adoption framework. This acknowledges the role of individual belief in their ability to use the technology successfully.

Second, the chapter develops a holistic model for CE tracking technology adoption. A more comprehensive model for understanding the absorption of green technologies, particularly in the context of CE reduction, has been proposed. This model integrates the extended TAM with contextual and individual-level factors, offering a more nuanced perspective than existing models. In addition, the chapter addresses the micro-level of CE initiatives. Most of the studies focus on macro-level policies and regulations for carbon reduction. However, this study uniquely examines at the micro-level, focusing on individual employee actions and their role in corporate-level CE reduction strategies. This micro-level perspective provides a unique contribution to the field of climate change mitigation and sustainability initiatives.

Finally, the chapter provides specific insights into carbon credit mechanisms for individuals employed. The work contributes valuable insights into the practical aspects of carbon credit implementation. It addresses the issues of data reliability, organizational integration of carbon tracking systems, and strategies for effective individual and corporate engagement. The research places significant emphasis on user experience and the critical role of cybersecurity concerns in technology adoption. This highlights the need for user-friendly and secure designs in green technologies.

6.6.2 Practical Implications

The study's practical contributions center on providing actionable insights and recommendations for the successful implementation and adoption of CE tracking technologies, particularly in the context of South Africa.

First, the chapter provides suggestions for the improvement of the CE tracking application design. Specifically, factors for CE tracker adoption have been provided such as PU, perceived ease of use, self-efficacy, PI, privacy and security, and organizational considerations. This allows developers to design and improve applications that are more user-friendly, effective, and readily adopted. Understanding the importance of security and privacy concerns allows for the development of safer and more trustworthy applications. In addition, the chapter provides actionable insights for developing targeted strategies to promote technology adoption among individuals and corporations. This includes tailoring interventions to address specific barriers identified in the study, such as the lack of self-efficacy, PI, or concerns about data privacy.

In order to harness the effective absorption of CO_2 tracking application by corporate organizations, the chapter provides actionable strategies for the absorption of CE tracker by undertaking organizational alignment, resource allocation, and adequate training in facilitating the successful integration of carbon tracking technologies within companies. The chapter enables corporates to develop more effective implementation plans, maximizing the benefits and minimizing disruptions. The research offers valuable insights into the practical challenges of carbon credit mechanisms, such as data reliability and organizational integration. This knowledge can be used to improve the design and management of carbon credit programs, making them more effective and equitable.

Finally, the chapter provides valuable data that can inform policy decisions aimed at promoting the adoption of green technologies. Understanding the factors that drive, or hinder adoption can help policymakers design more effective incentive programs, regulations, and support mechanisms. In essence, the study provides a practical guide for stakeholders involved in the development, implementation, and usage of CE tracking technologies. Its recommendations can improve the effectiveness of these technologies in mitigating climate change and promoting sustainability.

References

Adwan, A.S.A., Adwan, A.A., & Berger, H. (2018). Solving the mystery of mobile learning adoption in higher education. *International Journal of Mobile Communications*, *16*(1), 24.

Akdim, K., Casaló, L.V., & Flavián, C. (2022). The role of utilitarian and hedonic aspects in the continuance intention to use social mobile apps. *Journal of Retailing and Consumer Services, 66*, 102888.

Akour, I.A., Al-Maroof, R.S., Alfaisal, R., & Salloum, S.A. (2022). A conceptual framework for determining metaverse adoption in higher institutions of gulf area: An empirical study using hybrid SEM-ANN approach. *Computers and Education: Artificial Intelligence, 3*, 100052.

Al-Adwan, A.S., Li, N., Al-Adwan, A., Abbasi, G.A., Albelbisi, N.A., & Habibi, A. (2023). Extending the technology acceptance model (TAM) to Predict University students' intentions to use metaverse-based learning platforms. *Education and Information Technologies, 28*(11), 15381–15413.

Al-Rahmi, A.M., Al-Rahmi, W.M., Alturki, U., Aldraiweesh, A., Almutairy, S., & Al-Adwan, A.S. (2022). Acceptance of mobile technologies and M-learning by university students: An empirical investigation in higher education. *Education and Information Technologies, 27*(6), 7805–7826.

An, Y., Zhou, D., Wang, Q., Shi, X., & Taghizadeh-Hesary, F. (2022). Mitigating size bias for carbon pricing in small Asia-Pacific countries: Increasing block carbon tax. *Energy Policy, 161*, 112771.

Anjos, M.F., Feijoo, F., & Sankaranarayanan, S. (2022). A multinational carbon-credit market integrating distinct national carbon allowance strategies. *Applied Energy, 319*, 119181.

Bandura, A. (1986). *Social foundations of thought and action* (2nd ed., pp. 23–28). Englewood Cliffs, NJ: Prentice-Hall.

Berakon, I., Wibowo, M.G., Nurdany, A., & Aji, H.M. (2023). An expansion of the technology acceptance model applied to the halal tourism sector. *Journal of Islamic Marketing, 14*(1), 289–316.

Bhatti, M.A., & Juhari, A.S. (2023). Effects of block chain adoption on green innovation: Moderating the role of intention to use. *AgBioforum, 25*(1), 164–178.

Bubou, G.M., & Job, G.C. (2020). Individual innovativeness, self-efficacy and e-learning readiness of students of Yenagoa study centre, National Open University of Nigeria. *Journal of Research in Innovative Teaching & Learning, 15*(1), 2–22. https://doi.org/10.1108/JRIT-12-2019-0079

Calderón, S., Alvarez, A.C., Loboguerrero, A.M., Arango, S., Calvin, K., Kober, T., Daenzer, K., & Fisher-Vanden, K. (2016). Achieving CO_2 reductions in Colombia: Effects of carbon taxes and abatement targets. *Energy Economics, 56*, 575–586.

Cao, Q., Kang, W., Xu, S., Sajid, M.J., & Cao, M. (2019). Estimation and decomposition analysis of carbon emissions from the entire production cycle for Chinese household consumption. *Journal of Environmental Management, 247*, 525–537.

Chahal, J., & Rani, N. (2022). Exploring the acceptance for e-learning among higher education students in India: Combining technology acceptance model with external variables. *Journal of Computing in Higher Education, 34*(3), 844–867.

Chau, P.Y.K. (1996). An empirical assessment of a modified technology acceptance model. *Journal of Management Information Systems, 13*(2), 185–204.

Chen, S.-Y., & Lu, C.-C. (2016). Exploring the relationships of green perceived value, the diffusion of innovations, and the technology acceptance model of green transportation. *Transportation Journal, 55*(1), 51–77.

Chen, Y., & Tseng, C.-L. (2011). Inducing clean technology in the electricity sector: Tradable permits or carbon tax policies? *The Energy Journal, 32*(3), 149–174.

Chen, X., Gong, X., Li, D., & Zhang, J. (2019). Can information and communication technology reduce CO_2 emission? A quantile regression analysis. *Environmental Science and Pollution Research, 26*(32), 32977–32992.

Chen, L., Qie, K., Memon, H., & Yesuf, H.M. (2021). The empirical analysis of green innovation for fashion brands, perceived value and green purchase intention—Mediating and moderating effects. *Sustainability, 13*(8), 4238.

Christensen, R., & Knezek, G. (2016). Validating the Technology Proficiency Self-Assessment Questionnaire for 21st Century Learning (TPSA C-21). *Journal of Digital Learning in Teacher Education, 33*(1), 20–31. https://doi.org/10.1080/21532974.2016.1242391

Creswell, J.W., & Creswell, J.D. (2017). *Research design: Qualitative, quantitative, and mixed methods approaches*. Sage publications.

Davis, F.D., Bagozzi, R.P., & Warshaw, P.R. (1989). User acceptance of computer technology: A comparison of two theoretical models. *Management Science, 35*(8), 982–1003.

Di Pietro, R., & Cresci, S. (2021, December 13–15). Metaverse: Security and privacy issues. *2021 Third IEEE International Conference on Trust, Privacy and Security in Intelligent Systems and Applications (TPS-ISA)* (pp. 281–288). Atlanta, GA, USA.

Dou, K., Yu, P., Deng, N., Liu, F., Guan, Y., Li, Z., Ji, Y., Du, N., Lu, X., & Duan, H. (2017). Patients' acceptance of smartphone health technology for chronic disease management: A theoretical model and empirical test. *JMIR mHealth and uHealth, 5*(12), e7886.

El Khatib, S., & Galiana, F.D. (2019). Investigating emission regulation policy in the electricity sector: Modeling an oligopolistic electricity market under hourly cap-and-trade. *Energy Economics, 78*, 428–443.

Fan, W., Liu, J., Zhu, S., & Pardalos, P.M. (2020). Investigating the impacting factors for the healthcare professionals to adopt artificial intelligence-based medical diagnosis support system (AIMDSS). *Annals of Operations Research, 294*(1), 567–592.

Feijoo, F., Mignone, B.K., Kheshgi, H.S., Hartin, C., McJeon, H., & Edmonds, J. (2019). Climate and carbon budget implications of linked future changes in CO_2 and non-CO_2 forcing. *Environmental Research Letters, 14*(4), 044007.

Feijoo, F., Iyer, G., Binsted, M., & Edmonds, J. (2020). US energy system transitions under cumulative emissions budgets. *Climatic Change, 162*(4), 1947–1963.

Figliozzi, M., & Jennings, D. (2020). Autonomous delivery robots and their potential impacts on urban freight energy consumption and emissions. *Transportation Research Procedia*, *46*, 21–28.

Floros, N., & Vlachou, A. (2005). Energy demand and energy-related CO_2 emissions in Greek manufacturing: Assessing the impact of a carbon tax. *Energy Economics*, *27*(3), 387–413.

Gao, R., Zhang, H., Gong, C., & Wu, Z. (2022). The role of farmers' green values in creation of green innovative intention and green technology adoption behaviour: Evidence from farmers grain green production. *Frontiers in Psychology*, *13*, 980570.

Granić, A., & Marangunić, N. (2019). Technology acceptance model in educational context: A systematic literature review. *British Journal of Educational Technology*, *50*(5), 2572–2593.

Hai Yen, T.T., Wong, W.K., Ali Al-Abyadh, M.H., Muda, I., Julca-Guerrero, F., Hishan, S.S., & Islam, M.M. (2023). The impact of ecological innovation and corporate social responsibilities on the sustainable development: Moderating role of environmental ethics. *Economic research-Ekonomska istraživanja*, *36*(3).

Holappa, L. (2020). A general vision for reduction of energy consumption and CO_2 emissions from the steel industry. *Metals*, *10*(9). Article 9 https://doi.org/10.3390/met10091117

Hong, J.-C., Hwang, M.-Y., Tai, K.-H., & Lin, P.-H. (2019). Improving cognitive certitude with calibration mediated by cognitive anxiety, online learning self-efficacy and interest in learning Chinese pronunciation. *Educational Technology Research and Development*, *67*(3), 597–615.

Ji, L., Zhang, X.P., Huang, G.H., & Yin, J.G. (2015). Development of an inexact risk-aversion optimization model for regional carbon constrained electricity system planning under uncertainty. *Energy Conversion and Management*, *94*, 353–364.

Joo, Y.-J., Bong, M., & Choi, H.-J. (2000). Self-efficacy for self-regulated learning, academic self-efficacy, and internet self-efficacy in web-based instruction. *Educational Technology Research and Development*, *48*(2), 5–17.

Joo, Y.J., Lee, H.W., & Ham, Y. (2014). Integrating user interface and personal innovativeness into the TAM for mobile learning in Cyber University. *Journal of Computing in Higher Education*, *26*(2), 143–158.

Kogabayev, T., & Maziliauskas, A. (2017). The definition and classification of innovation. *HOLISTICA – Journal of Business and Public Administration*, *8*(1), 59–72.

Lange, S., Pohl, J., & Santarius, T. (2020). Digitalization and energy consumption. Does ICT reduce energy demand? *Ecological Economics*, *176*, 106760.

Lee, Z.H., Sethupathi, S., Lee, K.T., Bhatia, S., & Mohamed, A.R. (2013). An overview on global warming in Southeast Asia: CO_2 emission status, efforts done, and barriers. *Renewable and Sustainable Energy Reviews*, *28*, 71–81.

Lee, J., Kim, J., & Choi, J.Y. (2019). The adoption of virtual reality devices: The technology acceptance model integrating enjoyment, social interaction, and strength of the social ties. *Telematics and Informatics*, *39*, 37–48.

Li, K. (2023). Determinants of college students' actual use of AI-based systems: An extension of the technology acceptance model. *Sustainability*, *15*(6), 5221.

Li, M.W., Li, Y.P., & Huang, G.H. (2011). An interval-fuzzy two-stage stochastic programming model for planning carbon dioxide trading under uncertainty. *Energy*, *36*(9), 5677–5689.

Lise, W., Sijm, J., & Hobbs, B.F. (2010). The Impact of the EU ETS on prices, profits and emissions in the power sector: Simulation results with the COMPETES EU20 model. *Environmental and Resource Economics*, *47*(1), 23–44.

Liu, Y., Deng, S., Zhao, R., He, J., & Zhao, L. (2017). Energy-saving pathway exploration of CCS integrated with solar energy: A review of innovative concepts. *Renewable and Sustainable Energy Reviews*, *77*, 652–669.

Liu, J., Liu, L., Qian, Y., & Song, S. (2022). The effect of artificial intelligence on carbon intensity: Evidence from China's industrial sector. *Socio-Economic Planning Sciences*, *83*, 101002.

Lokuge, N., & Anders, S. (2022). Carbon-credit systems in agriculture: A review of literature (SSRN Scholarly Paper 4096174). *The School of Public Policy Publications*, *15*(1), 44.

Ma, Y.J., Gam, H.J., & Banning, J. (2017). Perceived ease of use and usefulness of sustainability labels on apparel products: Application of the technology acceptance model. *Fashion and Textiles*, *4*(1), 3.

Mailizar, M., Burg, D., & Maulina, S. (2021). Examining university students' behavioural intention to use e-learning during the COVID-19 pandemic: An extended TAM model. *Education and Information Technologies*, *26*(6), 7057–7077.

Mardones, C., & Flores, B. (2018). Effectiveness of a CO_2 tax on industrial emissions. *Energy Economics*, *71*, 370–382.

Martinho, D., Santos, E., Miguel, I., & Cordeiro, D. (2018). Factors that influence the adoption of postgraduate online courses. *International Journal of Emerging Technologies in Learning*, *13*(12), 123.

Mikusa, M. E. (2015). *The effect of technology self-efficacy and personal engagement on students' and teachers' attitudes toward technology use in education* [PhD Thesis, Appalachian State University].

Mortha, A., Taghizadeh-Hesary, F., & Vo, X.V. (2021). The impact of a carbon tax implementation on non-CO_2 gas emissions: The case of Japan. *Australasian Journal of Environmental Management*, *28*(4), 355–372.

Moyer, J.D., & Hughes, B.B. (2012). ICTs: Do they contribute to increased carbon emissions? *Technological Forecasting and Social Change*, *79*(5), 919–931.

Nardi, A., & Ranieri, M. (2019). Comparing paper-based and electronic multiple-choice examinations with personal devices: Impact on students' performance, self-efficacy and satisfaction. *British Journal of Educational Technology*, *50*(3), 1495–1506.

Natalwala, A.S., & Habibullah, M. (2024). Adaption of the tam to examine the influence of perceived ease of use and perceived usefulness on green banking practices. *Library of Progress-Library Science, Information Technology & Computer*, *44*(3), 20679–20684.

Polas, M.R.H., Kabir, A.I., Jahanshahi, A.A., Sohel-Uz-Zaman, A.S.M., Karim, R., & Tabash, M.I. (2023). Rural entrepreneurs behaviours towards green innovation: Empirical evidence from Bangladesh. *Journal of Open Innovation: Technology, Market, and Complexity*, *9*(1), 100020.

Qin, B., Liu, J., Wang, H., Wang, Z., Xiong, Z., Wang, M., & Qian, Q. (2025). Energy-efficient and reliable urban rail transit: A new framework incorporating underground energy storage systems. *iEnergy*.

van Raaij, E.M., & Schepers, J.J.L. (2008). The acceptance and use of a virtual learning environment in China. *Computers & Education*, *50*(3), 838–852. https://doi.org/10.1016/j.compedu.2006.09.001

Rafique, H., Ul Islam, Z., & Shamim, A. (2024). Acceptance of e-learning technology by government school teachers: Application of extended technology acceptance model. *Interactive Learning Environments*, *32*(6), 2970–2988.

Rocha, P., Das, T.K., Nanduri, V., & Botterud, A. (2015). Impact of CO_2 cap-and-trade programs on restructured power markets with generation capacity investments. *International Journal of Electrical Power & Energy Systems*, *71*, 195–208.

Rosli, M.S., Saleh, N.S., Ali, A.M., & Bakar, S.A. (2023). Factors determining the acceptance of E-wallet among gen Z from the lens of the extended technology acceptance model. *Sustainability*, *15*(7), 5752.

Sadriwala, M.F., & Sadriwala, K.F. (2022). Perceived usefulness and ease of use of artificial intelligence on marketing innovation. *International Journal of Innovation in the Digital Economy (IJIDE)*, *13*(1), 1–10.

Salahdine, F., & Kaabouch, N. (2019). Social engineering attacks: A survey. *Future Internet*, *11*(4), 89.

Schwartz-Shea, P., & Yanow, D. (2013). *Interpretive research design: Concepts and processes*. Routledge.

Shahzad, M., Qu, Y., Rehman, S.U., & Zafar, A.U. (2022). Adoption of green innovation technology to accelerate sustainable development among manufacturing industry. *Journal of Innovation & Knowledge*, *7*(4), 100231.

Song, J., & Wu, D. (2023). Modeling forest carbon sink trading with carbon credit using stochastic differential game. *Environmental Science and Pollution Research*, *30*(26), 68934–68950.

Sun, H., & Kim, G. (2021). The composite impact of ICT industry on lowering carbon intensity: From the perspective of regional heterogeneity. *Technology in Society*, *66*, 101661.

Taherdoost, H. (2018). A review of technology acceptance and adoption models and theories. *Procedia Manufacturing*, *22*, 960–967. https://doi.org/10.1016/j.promfg.2018.03.137

Tam, C., Santos, D., & Oliveira, T. (2020). Exploring the influential factors of continuance intention to use mobile apps: Extending the expectation confirmation model. *Information Systems Frontiers*, *22*, 243–257.

Tanaka, M., & Chen, Y. (2012). Market power in emissions trading: Strategically manipulating permit price through fringe firms. *Applied Energy*, *96*, 203–211. https://doi.org/10.1016/j.apenergy.2011.08.049

Tetik, G., Türkeli, S., Pinar, S., & Tarim, M. (2024). Health information systems with technology acceptance model approach: A systematic review. *International Journal of Medical Informatics*, *190*, 105556.

Trapero, H., Ilao, J., & Lacaza, R. (2020, December 22). An integrated theory for ChatBot use in air travel: Questionnaire development and validation. *2020 IEEE Region 10 Conference (TENCON)* (pp. 652–657), Osaka, Japan.

Trouwloon, D., Streck, C., Chagas, T., & Martinus, G. (2023). Understanding the use of carbon credits by companies: A review of the defining elements of corporate climate claims. *Global Challenges*, *7*(4), 2200158.

Usman, A., Ozturk, I., Ullah, S., & Hassan, A. (2021). Does ICT have symmetric or asymmetric effects on CO_2 emissions? Evidence from selected Asian economies. *Technology in Society*, *67*, 101692.

Venkatesh, V., & Davis, F.D. (1996). A model of the antecedents of perceived ease of use: Development and test. *Decision Sciences*, *27*(3), 451–481.

Venkatesh, V., Thong, J.Y., & Xu, X. (2012). Consumer acceptance and use of information technology: Extending the unified theory of acceptance and use of technology. *MIS Quarterly*, *36*, 157–178.

Wang, S., Wang, J., Li, J., Wang, J., & Liang, L. (2018). Policy implications for promoting the adoption of electric vehicles: Do consumer's knowledge, perceived risk and financial incentive policy matter? *Transportation Research Part A: Policy and Practice*, *117*, 58–69.

Wang, L., Chen, Y., Ramsey, T.S., & Hewings, G.J.D. (2021). Will researching digital technology really empower green development? *Technology in Society*, *66*, 101638.

Wang, C., Ahmad, S.F., Ayassrah, A.Y.B.A., Awwad, E.M., Irshad, M., Ali, Y.A., Al-Razgan, M., Khan, Y., & Han, H. (2023). An empirical evaluation of technology acceptance model for artificial intelligence in E-commerce. *Heliyon*, *9*(8), e18349.

Wang, Y., Su, Z., Zhang, N., Xing, R., Liu, D., Luan, T. H., & Shen, X. (2022). A survey on metaverse: Fundamentals, security, and privacy. *IEEE communications surveys & tutorials*, *25*(1), 319–352.

Wang, Y.-M., Chiu, W.-C., Lin, H.-H., Wang, Y.-S., Wang, Y.-Y., & Chen, I.-F. (2024). Determinants of students' adoption of virtual reality-based learning systems: An individual difference perspective. *Innovations in Education and Teaching International*, *61*(1), 143–153.

Wu, R., & Yu, Z. (2024). Investigating users' acceptance of the metaverse with an extended technology acceptance model. *International Journal of Human–Computer Interaction*, *40*(19), 5810–5826.

Wu, I.-L., Li, J.-Y., & Fu, C.-Y. (2011). The adoption of mobile healthcare by hospital's professionals: An integrative perspective. *Decision Support Systems*, *51*(3), 587–596.

Wu, Z., Li, Q., Wu, W., & Zhao, M. (2019). Crowdsourcing model for energy efficiency retrofit and mixed-integer equilibrium analysis. *IEEE Transactions on Industrial Informatics*, *16*(7), 4512–4524.

Yildiz Durak, H. (2018). Flipped learning readiness in teaching programming in middle schools: Modelling its relation to various variables. *Journal of Computer Assisted Learning*, *34*(6), 939–959.

Yin, R.K. (2009). *Case study research: Design and methods* (p. 282). Thousand Oaks, CA: Sage.

Zhang, B., Wang, Z., Yin, J., & Su, L. (2012). CO_2 emission reduction within Chinese iron & steel industry: Practices, determinants and performance. *Journal of Cleaner Production*, *33*, 167–178.

Zhang, X., Chen, Y., Hu, L., & Wang, Y. (2022). The metaverse in education: Definition, framework, features, potential applications, challenges, and future research topics. *Frontiers in Psychology*, *13*, 1016300.

Zhao, Y., Jiang, J., Chen, Y., Liu, R., Yang, Y., Xue, X., & Chen, S. (2022). Metaverse: Perspectives from graphics, interactions and visualization. *Visual Informatics*, *6*(1), 56–67.

Zhou, L., Xue, S., & Li, R. (2022). Extending the technology acceptance model to explore students' intention to use an online education platform at a University in China. *SAGE Open*, *12*(1), 21582440221085259.

Zhu, J., Sun, H., Liu, N., Zhou, D., & Taghizadeh-Hesary, F. (2020). Measuring carbon market transaction efficiency in the power industry: An entropy-weighted TOPSIS approach. *Entropy*, *22*(9), 973.

7

Toward Sustainable Business Travel: A Case of eco.mio Software Solution

Marcia Mkansi, Frank Ranganai Matenda, and Ntwanano Mboweni

Department of Operations Management, College of Economic and Management Sciences, University of South Africa, Pretoria, South Africa

7.1 Introduction

Reducing carbon emissions is now a shared objective globally as climate change is becoming increasingly severe with each passing day (Müller, 2024). Carbon dioxide emissions are a primary driver of global warming, which is a rise in the planet's average temperature. Increasing global temperatures are promoting the occurrence of extreme weather events like heatwaves, droughts, floods and wildfires, disruptions to ecosystems, and rising sea levels. Fundamentally, climate change threatens human life, environment, and development (Raihan & Tuspekova, 2023; Samour et al., 2022). Climate change has reached a critical point that requires instant and intensive consideration.

Business travel is an essential part of the economic panorama. A 2020 McKinsey report indicated that, pre-pandemic, business travel was responsible for approximately 15–20% of global travel (Borko et al., 2020). Business travel provides personal interface, reinforces client associations, promotes team collaboration, inspires a better comprehension of numerous global market conditions, and promotes trust between business associates. Nonetheless, business travel has consequences. Business travel is one of the main causes of environmental degradation. It significantly contributes to global carbon dioxide emissions (Müller, 2024 and references therein). This has been worsened by the fact that the question of who is answerable for decarbonization is still unsettled (Müller, 2024).

Green Innovations in Supply Chain Management: Case Studies and Applications, First Edition. Edited by Marcia Mkansi, Godfrey Mugurusi, Aaron L. Nsakanda, Temidayo Akenroye, and Frank Tietze.

Published 2026 by John Wiley & Sons, Inc.

Since the advent of the COVID-19 pandemic, digital options like virtual meetings have demonstrated that they can efficiently reduce business travel and the associated operational expenses and environmental impact. Virtual platforms substituted face-to-face meetings. Nevertheless, even though business travel significantly fell during the COVID-19 pandemic era, it has increased substantially in the post-pandemic era as business leaders highlighted the significance of face-to-face interactions. This shows that digital alternatives cannot entirely substitute business travel.

The transport sector is responsible for 37% of global carbon dioxide emissions (International Energy Agency, 2022) and is accountable for 30% of global energy usage (Energy Institute, 2022). In the same vein, Creutzig et al. (2015) stated that transport is accountable for roughly 23% of total energy-related carbon dioxide emissions globally, and the transport sector is growing faster than other sectors, with emissions anticipated to double by 2050. Nevertheless, the transport sector is one of the most problematic sectors to decarbonize (Creutzig et al., 2015). The Department for Business Energy and Industrial Strategy (2019) has indicated that in the United Kingdom, between 1990 and 2019, emissions from other end-use sectors were decreased by 44%, whereas those from transport were cut by less than 5%, leaving the transport sector as a problematic sector.

A huge chunk of national transport emissions emanates from passenger travel. Therefore, passenger travel needs to be examined in detail when designing any decarbonization strategy. Nowadays, numerous corporates are making efforts to include diverse facets of sustainability in their operations (Lozano & Garcia, 2020). They have identified the ecological influence of business travel and are subjected to regulatory pressure when it comes to the reduction of carbon emissions. Esty and de Arriba-Sellier (2023) and Black et al. (2021) proffered that the number of corporates that know that their activities contribute to global warming and are interested in plummeting emissions is increasing. This augments the attention of corporates toward sustainability issues.

Regulators have responded to the issue of corporate emissions by strengthening the reporting commitments of huge corporations within their dominions as part of nonfinancial reporting obligations. The European Union Corporate Sustainability Reporting Directive is one of the newest developments in this space and has been influencing a myriad of corporates since 2023. Under this directive, corporates are obligated to generate audit-ready reports that include emissions from their whole value chain, including Scope 3 business travel emissions. Scope 3 emissions incorporate all indirect emissions in the value chain, including downstream and upstream activities not dealt with in Scope 2 (Greenhouse Gas Protocol, 2013). Scope 2 emissions refer to indirect greenhouse gas (GHG) emissions related to the purchase of energy like steam, electricity, heat, or cooling.

Since business travel cannot be eradicated and the pressure is increasing on corporates to take more responsibility to reduce carbon emissions from their operations, the reduction of carbon emissions from business travel is now a priority. In practice, several tools have been proposed and adopted to reduce carbon emissions from business travel. These tools include encouraging sustainable transportation alternatives, carbon footprint calculators, offsetting emissions, emission impact dashboards, and encouraging eco-friendly practices among workers. However, most of these tools are nonbehavioral, they do not pursue the change in the behavior of travelers. Progress in business travel decarbonization therefore remains limited, raising questions concerning the efficiency of voluntarism in decarbonization attempts (Müller, 2024). This restricted progress is mainly due to the fact that (i) most tools that have been proposed and adopted to reduce carbon emissions from business travel are not directed on the critical subject of individual traveler behavior change; (ii) many corporates struggle to precisely trail and compute their business travel emissions, making it challenging to gazette profound targets and measure progress; (iii) business travel comprises numerous means of transportation and activities, making it sophisticated to analyze the environmental influence of each component; (iv) workers may not be entirely conscious of the environmental influence of their travel selections or may prioritize convenience over sustainability; and (v) many corporates use diverse online booking tools and travel management systems, making it challenging to integrate sustainability measures into their existing processes.

Since the environmental load of travel involves an extensive range of influences on climate, ecosystems, and human health (Poom et al., 2017), there is a dire need for a holistic approach to decarbonize business travel. Travelers need comprehensive tools that enable and motivate them to endorse behavioral changes. In practice, the application of decarbonization measures mainly centers on nonbehavioral methods, which are more broadly available instead of the behavioral methods that encourage changes in travelers' behaviors (Müller, 2024). Nonbehavioral measures are usually preferred due to the problems related to traveler behavior changes within the corporates where regular business travel is the norm (Müller, 2024). Adopting behavioral measures is generally challenging since organizations are usually resistant to change. Changing the business-travel behavior of the corporate for environmental motives can be viewed as even less essential from the perspective of the management (Müller, 2024). The author (Müller, 2024) further postulated that a changeover from nonbehavioral methods to modifications in business-travel behavior is required. This is essential since the demand for business travel is influenced by, among other things, individual preferences. Thus, comprehending passenger travel patterns is essential in the process of cutting carbon dioxide emissions (Wadud et al., 2024). Existing literature has indicated that corporates can meaningfully cut their environmental influence by encouraging behavioral change among workers and prioritizing sustainable practices (Müller, 2024; Sharpe et al., 2022). This indicates that even though the decarbonization attempts of several stakeholders are usually based on technology, technology alone is not adequate enough to decarbonize the transport sector (Martin et al., 2020). McLachlan and Callister (2022) and Peeters et al. (2016) also indicated that technology innovation alone cannot decrease emissions according to the decarbonization objectives. Moreover, implementing sustainable approaches that decrease the environmental impact of business travel without despising the advantages of in-person assemblies and global connectivity is vital. Thus, decreasing carbon emissions from business travel opens up opportunities for innovation and sustainable habits. Interestingly, a comprehensive assessment of business travel decarbonization is still lacking in the existing literature.

Generally, even though there are several studies that examined carbon emission reduction from business travel, there are some research gaps that need to be explored. There are inadequate approaches for obtaining and computing travel data, which is usually detailed, real time, and continuous (Zhang et al., 2024). This type of data is difficult to gather via traditional survey approaches. Also, there is a deficiency of research on the carbon emissions of business travel. Zhang et al. (2024) postulated that the existing literature is mostly devoted to the carbon emissions of commuting travel, neglecting carbon emissions of business travel. Further, the implementation of practical tools for reducing carbon emissions from business travel is restricted. To solve these challenges, more empirical research is required to reduce carbon emissions from business travel via technology, among other things. Behavior-based tools need to be designed to promote carbon emission reduction from business travel.

This study presents eco.mio software as a simple and "painless" behavioral measure for reducing carbon emissions emanating from business travel and examines its decarbonization potential for business travel-focused emissions. eco.mio software is a browser extension that assists corporates guiding their workers to greener business travel decisions by driving behavioral change through nudging and incentivizing corporate travelers (Advito, 2025; Cullen, 2023). The eco.mio solution drives behavioral change through its simplicity and by giving travelers an opportunity (to reduce their environmental impact) and personal motivation. This is in line with the COM-B model for behavior change, which states that capability (C), opportunity (O), and motivation (M) are the three major factors that are able to change behavior (B). eco.mio software addresses the problem of corporates striving to precisely compute (using a scientific methodology) and decrease their business travel carbon emissions by offering a solution that integrates the emission computation methodology into online booking tools, guiding workers toward greener choices. eco.mio software aims to reduce carbon emissions and travel costs, thereby taking conscious travel behavior to the new standard and encouraging corporates to attain their climate goals.

This study contributes to the existing literature on the discourse of sustainable business travel and corporate emission reduction. First, the study offers insights into innovation in eco-friendly travel technologies by detailing the development and application of the eco.mio software, which aims to encourage sustainable business travel practices. This includes understanding the challenges that are addressed by the eco.mio software and how it addresses them. The study offers empirical evidence on how technology-driven solutions enable reductions in carbon emissions connected to business travel, thus advancing the comprehension of sustainable technology applications in real-world settings. Gössling and Dolnicar (2023) and Erdoğan et al. (2022) are some of the authors who proposed that advancements in transportation technologies reduce carbon emissions. Second, since technology innovation alone cannot decrease emissions, the study provides new behavioral insights into business travel. By analyzing how corporate travelers respond to and engage with the eco.mio software, the study provides novel insights into behavioral patterns associated with sustainable travel. The chapter also focuses on consumer demand selections. This is vital since the extant literature (see, for instance, Gössling and Dolnicar 2023) has indicated that two principal solutions for the reduction in carbon emissions in the transport sector are the innovations of technology and measures to trigger behavioral change. This chapter focuses on technology innovations and travelers' behavior. Third, the study assesses both the economic benefits and environmental impacts of integrating the eco.mio software into corporate travel policies. This is vital since sustainability efforts are powered by cost considerations (Ekins & Zenghelis, 2021), and in most cases, the most efficient technologies in reducing carbon emissions in business travel are expensive. This dual analysis contributes to the wider discourse on the cost-effectiveness of sustainable practices in the business travel context, offering a balanced view of the trade-offs and synergies between economic objectives and environmental sustainability. The study is vital for the development of more effective corporate practices and environmental policies, and helps corporates to bridge the gap between benefits and costs in an attempt to promote more sustainable practices. Fundamentally, the study provides learning outcomes when it comes to the reduction of carbon emissions related to business travel using behavioral tools.

The rest of this chapter is structured as follows. Section 7.2 outlines the research methodology adopted in this study. Section 7.3 presents the challenges that are addressed by the eco.mio solution and how it addresses them. The impact of the innovation and the business model of the organization are given in Sections 7.4 and 7.5, respectively. Section 7.6 presents the learning outcomes of this study, and Section 7.7 outlines the plans to further develop the initiative.

7.2 Methodology

A qualitative case study research design was implemented in this study. Simons (2009) propounded that a "case study is an in-depth exploration, from multiple perspectives, of the complexity and uniqueness of a particular project, policy, institution, program, or system in a real-life context." The case study approach is one of the best methods for presenting evidence in a linear way (Yin, 2013). Further, the case study approach has the capacity to offer in-depth, context-rich perspectives into particular situations, investigate sophisticated issues, and apprehend the distinctions of real-life circumstances (Crowe et al., 2011). Therefore, the case study approach is suitable for achieving the objective of this study.

In this study, purposive sampling was adopted. Purposive sampling investigates only the population of interest (Andrade, 2021) and is associated with focused data collection. Campbell et al. (2020) propounded that "the reason for purposive sampling is the better matching of the sample to the aims and objectives of the research, thus improving the rigour of the study and trustworthiness of the data and results." Moreover, purposive sampling offers deeper insights and is related to efficient time and resource utilization. This makes it appropriate for qualitative and exploratory investigations.

The case study used eco.mio GmbH, a software company in Berlin, Germany. The company was established in 2021 and is on a mission to make conscious travel behavior the new standard. It works toward 3 of the 17 Sustainable Development Goals of the United Nations. This study is specifically focused on the eco.mio software solution, which is a product of eco.mio GmbH. The eco.mio software solution was incorporated into this study because it is one of the behavior-based tools that attempts to reduce carbon emissions from business travel and travel costs. It drives behavioral change through its simplicity and by giving travelers an opportunity to reduce their environmental impact and through personal motivation. Similarly, eco.mio software solution was selected for this case study as a representative example of all behavior-based tools that assist corporates in guiding their workers to greener business travel decisions.

In this study, web scraping was used to extract data from the website of eco.mio GmbH. Web scraping is a research technique that implements an automated software to unearth or collect and combine data from websites in a systematic manner or structured format (Dongo et al., 2021; Hillen, 2019). Ideally, it is a method to systematically collect data from the web by "scraping" information straight from webpages. Web scraping permits researchers to collect data at high speed and precision (Dongo et al., 2021), and allows a faster assessment of patterns and trends. Slamet et al. (2018) indicated that web scraping is one of the most practical methods of data harvesting.

7.3 Results

The results of the study are presented in this section.

7.3.1 Challenges and Solutions

This subsection outlines the challenges addressed by the eco.mio software solution (Table 7.1) and how it addresses them.

7.3.1.1 Challenges

Table 7.1 Challenges addressed by the eco.mio software solution

Challenge	Description
Tracking and reporting emissions	Challenging to track and report more accurate, real time, and continuous carbon emissions data. This includes tracking and reporting emissions reduction results. The challenge is amplified by the time-intensive nature of estimates.
Standardized reporting frameworks	Lack of standardized reporting frameworks, which are crucial for businesses aiming to measure and report on carbon emissions from business travel effectively. Standardized reporting frameworks ensure comparability, transparency, and precision in emission reporting. Business travel decarbonization attempts of corporates are diverse, making it difficult to understand them.
Data availability	Inconsistent data availability. The required data is difficult to gather via traditional methods like surveys and interviews.
Inadequate techniques for obtaining and computing business travel data.	Difficulties in obtaining and computing travel data due to inadequate techniques.
Cost considerations	Sustainability efforts are generally powered by high cost considerations.
Behavioral-based tools	Lack of behavioral-based tools that pursue the change in the behavior of travelers. Restricted progress in business travel decarbonization poses questions concerning the efficiency of nonbehavioral tools in decarbonization efforts.
Online booking systems and sustainability tools	Difficult to integrate online booking systems and sustainability tools.

(*Continued*)

Table 7.1 (Continued)

Challenge	Description
Supply chains	Challenging to estimate emissions throughout the whole value chain and sophisticated supply chains.
Awareness and sustainable business travel practices	Lack of awareness and sustainable business travel practices. For instance, travelers are booking according to their personal comfort and time, instead of corporates' environmental objectives.
Structural barricades	Structural barricades like strategy and availability of infrastructure can obstruct the change of traveler behavior. Travelers may have restricted control over this. There is no guidance and opportunities that make it easier for workers to select more environmentally friendly travel engagements.

Source: Authors' compilation (2025).

7.3.1.2 Overcoming the Challenges

eco.mio software impeccably interfaces with recognized online booking tools such as Egencia, Cytric, and SAP Concur. It functions with the customer's present online booking tools and needs no modifications to the environment or booking process. Interestingly, eco.mio software addresses a service accessible market of €15–20 billion.

To analyze up-to-the-minute company pricing and availability, past travel patterns, customer's policy, and travel culture to estimate carbon emissions and cost yardsticks, eco.mio software implements robust customized machine learning algorithms. Travel options that are more carbon dioxide efficient than the normal default alternative are indicated in the booking system with striking enticements directing workers to select more intentionally. Interestingly, eco.mio software can be used to inspire and rate behavior variations by underscoring greener and more cost-effective options within the online booking tools and providing custom-made enticements.

The eco.mio software solution addresses the business travel decarbonization challenges faced by corporates and employees through the following innovative approaches:

- **Promotion of greener travel choices**: The software offers tools and perspectives that inspire business travelers to choose more sustainable travel alternatives or make greener choices without harsh travel policies, which assist their corporates to reduce overall carbon footprints (Figure 7.1). Travelers cannot book as if they are "on autopilot" and mindless of the environmental effects and budgets.

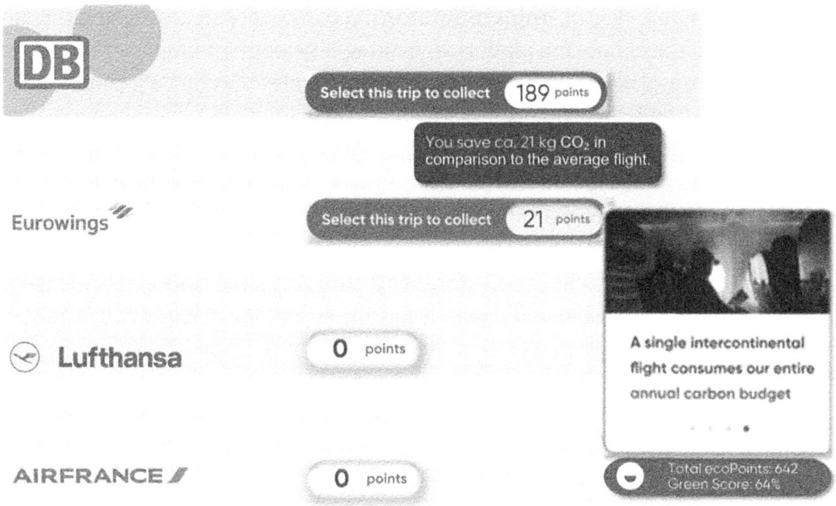

Figure 7.1 Features within the booking tool. *Source:* eco.mio GmbH/https://ecomio.com/how-it-works/ (accessed May 08, 2025).

- **Awareness and behavioral change**: The eco.mio solution strives to increase awareness concerning the environmental effects of business travel choices by giving travelers all the required information when booking so that they make well-versed decisions. It includes educational content and insights that advise employees regarding carbon emissions related to diverse travel alternatives and the benefits of plummeting them. In guiding employees to make the right decisions, the innovation uses green nudging, which is a psychological notion that functions by delicately guiding people in the direction of particular verdicts or behaviors while giving them the autonomy in their selections. Since individuals make suboptimal verdicts as a result of issues like personal bias and deficiency of information, nudging in the form of, for example, proposed travel alternatives, supplementary information (pop-ups, icons), and peer comparisons, guide workers so that they make right verdicts. This awareness advances a culture of responsibility and promotes behavioral changes with regard to sustainability.
- **Cost reduction via sustainable practices**: eco.mio software assists corporates in lessening costs by encouraging the selection of more sustainable travel alternatives. Greener travel alternatives are not only better for the environment but usually lead to reduced expenses, particularly in the long-term, via efficacies and compliance with regulatory prerequisites for the reduction of emissions. Further, eco.mio software automates numerous processes, further reducing operational costs associated with managing business travel logistics in a manual manner. Corporates implementing eco.mio software usually experience a 20% fall in cost savings and travel emissions.

- **Integration and ease of implementation**: eco.mio software assimilates directly into existing travel booking platforms, making it smooth for employees to access sustainable travel alternatives. This integration enables selecting a greener option that is as easy and direct as any other booking process.
- **Data analytics and reporting**: The eco.mio solution offers analytics and reporting tools that enable corporates to track their business travel emissions and savings over time. Specifically, route and traveler data are assessed using machine learning algorithms, and sustainability components are clearly exhibited for every booking choice. eco.mio software also makes carbon dioxide emission savings quantifiable and crystal clear for everybody (Figure 7.2). The new scientific methodology (GATE4) is used to calculate carbon emissions and savings accurately. GATE4 refers to Global Analytics for Travel Emissions, with the number "4" symbolizing both the four business travel sectors it addresses—air, hotel, rail, and car—as well as the Intergovernmental Panel on Climate Change's call to cut GHG emissions by a factor of four by 2050 to avert the most severe impacts of climate change. The resulting data is vital for assessing progress in the direction of corporate sustainability objectives and can be adopted in sustainability reports and other pieces of communication to different stakeholders.
- **Gamification and incentives**: The eco.mio solution includes gamification components such as rewards, scores, and climate facts within the booking tool, providing rewards and points for adopting greener travel alternatives (Figures 7.3 and 7.4). These topographies are developed to engage workers and inspire them

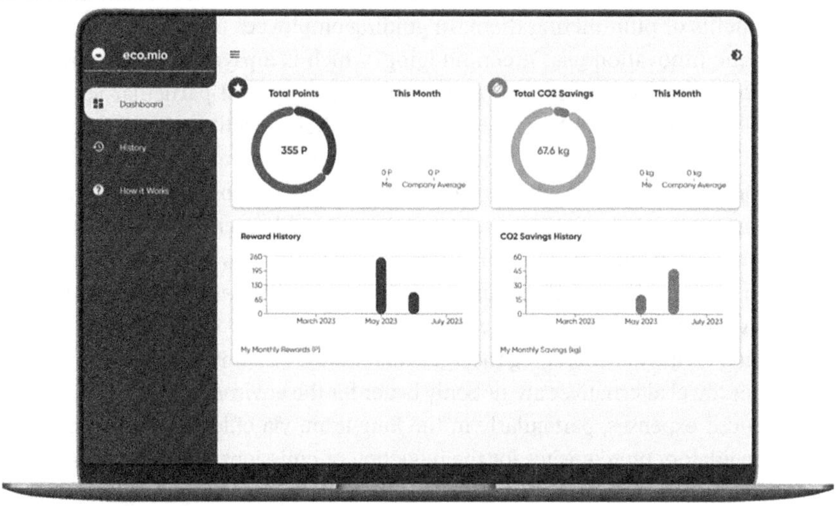

Figure 7.2 Impact measurement. *Source:* eco.mio GmbH/https://ecomio.com/how-it-works/ (accessed May 08, 2025).

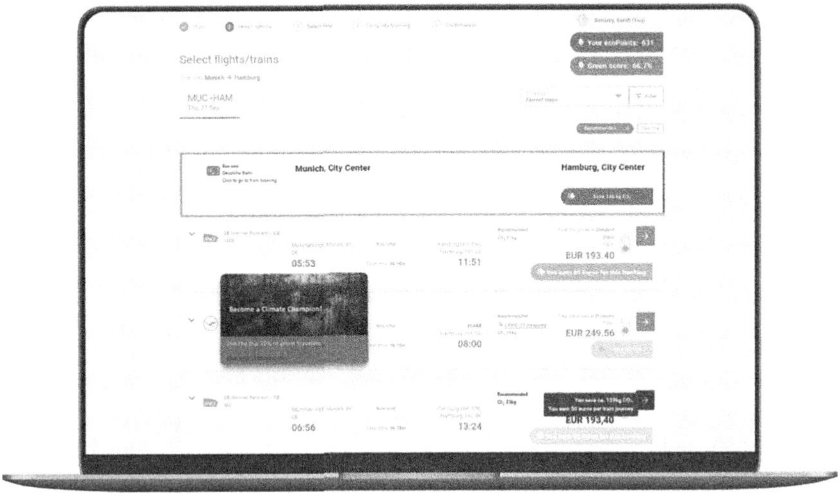

Figure 7.3 Gamification. *Source:* eco.mio GmbH/https://ecomio.com/how-it-works/ (accessed May 08, 2025).

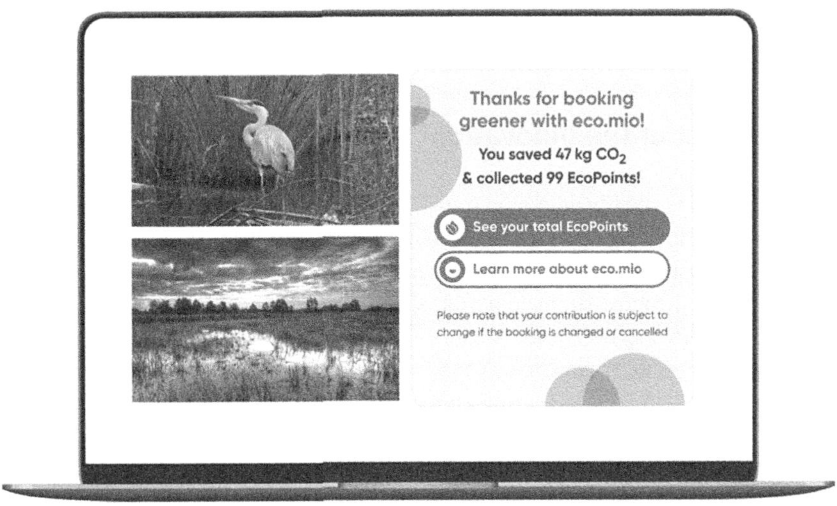

Figure 7.4 Rewards. *Source:* eco.mio GmbH/https://ecomio.com/how-it-works/ (accessed May 08, 2025).

to select more sustainable alternatives by making the process fun, interactive, competitive, and rewarding. For instance, workers can get points for selecting lower-emission flights, which can be converted to rewards within the corporate. The logic here is that, the greener the option, the more the points are awarded.

When it comes to the use of points, the innovation has four options, which can be tailored according to the culture of the corporation and structural and financial possibilities:
- Corporate donation to a nonprofit (climate protection) institution
- Vouchers through the voucher shop
- Raffles
- Pay-out to the salary account

- **Customization and personalization**: eco.mio software permits corporates to arrange the software's functionalities such as informational and gamification facets to align with their particular requirements and company policies. This implies that corporates can configure the eco.mio software topographies to align them with their corporate culture and sustainability objectives, thereby promoting employee's adoption and effectiveness. The innovation provides data-driven perspectives to assist corporates in identifying areas for improvement and tracking their progress toward sustainability goals.
- **Real-time feedback and visualization**: The eco.mio solution transparently offers immediate feedback on the possible environmental effects of business travel selections and visualizes the benefits of alternative selections. This real-time feedback assists travelers in comprehending the direct effects of their choices, enabling them to make more responsible decisions.

7.4 Impact of the Innovation

eco.mio GmbH is spearheading the transfiguration of the business travel landscape with a forward-looking approach. The start-up has assimilated contemporary technologies and sustainability schemes into online booking systems of huge companies. It has developed robust strategic partnerships with some of the industry's biggest booking tool providers such as Amadeus, Lufthansa City Center, SAP, Deutsche Bahn, and Advito (BCD Travel).

As an example, Advito, a leading global travel consultancy, and eco.mio GmbH have publicized their partnership to assimilate Advito's emissions data into eco.mio software's gamification platform in an attempt to deal with the growing need for precise pre-trip emission insights. This association takes advantage of eco.mio software's strong integration abilities and engaging gamification interface together with Advito's ISO-certified GATE4 methodology, permitting travel managers to enhance air travel booking patterns in real time and inspire workers to choose more sustainable travel alternatives. GATE4 is the only ISO-certified estimation methodology for business travel emissions in the sector.

The associations between eco.mio GmbH and these huge corporates reinforce eco.mio GmbH's market position and assist the corporates in using its solutions via interfaces on a wide scale. Consequently, eco.mio GmbH has expanded its user base to 40,000 and cut its customers' carbon dioxide emissions by 26%. Reference corporate clients like Kearney and Roland Berger are also benefiting from the eco.mio innovation. Northstar Travel Group (2025) indicated that eco.mio GmbH deals with a service existing market of €15–20 billion, considering heavy travel industries that implement supported booking tools.

The quotes to support this argument are as follows:

Erin Singleton, Senior Director at Advito:

> "We are beyond happy to be working with eco.mio to integrate our emission calculation methodology, GATE4, into online booking tools and have a direct impact on travel behaviour. Together, we pave the way to sustainable choices."

Axel Mueller, Regional Vice President at Amadeus:

> "We see great potential in working with start-ups. With eco.mio in particular, we have seen how well we complement each other and how we can benefit from their innovative strength and time-to-market. We are proud of our collaboration on the road to more sustainable business travel."

Lutz Nauert, Senior Vice President—Global Program Management EMEA at BCD Travel:

> "We are increasingly seeing the need among our customers for sustainable initiatives that go beyond pure reporting. I am excited about the partnership with eco.mio, which enables us not only to recognise these needs, but also to offer companies concrete solutions."

Stefan Pagel, Division Manager at Lufthansa City Centre:

> "We are delighted to have gained eco.mio as a young partner in the industry to help us complement our sustainability portfolio for our customers with a seamless POS solution."

Maximilian Meurer, ESG Team at Roland Berger:

> "I am glad, we found a strong technical partner to close the gap of our sustainable travel roadmap helping us to meet our sustainability targets through the facilitation of behavioral change."

eco.mio GmbH has managed to win the hearts of investors. It has raised a low seven-figure seed funding round. The financing is spearhead by IBB Ventures, the Hans(wo)men Group, and identity.vc with support from Partners in Clime, N&V

Capital, Robert Jacobi, Volker Rofalski, Alfred Luger, Patrick Diemer, Nicole Büttner, and Lea-Sophie Cramer. The founders (i.e., these individuals) of eco.mio GmbH possess insider knowledge from the software development and travel industries. Also, the founders work with a diverse, international, and determined group that oozes inherent motivation. This played a vital role in fundraising. This argument is supported by the following quotes from the investors:

Til Klein, co-founder of Identity.vc,

> "The ability to integrate directly into the existing systems of business customers makes eco.mio's solution extremely valuable. This is a crucial factor for our investment decision."

Hansi Hansmann, Hans(wo)men Group, underscores:

> "As an investor, I look for teams that not only have a ground-breaking idea but also the ability and passion to implement it. eco.mio has shown both, and I am proud to support their vision."

Cindy Beckel from IBB Ventures says:

> "The strong founding team of eco.mio impressively combines sustainability with economic viability, making it a valuable addition to our Impact Fund. We are firmly convinced of their potential and look forward to supporting eco.mio on their path to success."

eco.mio GmbH has received several awards and recognitions, indicating its impact in the business travel landscape. In 2023, eco.mio GmbH was awarded the prestigious Business Travel Sustainability Award. Also, eco.mio GmbH has been named one of PhocusWire's Hot 25 Travel Startups for 2025. The Hot 25 recognizes bold ideas and highpoints emerging corporates, which are addressing problems in the travel industry. Moreover, eco rewards have been named the number 1 business travel trend for 2025 by the Business Travel Magazine.

Users and supporters of eco.mio GmbH have been providing positive feedback concerning the impact of the innovation. These positive reviews are a strong indicator of its impact. They show user satisfaction, market acceptance, and product validation, enhancing the reputation of the corporate behind the innovation, and act as a feedback loop. Positive quotes from the users and supporters include the following:

Katja Firat, Travel Manager at adesso:

> "Reducing emissions from necessary business travel is a challenge. We do not want to impose travel bans, but instead strive for a sustainable change in the behaviour of our employees with the help of eco.mio"

Moritz, Consultant at adesso:

> "The additional information when booking helps me a lot to compare options. Being rewarded for taking the train makes it even more worth it for me."

Ben Lenk-Ostendorf, Behavioral Science Expert, Technical University Munich:

> "Climate change is the most important challenge of our time. To tackle this problem, my research simply is not enough. That's why I am proud to support eco.mio in merging scientific insights with practical solutions to combat climate change in business travel."

Fabian Kienbaum, CEO at Kienbaum:

> "Through eco.mio's innovative plug-in, we've transformed our approach to travel, engaging our employees as partners in our sustainability journey, making a real impact on the environment while maintaining our business operations."

7.5 Business Model of the Organization

eco.mio GmbH targets corporates associated with ambitious objectives to decrease their carbon emissions. eco.mio software promotes behavioral change within corporates and makes corporates appealing to prospective workers. The eco.mio solution functions on a SaaS (software-as-a-service) basis. Interestingly, the pricing is a fixed fee per booking and is dependent on the possible savings and transaction volume of the corporate.

The organization has a forward-looking business model. It has a proactive mindset, where it aggressively looks ahead to recognize prospects and alleviate probable dangers. The company meticulously plans for potential future market developments, technological advancements, and changes in customer needs. It seeks to be a leader in the market by being proactive in product, service, and operations development in order to deal with emerging market demands, instead of dealing with present market demands only. The organization is fully engaged in strategic product diversification and expansion.

The organization's urgency for immediate action is motivated by the European Union Corporate Sustainability Reporting Directive, which mandates companies to disclose their carbon dioxide emissions, reduction goals, and transition strategies. Since corporates are now forced to consider business travel carbon emissions in their emissions reduction plans, eco.mio solution provides a vital answer to satisfy the Corporate Sustainability Reporting Directive requirements.

7.6 Learning Outcomes

Business travel is the main contributor to the GHG emissions from commercial and noncommercial corporates. GHG emissions are the principal driver of climate change. Nonetheless, the lack of suitable and user-friendly behavior-based tools and strategies for sustainable business travel can hamper corporates' efforts to reduce their environmental impact. It is challenging to compute and reduce carbon emissions emanating from business travel. This motivated the development of behavior-based tools for sustainable business travel that corporates require to measure, reduce, and mitigate their climate impacts. One such tool designed to assist corporates attain their climate targets by promoting sustainable travel is the software solution called eco.mio.

eco.mio GmbH refers to a business-to-business software-as-a-service business that assists corporates in reducing business travel carbon dioxide emissions and cut travel costs without announcing harsher travel policies. The eco.mio platform combines technology and behavioral changes to make sure that corporates make sustainable decisions and chase their climate targets. Workers can select greener travel alternatives and change their behavior. eco.mio solution can reduce emissions of business travel by 26%.

It has been indicated that when dealing with carbon emissions from business travel, it is important to consider pertinent systemic interdependencies between corporates, travelers, and the environment. Inclusive mechanisms should be designed to inspire employees to reduce high-carbon business travel and elevate the consciousness of green travel. Also, measures, tools, and strategies must focus on behavioral change. Travelers need to adopt more environmentally sustainable behaviors. Generally, since it is challenging to make sweeping changes when it comes to traveler behaviors, travelers require comprehensive measures, tools, and strategies to allow and motivate them to endorse behavioral changes. Basically, there is a need to adopt an explicit behavior change perspective on the issue of carbon emissions from business travel. In terms of research, behavior change models should be incorporated when analyzing carbon emissions from business travel.

For start-up organizations like eco.mio GmbH to flourish, there is a need for trust-premised strategic relationships with investors. Investors need to have deep confidence in the organization's vision and team. In fundraising, perseverance is the key. Strategic partnerships with booking tool providers are vital. eco.mio GmbH has developed strong strategic partnerships with some of the industry's major booking tool providers such as Amadeus, Lufthansa City Center, SAP, Deutsche Bahn, and Advito (BCD Travel). Interestingly, eco.mio software works with the client's current online booking tool and requires no modifications to the environment or booking process.

In dealing with business travel emissions, clear and effective communication is critical in getting extensive support and promoting a robust collective commitment from stakeholders. For instance, eco.mio GmbH hosts webinars, is present on social media handles like LinkedIn, participates in personalized mailing campaigns, and attends numerous international conferences to get recognition, resulting in inbound leads. Recognition is precious for start-up organizations. It brings visibility to the organization and assists to link with partners, supporters, and prospective customers who support the cause.

7.7 Plans to Further Develop the Initiative

The vision of the company is to go international beyond Germany, strategically diversifying its basket of products and services, and become a European leader in decarbonizing business travel. Of late, the company has been involved in some fundraising activities to raise fresh capital. The fresh capital is expected to be used to expand eco.mio GmbH's global presence. Catering for international routes increases usage of products and services. Further, new applications for rental cars, corporates without booking tools and hotels, and innovative solutions in the field of sustainability governance taking into account internal carbon dioxide taxation and budgets are anticipated to be announced. The quote to support this is as follows:

Katharina Riederer, Co-Founder and CEO, comments:

> *"This funding allows us to scale our solutions to become a European leader and significantly expand our product range. With the support of our investors, we will significantly accelerate our vision of making sustainability a core element of every business trip."*

eco.mio GmbH can improve its software along the lines discussed below. Improving eco.mio software can involve the enhancement of its functionality, user experience, and integrating features that support its eco-friendly goal. In terms of user experience, the software interface can be enhanced in terms of its user-friendliness and intuitiveness, and the accessibility of its features can also be improved to accommodate all users, including those with disabilities. This could include simplifying navigation, adding screen reader support, and color contrast options.

Feature expansion is essential to permit users to customize facets of the software to fit their particular needs in a better way. This can include customizable reports and notification settings. Integration abilities can be enhanced through the design of plugins that permit integration with more other tools and systems that corporates usually apply in their operations. This promotes the software's versatility.

Advanced analytics and machine learning algorithms can be incorporated to offer deeper perspectives into sustainability metrics. This assists users to make more informed decisions.

Sustainability-specific features can be enhanced. These include carbon footprint tracking features that assist users in tracking and managing their carbon footprint via the software, including strategies and tips for reduction. Sustainability reporting features can be enhanced to assist users to generate reports that are in line with the universal standards. Resource management tools can be designed to help in the management of resources such as water, offering analytics and propositions for efficiency enhancements.

An online community platform can be developed within the software where users can discuss strategies, ideas, and best practices for sustainability. Support can be enhanced by offering support via webinars, tutorials, and live chats, among other ways. This assists users to maximize the software's abilities. Security measures should be continuously updated to protect user data and observe international data protection rules and regulations. In terms of compliance, the software should be continuously improved to meet pertinent environmental and data privacy rules and regulations.

Feedback mechanisms for gathering user feedback can be improved. Moreover, premised on user feedback and evolving technology, a flexible development approach that permits recurrent updates and improvements should be adopted.

References

Advito. (2025). *Advito partners with eco.mio to bring GATE4 emissions methodology into multiple online booking tools*. https://www.advito.com/resources/advito-partners-with-eco-mio-to-bring-gate4-emissions-methodology-into-multiple-online-booking-tools/

Andrade, C. (2021). The inconvenient truth about convenience and purposive samples. *Indian Journal of Psychological Medicine, 43*(1), 86–88.

Black, R., Cullen, K., Fay, B., Hale, T., Lang, J., Mahmood, S., & Smith, S. M. (2021). *Taking stock: A global assessment of net zero targets*. https://ca1-eci.edcdn.com/reports/ECIU-Oxford_Taking_Stock.pdf

Borko, S., Wouter, G., & Haixia, W. (2020). *The travel industry turned upside down: Insights, analysis and actions for travel executives*. McKinsey & Company with Skift Research. https://www.mckinsey.com/industries/travel-logistics-and-infrastructure/our-insights/the-travel-industry-turned-upside-down-insights-analysis-and-actions-for-travel-executives#/

Campbell, S., Greenwood, M., Prior, S., Shearer, T., Walkem, K., Young, S., Bywaters, D., & Walker, K. (2020). Purposive sampling: Complex or simple? Research case examples. *Journal of Research in Nursing, 25*(8), 652–661.

Creutzig, F., Jochem, P., Edelenbosch, O.Y., Mattauch, L., Vuuren, D.P.V., McCollum, D., & Minx, J. (2015). Transport: A roadblock to climate change mitigation? *Science*, *350*(6263), 911–912.

Crowe, S., Cresswell, K., Robertson, A., Huby, G., Avery, A., & Sheikh, A. (2011). The case study approach. *BMC Medical Research Methodology*, *11*, 100.

Cullen, P. (2023). eco.mio. *The Business Travel Magazine*. https://thebusinesstravelmag.com/eco-mio/

Department for Business Energy and Industrial Strategy. (2019). *UK greenhouse gas emissions, final figures*. https://assets.publishing.service.gov.uk/government/uploads/system/uploads/attachment_data/file/957887/2019_Final_greenhouse_gas_emissions_statistical_release.pdf

Dongo, I., Cardinale, Y., Aguilera, A., Martinez, F., Quintero, Y., Robayo, G., & Cabeza, D. (2021). A qualitative and quantitative comparison between web scraping and API methods for Twitter credibility analysis. *International Journal of Web Information Systems*, *17*(6), 580–606.

Ekins, P., & Zenghelis, D. (2021). The costs and benefits of environmental sustainability. *Sustainability Science*, *16*, 949–965.

Energy Institute. (2022). *Transport sector accounts for almost 30% of all energy consumption globally*. https://www.energyinst.org/exploring-energy/topic/transport

Erdoğan, S., Gedikli, A., Cevik, E.I., & Erdoğan, F. (2022). Eco-friendly technologies, international tourism and carbon emissions: Evidence from the most visited countries. *Technological Forecasting and Social Change*, *180*, 121705.

Esty, D.C., & de Arriba-Sellier, N. (2023). Zeroing in on net-zero: From soft law to hard law in corporate climate change pledges. *University of Colorado Law Review*, *94*, 635.

Gössling, S., & Dolnicar, S. (2023). A review of air travel behavior and climate change. *Wiley Interdisciplinary Reviews: Climate Change*, *14*(1), e802.Greenhouse Gas Protocol. (2013). *Technical guidance for calculating scope 3 emissions*. https://ghgprotocol.org/sites/default/files/standards/Scope3_Calculation_Guidance_0.pdf

Hillen, J. (2019). Web scraping for food price research. *British Food Journal*, *121*(12), 3350–3361.

International Energy Agency. (2022). *Transport: Improving the sustainability of passenger and freight transport*. https://www.iea.org/topics/transport

Lozano, R., & Garcia, I. (2020). Scrutinizing sustainability change and its institutionalization in organizations. *Frontiers in Sustainability*, *1*, 1. https://doi.org/10.3389/frsus.2020.00001

Martin, B., Pestiaux, J., Schobbens, Q., Emmrich, J., & Hagemann, M. (2020). *A radical transformation of mobility in Europe: Exploring the decarbonisation of the transport sector by 2040*. NewClimate Inst.

McLachlan, R., & Callister, P. (2022). *Managing New Zealand's greenhouse gas emissions from aviation*. Institute for Governance and Policy Studies.

Müller, A. (2024). Decarbonizing business travel: A qualitative exploration of the (mis-)alignment between knowledge organizations' climate strategies and travel practices. *Journal of Sustainable Tourism, 32*(8), 1557–1583.

Northstar Travel Group. (2025). *Startup stage: eco.mio helps companies guide employees to choose greener travel options.* https://www.phocuswire.com/startup-stage-eco-mio-helps-companies-guide-employees-green-travel

Peeters, P., Higham, J., Kutzner, D., Cohen, S., & Gössling, S. (2016). Are technology myths stalling aviation climate policy? *Transportation Research Part D: Transport and Environment, 44*, 30–42.

Poom, A., Orru, K., & Ahas, R. (2017). The carbon footprint of business travel in the knowledge-intensive service sector. *Transportation Research Part D: Transport and Environment, 50*, 292–304.

Raihan, A., & Tuspekova, A. (2023). The role of renewable energy and technological innovations toward achieving Iceland's goal of carbon neutrality by 2040. *Journal of Technology Innovations and Energy, 2*(1), 22–37.

Samour, A., Moyo, D., & Tursoy, T. (2022). Renewable energy, banking sector development, and carbon dioxide emissions nexus: A path toward sustainable development in South Africa. *Renewable Energy, 193*, 1032–1040.

Sharpe, E., Ruepert, A., van der Werff, E., & Steg, L. (2022). Corporate environmental responsibility leads to more pro-environmental behavior at work by strengthening intrinsic pro-environmental motivation. *One Earth, 5*(7), 825–835.

Simons, H. (2009). *Case study research in practice*. Sage Publications.

Slamet, C., Andrian, R., Maylawati, D., Suhendar, D., & W., & Ramdhani, M. (2018). Web scraping and Naïve Bayes classification for job search engine. *IOP Conference Series: Materials Science and Engineering, 288*, 012038.

Wadud, Z., Adeel, M., & Anable, J. (2024). Understanding the large role of long-distance travel in carbon emissions from passenger travel. *Nature Energy, 9*(9), 1129–1138.

Yin, R.K. (2013). *Case study research: Design and methods*. Sage Publications.

Zhang, L., Bai, Y., Zhang, R., Ma, Y., & Shen, C. (2024). Carbon emission characteristics and carbon reduction analysis of employee travel-taking a research institute as an example. *Energy Informatics, 7*(1), 95.

8

Greening the Seafood Supply Chain Through Circular Economy and Digital Transformation: A Case Study of the UNIDO Indonesian Global Quality and Standards Program

Ita Sualia[1], *Eko Ruddy Cahyadi*[2], *and Anthony Halog*[1]

[1] *School of Environment, Faculty of Science, The University of Queensland, Brisbane, Australia*
[2] *Department of Management, Faculty of Economics and Management, IPB University, Bogor, Indonesia*

8.1 Introduction

8.1.1 Background

As an archipelagic country with over 17,000 islands, Indonesia's seafood industry is vital to its national economy and global food security, generating over USD 6.4 billion in export value annually (MMAF, 2022) and providing jobs for 4.8 million Indonesian people (OECD, 2021). Globally, Indonesia ranks third in seafood production, supplying 7% of the world's seafood consumption (FAO, 2024). Despite its critical economic role, the rapid growth of the Indonesian aquaculture-based seafood raises major environmental concerns (Adyasari et al., 2021; Kunzmann et al., 2023). The expansion of aquaculture, largely driven by resource-intensive production practices such as feed and chemical overuse and land conversion, has led to water pollution and habitat degradation, particularly in sensitive ecosystems such as mangrove ecosystem (Adyasari et al., 2021; Bull et al., 2021; Ilman et al., 2016; Kunzmann et al., 2023).

The environmental footprint of Indonesia's seafood industry, which includes water pollution and habitat degradation, particularly in the production stage, underscores the limitations of the traditional linear economy model, where resources are extracted, processed, and disposed of without considering ecological boundaries and resource regeneration (Cooney et al., 2023; Laso et al., 2018).

Green Innovations in Supply Chain Management: Case Studies and Applications, First Edition. Edited by Marcia Mkansi, Godfrey Mugurusi, Aaron L. Nsakanda, Tomidayo Akenroye, and Frank Tietze.

Published 2026 by John Wiley & Sons, Inc.

Therefore, there is an urgent need to shifting aquaculture-based seafood supply chain into the circular economy model, where circular principles—such as minimizing waste, repurposing by-products of seafood or non-seafood product, and optimizing resource use—can reduce ecological strain and promote long-term sustainability (Borrello et al., 2020; Cooney et al., 2023). Emerging digital technologies, particularly the Internet of Things (IoT), offer effective solutions for enhancing sustainability in aquaculture. IoT enables real-time water quality monitoring, data collection, and supply chain transparency, fostering "greening" initiatives that support data-driven decision-making to optimize resource use and reduce environmental impact from farm to plate (Aithal et al., 2024; Rowan, 2023). In this chapter, "greening supply chain" refers to the implementation of sustainable practices aimed at reducing resource use and environmental impacts throughout seafood supply chain. Thus, the integration of the IoT technology will further support these circular processes by enhancing resource efficiency and enabling precise decision-making, ultimately improving environmental, social, and economic sustainability (Aithal et al., 2024; Fletcher et al., 2021; Rowan, 2023).

8.1.2 Objective

This chapter aims to analyze key challenges in greening Indonesia's aquaculture-based seafood supply chain and examines the role of IoT in supporting circular economy through enhanced operational efficiency and precise decision-making. To support this analysis, the chapter presents key lessons from the GQSP initiative implemented in Indonesia between 2014 and 2023 aimed at increasing productivity and expanding Indonesia's seafood global market access. These initiatives align with the global agenda, particularly Sustainable Development Goal (SDG) 1: End Hunger, SDG 14: Life Below Water, and the Blue Economy Development Framework.

8.2 State-of-the-Art Greening Seafood Supply Chains Concept

8.2.1 Green Supply Chain Management (GSCM) in the Seafood Supply Chains

Green Supply Chain Management (GSCM) has emerged as a prominent approach in the seafood industry for enhancing environmental and economic sustainability (Anugrah et al., 2024; Pintuma & Aunyawong, 2021). In this chapter, "greening" the supply chain refers to sustainable practices that reduce resource consumption and environmental degradation (Puglieri & Saavedra, 2021; Srivastava, 2007). This approach integrates environmental considerations across the supply chain

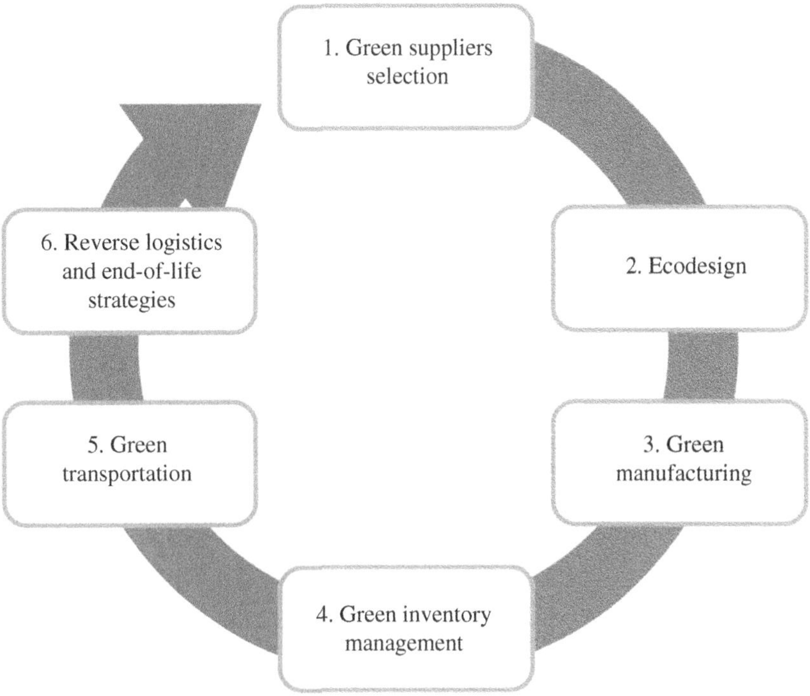

Figure 8.1 The main component of Green Supply Chain Management (Puglieri & Saavedra, 2021/Springer Nature).

and emphasizes collaboration among supply chain actors (Puglieri & Saavedra, 2021), of which, in the context of aquaculture-based seafood, includes hatcheries, farmers, collectors, processors, retailer, and exporters.

Figure 8.1 illustrates GSCM's six key components: (i) Green supplier selection involves choosing suppliers based on their environmental performance; (ii) Green design (eco-design) integrates environmental considerations into product development; (iii) Green manufacturing focuses on reducing environmental impacts during production; (iv) Green inventory management optimizes material handling and storage to minimize waste; (v) Green transportation aims to reduce emissions and energy use in logistics; (vi) reverse logistics deals with the recovery and recycling of products at the end of their lifecycle. Adopting GSCM will improve company's environmental performance and generate value through enhanced efficiency and regulatory compliance (Puglieri & Saavedra, 2021).

By adopting key components of GSCM tailored for the aquaculture-based seafood supply chain, as illustrated in Figure 8.2, starting with Green Suppliers for sustainable feed, fertilizer, and fuel, the model progresses to eco-design for fishers, hatcheries,

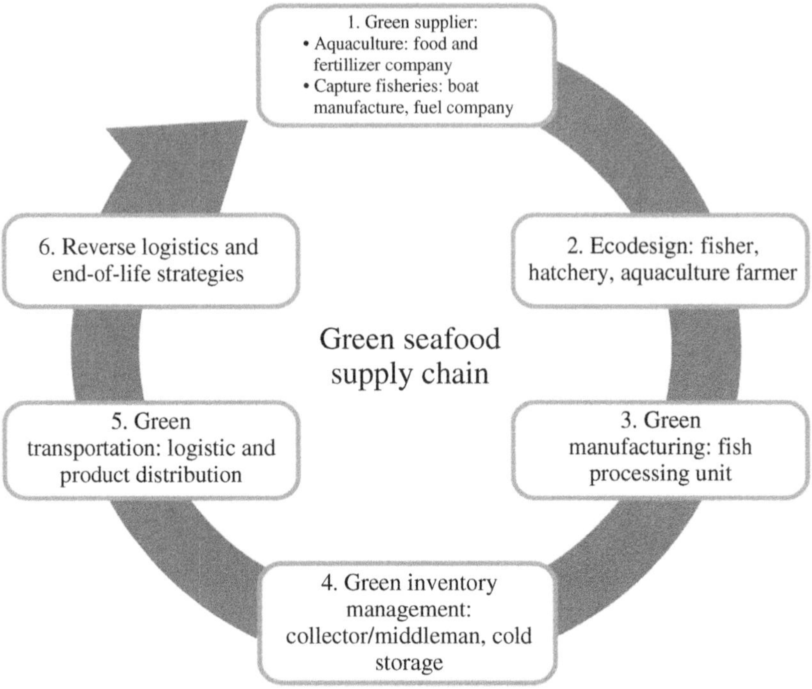

Figure 8.2 Green seafood supply chain management adopting GSCM concept.

and farmers. Green Manufacturing focuses on environmentally responsible fish processing, Green Inventory Management reduces waste through efficient fish handling and storage, and Green Transportation emphasizes low-emission logistics. Finally, Reverse Logistics and End-of-Life Strategies ensure recycling and sustainable disposal, minimizing environmental impact across the supply chain. Implementing the above GSCM components is key to reducing the aquaculture-based seafood supply chain's environmental footprint, addressing critical issues such as water pollution and habitat degradation (Anugrah et al., 2024; Pintuma & Aunyawong, 2021). By implementing these practices, seafood companies contribute to global sustainability initiatives, including SDG 1 (End Hunger) and SDG 14 (Life Below Water) increasing seafood production while protecting the environment.

8.2.2 Circular Economy in Seafood Supply Chains

The traditional linear economy model, characterized by the extract–produce–dispose approach (Cooney et al., 2023; Laso et al., 2018) is increasingly incompatible with planetary ecological limits. In seafood production, this model results in significant by-product generation throughout production, processing, distribution, and consumption stages (Cooney et al., 2023; Laso et al., 2018; Venugopal, 2022), with up to 36% seafood potentially either lost or wasted and disposed into aquatic environments, causing environmental harm (FAO, 2018; Gustavsson et al., 2013).

Shifting toward a circular economy can address these issues by promoting the reuse of by-products, enhancing resource efficiency, and reducing pressure on natural resources and ecosystems (Cooney et al., 2023); Fletcher et al. (2021) emphasize that circular economy principles, such as waste reduction and resource optimization, can improve the resilience of seafood businesses to external challenges. Integrating these principles throughout the seafood supply chain can contribute to more sustainable food systems and generate long-term environmental and economic benefits.

8.2.3 Key Actors in Aquaculture-Based Seafood Supply Chains

Identifying the roles of seafood supply chain actors is essential to effectively implement the GSCM principles (Puglieri & Saavedra, 2021). The aquaculture-based seafood supply chain is highly complex, with numerous stakeholders involved across production, processing, and logistic distribution (Anugrah et al., 2024). The key actors include hatchery, nursery operators, farmers, fish collectors, feed suppliers, processors, and distributors as illustrated in Figure 8.3.

Each actor plays a crucial role in promoting sustainability and aligning with SDGs. Hatcheries supply juvenile/post larva fish, while nursery operators prepare juveniles for grow-out stages. Fish farmers then raise the fish to market size, implementing sustainable practices such as efficient feed use and water quality management to minimize waste and pollution. Fish collectors aggregate the harvested fish to processors, who maintain quality and safety standards with minimal resource use. The supply chain then branches into local and export markets: domestic distributors and sellers serve local consumers, while exporters reach international markets that typically have stricter quality and safety standards compared to local markets in Indonesia (FAO, 2020). This integrated approach fosters an eco-friendly, resilient seafood industry, aligning with GSCM principles.

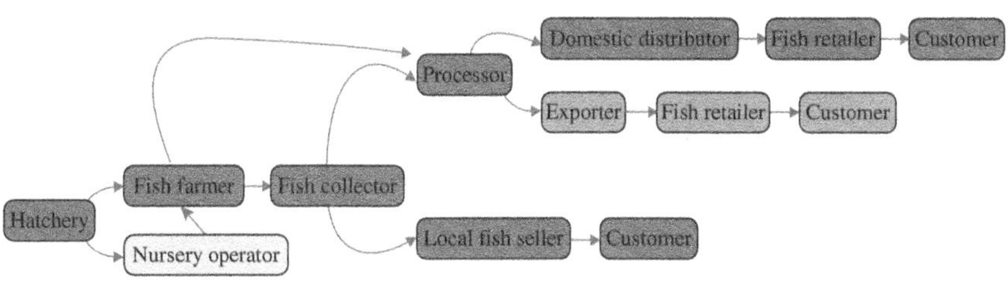

Figure 8.3 Aquaculture-based seafood supply chain actors.

8.2.4 Digital Transformation in Seafood Supply Chains

Digital transformation refers to the integration of digital technologies into all areas of a business, fundamentally changing how it operates and delivers value to customers (Verhoef et al., 2021; Vial, 2019). In the context of the seafood supply chain, digital transformation encompasses the adoption of advanced technologies that improve efficiency, traceability, and sustainability. Significant advancements have been made in digitalizing the seafood supply chain, particularly by adopting IoT and Big Data technologies (Aithal et al., 2024; Belaud et al., 2019; Pratiwy et al., 2022). IoT technologies facilitate real-time data collection and monitoring, enabling precise control over environmental conditions and resource use, which are crucial for optimizing production while minimizing environmental impact (Aithal et al., 2024; Pratiwy et al., 2022; Rowan, 2023). There are at least two digitalized aquaculture production tools that have gained popularity, namely automatic feeders and automatic water quality monitoring, as explained below.

8.2.4.1 Automatic Fish Feeders

The introduction of automatic feeders by David C. Smeltzer in 1985 marked a significant transformation in aquaculture feeding practices (Uddin et al., 2016). In Indonesia, automatic fish feeders were further developed in 2013 (Figure 8.4). Automatic feeders enable precise control over feeding schedules, quantities, and methods, allowing adjustments based on water temperature and fish behavior. The use of automatic feeders reduces feed waste, improves FCRs, and enhances overall growth rates in fish. This efficiency is crucial as feed accounts for

Figure 8.4 Automatic feeder (*Source:* e-Fishery.com).

approximately more than 50% of production costs in semi-intensive and intensive aquaculture systems (Utomo & Hermawan, 2018)

8.2.4.2 Automatic Water Quality Monitoring

Automatic water quality monitoring systems provide real-time data on critical water parameters, including dissolved oxygen (DO), pH, temperature, ammonia, and nitrate levels, (Figure 8.5) which are vital for maintaining the health of aquaculture systems (Rowan, 2023). Continuous monitoring enables immediate adjustments, ensuring that water quality remains optimal for the healthy growth of fish or shrimp. Additionally, this technology helps prevent mass mortality events, such as those caused by low DO levels.

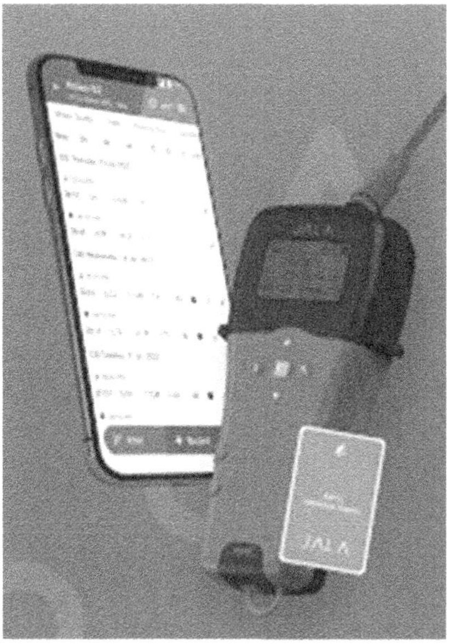

Figure 8.5 Internet-based water quality monitoring (*Source:* https://jala.tech).

To fully unlock the potential of digitalization in transforming the seafood supply chain, it is essential to address the challenges of limited infrastructure, especially in remote areas, and the high initial costs associated with acquiring and maintaining digital technologies. Moreover, improving the knowledge and skills of small-scale farmers is crucial. Overcoming these barriers will be critical for adopting more inclusive and effective digital technologies, thereby enhancing sustainability and efficiency throughout the seafood supply chain (Rowan, 2023).

8.3 The Case Study of Integrating GSCM and Digital Transformation in Aquaculture-Based Seafood Supply Chains

Integrating GSCM principles with IoT creates a robust framework for enhancing sustainability and efficiency in aquaculture-based seafood supply chains. GSCM practices reduce environmental impact through sustainable sourcing, waste reduction, and efficient resource management, while IoT enhances transparency, traceability, and operational efficiency (Belaud et al., 2019; Pratiwy et al., 2022). This approach could support Indonesia in achieving SDG 1 and 14 and aligns with the Blue Economy Development Framework by promoting sustainable and traceable seafood production. This chapter draws on lessons from a case study of the UNIDO Global Quality and Standards Programme (GQSP) in Indonesia to illustrate these integrations in the aquaculture-based seafood supply chain.

8.3.1 Method

This case study examines the GQSP Indonesia program funded by the Swiss State Secretariat for Economic Affairs (SECO), implemented from 2014 to 2023. The GQSP program was aimed at strengthening the capacities of Micro, Small & Medium Enterprises (MSMEs) to meet product quality, food safety standards, and sustainability criteria. Key interventions included implementing GAP, circular economy on feed production, and digital transformation in farming practices.

8.3.1.1 Data Collection

The impact of assessment of the greening supply chain in the case study utilized a triangulation approach, combining quantitative production data, qualitative feedback from farmers, and regression analysis. Quantitative data provided insights into production metrics, qualitative feedback captured farmers' experiences and perceptions, and regression analysis assessed the relationship between GAP adoption and key performance indicators such as feed conversion ratio (FCR), survival rates (SR), and cost savings. Data were gathered from pangasius and shrimp farmers across different regions as shown in Figure 8.6:

- pangasius farming in Tulungagung (East Java) and Muaro Jambi and Batanghari (Jambi Province)
- shrimp farming in South Lampung (Lampung Province).

a) **Primary Data**

   Primary data were gathered through structured interviews and field observations detailed as follows:

1) **Pangasius Supply Chain Evaluation**

   Data were collected between March and April 2019 using production metrics such as feed conversion ratios, production costs, and production quality. The target population included three distinct groups of farmers:

   - **Demo Farm/Full Adopters**: 12 farmers who utilized the UNIDO-formulated least cost feed (LCF) and fully implemented GAP Standard Operating Procedure (SOP), providing optimal conditions for pangasius growth.
   - **Partial Adopters**: 156 farmers who independently applied the UNIDO LCF formula using locally sourced ingredients to reduce costs but did not fully adhere to GAP SOP.

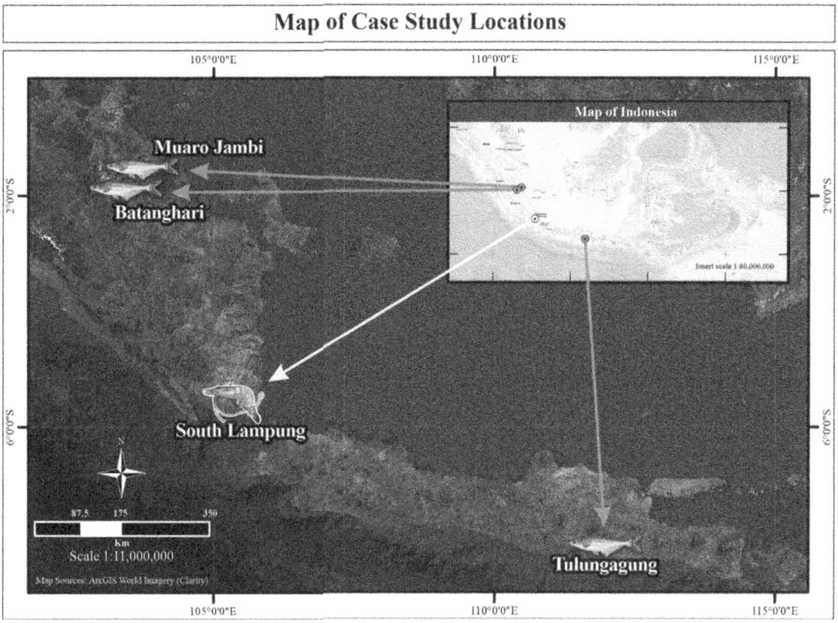

Figure 8.6 Data collection locations in East Java, Jambi, and Lampung Province, Indonesia: shrimp icons and pangasius icons.

- **Non-adopters**: 53 farmers who relied on either commercial feed or self-made feed without incorporating the LCF formula, missing the cost-efficiency benefits of the UNIDO approach.

2) **Shrimp Supply Chain Evaluation**
Data were collected from 30 shrimp farmers implementing the UNIDO GAP SOP. Key performance indicators (KPIs) evaluated included feed conversion ratio (FCR) to measure feed efficiency, survival rates (SR) of shrimp populations and cost savings achieved through improved farming practices across two distinct timeframes:
- Pre-adoption (March 2022): Before the implementation of GAP SOP.
- Post-adoption (March 2023): After the implementation of GAP SOP.

8.3.1.2 Data Collection Methods

1) **Structured Interviews**
Structured interviews were conducted with both UNIDO Good Aquaculture Practices SOP adopters (pangasius and shrimp farmers) and non-SOP adopter farmers to evaluate the program's effectiveness. For Pangasius farmers, the sampling process involved two steps. First, locations were purposively selected based on their operational length (with at least two years of project implementation)

and their geographic and cultural diversity. The selected locations included Tulungagung (East Java), Muaro Jambi, and Batanghari (Jambi). Second, respondents from these locations were randomly selected representing full adopters, partial adopters, and non-UNIDO SOP adopters. A modified Cochran formula for small populations was applied to determine a sample size of 251 farmers from a population of 511, accounting for a 5% margin of error, a 95% confidence level, and a 13% additional allowance for attrition.

For shrimp farmers, the sample was drawn from a population of 131 farmers adopting the UNIDO Good Aquaculture Practices SOP. A sample size of 30 was determined sufficient for statistical inference, following the central limit theorem (CLT), which suggests that sample sizes of 30 or more are generally sufficient for statistical inference. The sampling approach ensured a balance between UNIDO SOP adopters and non-adopters. Structured interviews were conducted to assess key performance indicators pre- and post-adoption of GAP SOP.

2) **Field Observations**

Field visits to aquaculture ponds were conducted to:

- Assess the application of GAP SOP in daily farming activities.
- Evaluate the usage of automated feeders and water quality monitoring systems.

b) **Secondary Data**

Secondary data were reviewed from various GQSP Indonesia reports, including:

- The Impact Evaluation for the National Upscaling Strategy (2019).
- Regular quarterly monitoring data provided by UNIDO field enumerators.

8.3.1.3 Data Analysis

1) **Descriptive Statistics**

- **Pangasius Production**: Descriptive statistics were employed to examine significant differences in characteristics between adopters and non-adopters, providing insights into the impact of interventions.
- **Shrimp Production**: Descriptive statistics were used to compare changes in adoption scores and outcomes between baseline and impact surveys, focusing on the same group of farmers to measure individual progress.

2) **Regression Model**

Regression analysis was applied to analyze the relationship between adoption score changes and outcome variables. Outcome changes served as the dependent variable, while adoption score changes acted as the independent variable.

$$Y_1 - Y_0 = a + b(X_1 - X_0) + c$$

$$\Delta Y = a + b\Delta X + c$$

Explanation:
Y_1 is the outcome after intervention; Y_0 is the outcome before intervention; X_1 is the adoption score after intervention; X_0 is the adoption score before intervention; a is the intercept term, representing the baseline outcome; b is the regression coefficient, measuring the effect of changes in adoption score; c is the error term or residual.

3) **Triangulation**
   The triangulation method ensured a comprehensive analysis by integrating quantitative data, qualitative feedback, and regression modelling. This approach provided an understanding of the impact of the UNIDO GQSP interventions on aquaculture practices in Indonesia.

8.3.2 Result and Discussion

This section highlights circular economy practices and digital transformation through the GQSP program in Indonesia's aquaculture-based seafood supply chain.

8.3.2.1 Circular Economy: Least Cost Feed Formulation

The UNIDO program applied circular economy principles by developing self-made fish feeds from alternative by-products and local ingredients, including palm kernel meal, leucaena leaves, water hyacinth, coconut meal, duckweed, bycatch fish, and maggots, to reduce pangasius farms' reliance on conventional/factory-made fishmeal and fish oil. The rapid expansion of pangasius (*Pangasianodon hypophthalmus*) farming in Indonesia faces challenges, especially high production costs and quality issues like yellowish tint and muddy odors in the fish flesh. Additionally, feed expenses, making up over 80% of production costs, push farmers to seek affordable, high-quality feed that meets market demands for odorless, white, lean fish (Utomo & Hermawan, 2018).

This study compared feed conversion, production costs, and production quality between ponds using LCF and conventional feeds, both with equal protein content (25–27%) to ensure iso-nitrogen conditions, with LCF formulated according to the Indonesian National Standard (SNI), while the comparison feed was either commercial or homemade without using LCF principles as detailed in the SOP.

Production data from the pangasius farming in Tulungagung, East Java, and Muaro Jambi, Jambi Province (Table 8.1) demonstrate the effectiveness of the LCF formulation. Farms using LCF formulation achieved a notably lower FCR of 1.56 compared to 1.9 in non-LCF ponds, indicating a more efficient feed use. Additionally, the LCF demo farm showed higher productivity, reaching 14.1 kg/m^2 compared to 9.7 and 9.0 kg/m^2 for adopters and non-adopters, respectively. The demo farm also yielded 3.95 kg per pond, significantly surpassing the 2.97 and 2.42 kg harvest of

Table 8.1 Comparison of pangasius farming: GQSP feed optimization adopters versus non-adopters in Tulungagung, East Java, and Muaro Jambi, Jambi

Performance	Full adopter/ demonstration farm Mean (SD)	Partial adopter pond Mean (SD)	Non-adopter pond Mean (SD)
Juvenile (stock/m^2)	18.6 (4.3)	18.8 (20.0)	16.2 (4.9)
Feed (kg/m^2)	20.8 (8.5)	17.3 (21.6)	15 (6.3)
Harvest (kg/pond)	3951.3 (2412.9)	2972.4 (1782.0)	2420.7 (984.6)
FCR	1.56 (0.2)	1.9 (1.3)	1.9 (0.7)
Pond size (m^2)	282.2 (162.7)	359.3 (217.6)	306.5 (106.7)
Productivity (kg/m^2)	14.1 (6.5)	9.7 (6.3)	9.0 (5.0)
Fish size (gram)	700	700	700
Mortality (%)	12	15.9	15.1
Number of samples	12	156	53
Flesh color	Brighter white	Slightly white	Yellowish tint
Smell	Odorless	Odorless	Muddy smell

adopters and non-adopters. Furthermore, the LCF demo farm managed to maintain a lower mortality rate of 12% compared to 15.9% and 15.1% in other ponds, with fish displaying a brighter flesh color and lacking undesirable odors. These results underscore the LCF formula's effectiveness in enhancing feed efficiency, productivity, fish flesh quality, and overall profitability in pangasius farming.

8.3.2.2 Good Aquaculture Practices and Digital Technology Transformation

The integration of SOPs for GAPs with digital technology, including the use of automatic feeders and real-time water quality monitoring, has demonstrated improvements in productivity, sustainability, and responsible resource use in shrimp aquaculture, compared to non-SOP-adopting farms. These SOPs were developed based on SNI and the FAO Code of Conduct for Responsible Fisheries, with extensive consultation across the seafood industry to ensure practical applicability.

The study shows that shrimp farmers in Lampung Province who implement the SOPs experience higher production and survival rates compared to the pre-SOP harvests, as illustrated in Figures 8.7 and 8.8. For shrimp production, every 10-point (out of 100 points) increase in the SOP level adoption corresponded to an additional 54 kg in yield per shrimp grow out cycle. The survival rate also improved with SOP adoption, addressing prior inconsistencies often caused by external

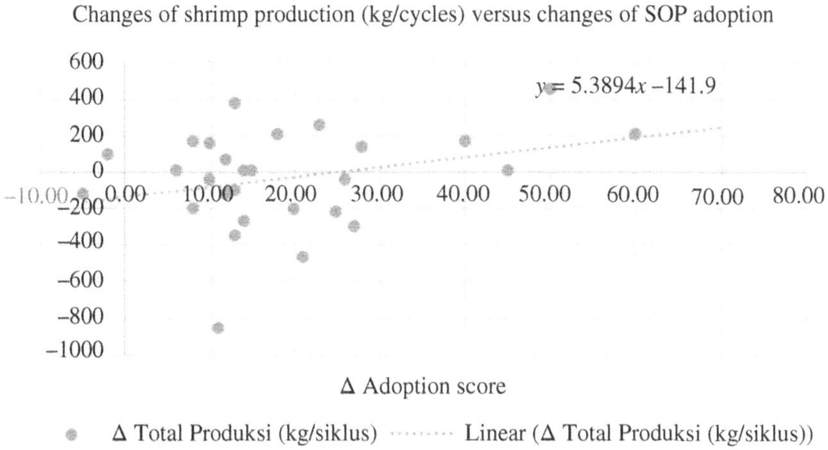

Figure 8.7 Shrimp production improvement.

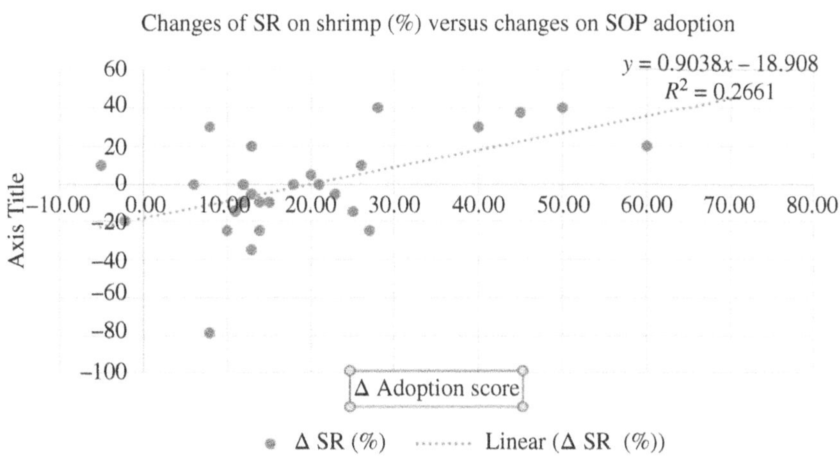

Figure 8.8 The changes of SR on shrimp farming after SOP adoption.

factors. After SOP implementation, a 10-point rise in the adoption score was associated with a 9% increase in the survival rate, demonstrating the SOPs' role in enhancing shrimp, illustrating how the SOP can boost overall production outcome, resilience, and overall farm productivity.

In terms of feed efficiency, the FCR (Figure 8.9) and revenue (Figure 8.10) showed significant improvement following the SOP adoption. Before implementation, farmers struggled with high FCRs due to inefficient feeding practices and inconsistent feed quality. However, post-adoption data indicated that a 10-point

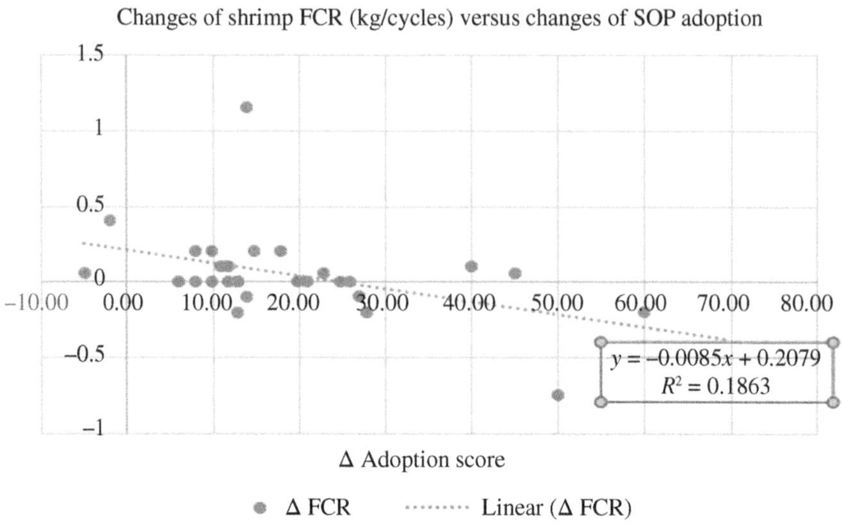

Figure 8.9 FCR and SOP adoption score.

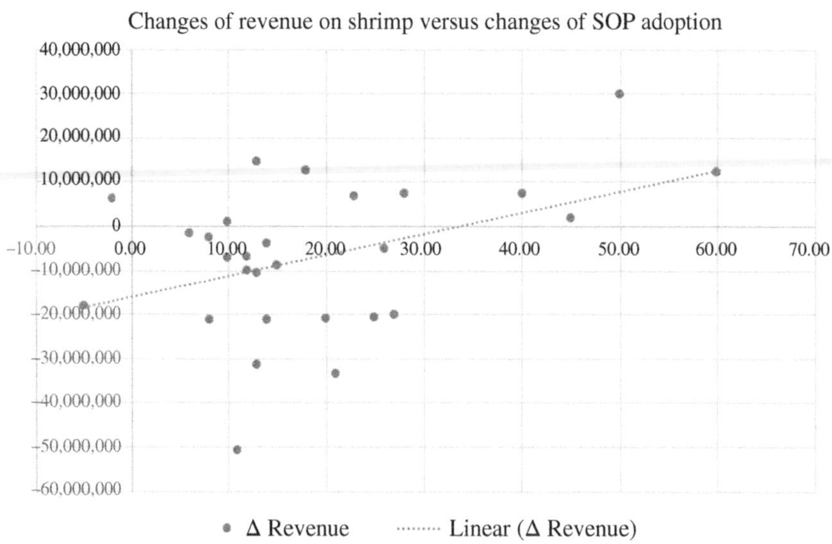

Figure 8.10 Revenue in each level of SOP adoption.

improvement in the SOP score resulted in a 0.085 reduction in FCR, highlighting better feed utilization and reduced feed costs. This improvement not only supports profitability but also aligns with sustainable practices, promoting a more efficient and eco-friendly approach to shrimp farming.

Table 8.2 Performance of vannamei shrimp ponds before (2022) and after (2023) the use of real-time water quality monitoring

Parameters	2022	2023
Average productivity (kg/m^2/cycle)	0.89	0.93
FCR	1.39	1.46
SR (%)	73.93	77.86
Average received price per kg (IDR)	62,500	50,643
Average sales (revenue)/farmer (IDR)/cycle	28,966,760	22,057,321

8.3.2.3 Digital Transformation: Automatic Feeders

In South Lampung District, Lampung Province, shrimp farmer partners utilized automatic feeders integrated with data analytics platform, alongside the adoption of SOPs to enhance feed efficiency in vannamei shrimp farming. This system delivers precise amounts of feed based on real-time data, optimizing feeding schedules and quantities. Our study showed that the automatic feeders improved FCR (Figure 8.9), reduced feed waste by 20%, and lowered production costs by IDR 6000 per m^2 compared to traditional feeding methods. This efficiency reduced the environmental impact of overfeeding and mitigated revenue losses due to the feed price. Although average gross revenue per pond per cycle decreased from IDR 28,966,760 per pond per cycle in 2022 to IDR 22,057,321 in 2023 due to external market factors, the use of automatic feeders mitigated these effects by enhancing feed efficiency and reducing costs.

8.3.2.4 Digital Transformation: Real-Time Water Quality Monitoring

In South Lampung District, Lampung Province, shrimp farmer partners utilized JALA, online water quality monitoring tools alongside the adoption of SOPs to monitor critical water quality parameters, such as pH, temperature, dissolved oxygen, and salinity. The data, accessible through the JALA web and mobile phone-based application platform, enable farmers to make informed decisions quickly, optimizing their farming practices. This real-time monitoring helped maintain better water quality, reduce disease outbreaks, and improve shrimp health, increasing the average survival rate from 73.93% in 2022 to 77.86% in 2023 and increasing productivity from 0.89 (kg/m^2/cycle) in 2022 to 0.93 in 2023 (Table 8.2).

8.4 The Learning Outcomes and Further Development

Indonesia's seafood supply chain, as evidenced by the experience of the UNIDO's GQSP Indonesia project (2014–2023), faces critical challenges that impact productivity and sustainability. Inefficiencies in production and management, a fragmented supply chain, and high logistics costs contribute to reduced productivity and efficiency. Additionally, multiple intermediaries decrease farmers' profit margins, diminish product freshness, and hinder the adoption of innovative practices. Addressing these challenges requires a holistic approach that integrates circular economy, digital transformation, and capacity building to create a more sustainable and resilient supply chain. The key learning outcomes from the GQSP and the plans for further developments in greening Indonesia's seafood supply chains are:

a) The implementation of SOPs is important to ensure consistent application of green practices. The GAP's SOP should be applied by adjusting to the local context to improve resource efficiency, reduce environmental impact, and enhance overall sustainability in shrimp aquaculture.
b) A collaboration for Green Innovation between government agencies, industry stakeholders, international organizations, and local communities is an important factor for driving the successful implementation of the SOP and other green initiatives of the Program. The Program also highlighted the necessity of inclusive approaches to drive the adoption of sustainable practices, particularly among small-scale farmers. In the future, business linkages among the supply chain actors from production input suppliers (hatchery, feed), farmers traders/industrial markets (processors), processors to buyers (local and international buyers), and financial institutions should also be an integral part of the program.
c) Monitoring and Evaluation (M&E) framework needs a combination of quantitative and qualitative assessments to measure a program's impact on environmental sustainability, economic performance, and social outcomes. In tracking the progress in the future, the M&E framework should involve external reviewers and measure the impact on the non-program farms to ensure accountability of the program.
d) Sustainability of green and circular innovations through continuous capacity-building and education is the key learning from the program. Future programs should equip farmer partners with the necessary skills and knowledge to maintain and develop sustainable practices, supported by digital platforms for ongoing training and monitoring.
e) Scalability and replicability are dependent on the project's SOP adaptability across different regions and aquaculture systems within Indonesia. Future programs should design innovations not just to address the local context but also to anticipate the project's potential to influence broader regional and global sustainability efforts. The adaptability of the innovation will ensure its replicability in other areas or countries with similar environmental and economic conditions.

References

Adyasari, D., Pratama, M.A., Teguh, N.A., Sabdaningsih, A., Kusumaningtyas, M.A., & Dimova, N. (2021). Anthropogenic impact on Indonesian coastal water and ecosystems: Current status and future opportunities. *Marine Pollution Bulletin, 171*, 112689. https://doi.org/10.1016/j.marpolbul.2021.112689

Aithal, P.S., Aithal, S., & Dillon, M. (2024). Challenges and opportunities for the convergence of blue economy, agriculture, and environmental information systems aligned with the principles of Society 5.0. In Paul, P.K., Lingam, G.D.S., & Kumar, N.S. (Eds.), *Agricultural and environmental information systems in society 5.0*. New Delhi: New Delhi Publishers. https://doi.org/10.5281/zenodo.10865385

Anugrah, R., Situmorang, M.L., & Suantika, G. (2024). Development of a green supply chain management strategy for the aquaculture industry: The case of the common carp (Cyprinus carpio L.) hatchery industry in Ciparay District, West Java, Indonesia. *Journal of Cleaner Production*, 453, 142160. https://doi.org/10.1016/j.jclepro.2024.142160

Belaud, J.P., Prioux, N., Vialle, C., & Sablayrolles, C. (2019). Big data for agri-food 4.0: Application to sustainability management for by-products supply chain. *Computers in Industry, 111*, 41–50. https://doi.org/10.1016/j.compind.2019.06.006

Borello, M., Pascucci, S., & Cembalo, L. (2020). The propositions to unify circular economy research: A review. *Sustainability, 12*(10), 4069. https://doi.org/10.3390/su12104069

Bull, E.G., Cunha de, L., & da N., & Scudelari, A. C. (2021). Water quality impact from shrimp farming effluents in a tropical estuary. *Water Science and Technology, 83*(1), 123–136. https://doi.org/10.2166/wst.2020.559

Cooney, R., de Sousa, D.B., Ríos, A.F., Mellett, S., Rowan, N., Morse, A.P., Hayes, M., Laso, J., Regueiro, L., Wan, A.H.L., & Clifford, E. (2023). A circular economy framework for seafood waste valorization to meet challenges and opportunities for intensive production and sustainability. *Journal of Cleaner Production, 392*, 136283. https://doi.org/10.1016/j.jclepro.2023.136283

European Commission. (2022). *Discarding in fisheries*. European Commission https://oceans-and-fisheries.ec.europa.eu/fisheries/rules/discarding-fisheries.en

Fletcher, A.C., Clair, R.S., & Sharmina, M. (2021). Seafood businesses' resilience can benefit from circular economy principles. *Nature Food, 2*, 228–232. https://doi.org/10.1038/s43016-021-00262-4

Food and Agriculture Organization of the United Nations (2018). *The state of world fisheries and aquaculture 2018: Meeting the sustainable development goals*. FAO. https://www.fao.org/3/i9540en/i9540en.pdf

Food and Agriculture Organization of the United Nations (2020). *The state of world fisheries and aquaculture 2020: Sustainability in action*. FAO. https://doi.org/10.4060/ca9229en

Food and Agriculture Organization of the United Nations (2024). *The state of world fisheries and aquaculture 2024: Blue transformation in action*. Rome: FAO. https://doi.org/10.4060/cd0683en

Gustavsson, J., Cederberg, C., Sonesson, U., & Emanuelsson, A. (2013). *The methodology of the FAO study: Global food losses and food waste—Extent, causes and prevention (FAO, 2011)*. SIK – The Swedish Institute for Food and Biotechnology. https://www.diva-portal.org/smash/get/diva2:944159/FULLTEXT01.pdf

Hayes, M., & Gallagher, M. (2019). Processing and recovery of valuable components from pelagic blood-water waste streams: A review and recommendations. *Journal of Cleaner Production, 215*, 410–422. https://doi.org/10.1016/j.jclepro.2019.01.028

Ilman, M., Dargusch, P., Dart, P., & Onrizal. (2016). A historical analysis of the drivers of loss and degradation of Indonesia's mangroves. *Land Use Policy, 54*, 448–459. https://doi.org/10.1016/j.landusepol.2016.03.010

Kumar, R., & Rahman, Z. (2015). Sustainability adoption through buyer-supplier relationship across supply chain: A literature review and conceptual framework. *International Strategic Management Review, 3*(1–2), 110–127. https://doi.org/10.1016/j.ism.2015.04.002

Kunzmann, A., Todinanahary, G., Msuya, F.E., & Alfiansah, Y. (2023). Review: Comparative environmental impacts and development benefits of coastal aquaculture in three tropical countries: Madagascar, Tanzania and Indonesia. *Tropical Life Sciences Research, 34*(3), 279–302. 10.21315/tlsr2023.34.3.15

Laso, J., Herrero, I.G., Margallo, M., Rowe, I.V., Fullana, P., Bala, A., Gazulla, C., Irabien, A., & Aldaco, R. (2018). Finding an economic and environmental balance in value chains based on circular economy thinking: An eco-efficiency methodology applied to the fish canning industry. *Resources, Conservation and Recycling, 133*, 428–437. https://doi.org/10.1016/j.resconrec.2018.02.004

Ministry of Marine Affairs and Fisheries. (2022). *Marine data release for the fourth quarter of 2022*. Center for Data, Statistics, and Information, Secretariat General of the Ministry of Marine Affairs and Fisheries.

Muslimin, M., Mangun, N., Rombe, E., Taqwa, E., Sutomo, M., & Hadi, S. (2021). AHP structure for determining sustainable performance of Indonesian seafood supply chain from stakeholders' perspective. *Journal of Management Information and Decision Sciences, 24*(7), 1–10.

Organisation for Economic Co-operation and Development (2021). *Sustainable ocean economy country diagnostics of Indonesia (OECD development perspectives No. 5)*. OECD Publishing. https://doi.org/10.1787/9bc36234-en

Pintuma, S., & Aunyawong, W. (2021). The Effect of Green Supply Chain Management Practices on Environmental, Operational And Organizational Performances of Seafood Manufacturers In Thailand. *International Journal of eBusiness and eGovernment Studies*, 13(2), 33–48. https:// doi.org/10.34111/ijebeg.202113203

Pratiwy, F.M., Cahya, M.C., & Andriani, Y. (2022). Digitalization of aquaculture: A review. *International Journal of Fisheries and Aquatic Studies, 10*(1), 18–22. 10.22271/fish.2022.v10.i1a.2623

Puglieri, F.N., & Saavedra, Y.M.B. (2021). Green supply chain management. In de Oliveira, J.A., Lopes Silva, D.A., Puglieri, F.N., & Saavedra, Y.M.B. (Eds.), *Life cycle engineering and management of products* (pp. 283–307). Springer. https://doi.org/10.1007/978-3-030-78044-9_12

Regueiro, M., & Jimenez, A.A. (2021). Minerals in the future of Europe. *Mineral Economics, 34*, 209–224. https://doi.org/10.1007/s13563-021-00254-7

Rowan, N.J. (2023). The role of digital technologies in supporting and improving fishery and aquaculture across the supply chain—Quo vadis? *Aquaculture and Fisheries, 8*, 365–374. https://doi.org/10.1016/j.aaf.2022.06.003

Seuring, S., & Müller, M. (2008). From a literature review to a conceptual framework for sustainable supply chain management. *Journal of Cleaner Production, 16*(15), 1699–1710. https://doi.org/10.1016/j.jclepro.2008.04.020

Srivastava, S.K. (2007). Green supply-chain management: A state-of-the-art literature review. *International Journal of Management Reviews, 9*(1), 53–80. https://doi.org/10.1111/j.1468-2370.2007.00202.x

Uddin, M.N., Rashid, M., Mostafa, M., Belayet, H., Salam, S., Nithe, N., Rahman, M., & Aziz, A. (2016). Development of automatic fish feeder. *Global Journal of Researches in Engineering: A Mechanical and Mechanics Engineering, 16*(2), 15–23.

Utomo, N.B.P., & Hermawan, I. (2018). Standar prosedur operasional aplikasi teknik least cost formulation (LCF) untuk produksi pakan ikan patin skala kecil. In Prawiro, S. (Ed.), *SMART-Fish Indonesia*. United Nations Industrial Development Organization (UNIDO). ISBN: 978-602-9156-44-7. https://drive.google.com/file/d/1nZNh9XlwKlZRBKWSWkedx-9u4gwF6hyf/view

Venugopal, V. (2022). Green processing of seafood waste biomass towards blue economy. *Current Research in Environmental Sustainability, 4*, 100164. https://doi.org/10.1016/j.crsust.2022.10.164

Verhoef, P.C., Broekhuizen, T., Bart, Y., Bhattacharya, A., Qi Dong, J., Fabian, N., & Haenlein, M. (2021). Digital transformation: A multidisciplinary reflection and research agenda. *Journal of Business Research, 122*, 889–901. https://doi.org/10.1016/j.jbusres.2019.09.022

Vial, G. (2019). Understanding digital transformation: A review and a research agenda. *The Journal of Strategic Information Systems, 28*(2), 118–144. https://doi.org/10.1016/j.jsis.2019.01.003

9

Business Model Innovation and Technological Advancement for Sustainable E-Waste Management: A Case Study of EWaste Africa in South Africa

Ernest Marfo Asiedu[1], *Frank Ranganai Matenda*[1], *and Pravashen Naidoo*[2]

[1] *Department of Operations Management, University of South Africa, Pretoria, South Africa*
[2] *Electro-Mechanical Engineering, University of Cape Town, Cape Town, South Africa*

9.1 Introduction

Electronic waste, commonly known as e-waste, refers to discarded electronic and electrical equipment nearing the end of its useful life. This comprises any discarded product with a battery or plug, containing toxic and hazardous substances such as mercury, which can pose severe risks to human and environmental health (Geneva Environment Network, 2024; United Nations Environment Management Group, n.d.). This complex waste stream, comprising a mixture of metals, plastics, and composites, originates from a wide range of unwanted electronic and electrical objects, including computers, televisions, smartphones, speakers, music players, keyboards, and lithium car batteries (The global E-waste Monitor, 2024).

The proliferation of e-waste presents an escalating global challenge. The rate and volume at which e-waste is generated have amplified rapidly in recent years, reaching extraordinary levels (Balde et al., 2024). In 2022, a record 62 million metric tons (Mt) of e-waste was produced globally (equivalent to an average of 7.8 kg per capita per year), representing an 82% increase since 2010 (Balde et al., 2024; ITU, 2024). Projections indicate this figure is on track to rise another 32% to 82 Mt by 2030, underscoring an annual increase of 2.6 Mt (Balde et al., 2024). This growth in e-waste generation is significantly outpacing formal collection and recycling efforts, by almost a factor of five since 2010 (ITU, 2024). The sheer

Green Innovations in Supply Chain Management: Case Studies and Applications, First Edition. Edited by Marcia Mkansi, Godfrey Mugurusi, Aaron L. Nsakanda, Temidayo Akenroye, and Frank Tietze.

Published 2026 by John Wiley & Sons, Inc.

volume of e-waste generated in 2022, for instance, would fill 1.55 million 40-tonne trucks, enough to form a bumper-to-bumper line encircling the equator (Balde et al., 2024). This escalating volume places considerable pressure on disposal facilities (Althaf et al., 2021).

The exponential increase in e-waste generation is driven by several interconnected factors. Key among these are rapid technological advancements, leading to shorter product lifespans and amplified consumer demand for newer devices, often motivated by consumerism (Saha et al., 2021; United Nations Environment Management Group, n.d.). For example, studies have shown a progressive decline in the average life expectancy of computers, with some reporting a reduction from 4–6 years in 1997 to just two years in 2005 (Babbitt et al., 2009; Needhidasan et al., 2014). This trend is compounded by a lack of repair alternatives for older gadgets. Furthermore, the problem is exacerbated by the unlawful exportation and unsuitable donation of electronic devices from developed to developing economies, thereby shifting the burden of managing complex waste to regions with limited infrastructure and regulatory frameworks (Saha et al., 2021).

Despite containing valuable materials such as copper, silver, and gold, e-waste also harbors hazardous toxins like mercury, lead, arsenic, and cadmium. These dangerous substances can result in severe air, soil, and water pollution, posing significant health hazards including cancers, neurological damage, and respiratory issues (Saha et al., 2021). Beyond direct health impacts, improper waste management contributes to environmental degradation and greenhouse gas emissions, exacerbating global warming, which in turn leads to extreme weather conditions and disturbances of ecosystems and food production. Annually, billions of dollars' worth of strategically valuable resources contained within e-waste are squandered and dumped, with just 1% of rare earth element demand currently being met by e-waste recycling (Balde et al., 2024).

The majority of e-waste worldwide is currently recycled in the unregulated informal sector (Saha et al., 2021), predominantly located in developing countries. For instance, in India, approximately 95% of e-waste is processed informally (Henzler et al., 2017). Workers in this sector are often exposed to dangerous substances due to rudimentary methods like open burning and a general lack of safety measures, leading to severe environmental and health impacts (United Nations Environment Management Group, n.d.). Globally, only 22.3% of the total e-waste generated in 2022 (13.8 Mt) was documented as being properly collected and recycled (Balde et al., 2024; ITU, 2024). Documented formal collection and recycling rates vary significantly across regions; Europe had the highest rate, recycling 42.8% (7.5 kg per capita) of its e-waste in 2022, whereas African countries reported the lowest rate, with less than 1% formally collected and recycled (ITU, 2024). Alarmingly, the United Nations Environment Programme (UNEP) estimated that 60–90% of the world's electronic waste, worth nearly USD 19 billion, is illegally traded or dumped each year (United Nations Environment Management Group, n.d.).

This discrepancy in the amount of e-waste produced and the amount that is properly recycled reflects an urgent need for all stakeholders, including the youth, to address this issue. Recognizing this, an increasing number of initiatives from international actors, including the United Nations, have emerged (United Nations Environment Management Group, n.d.). The UN's System-wide Response to Tackling E-waste (2017) highlighted the need for strengthened collaboration among over 20 United Nations organizations active in e-waste, with more than 150 initiatives undertaken since 2004, offering recommendations for maximizing system-wide coherence towards a life-cycle approach (United Nations Environment Management Group, n.d.).

South Africa is no exception to this global challenge, being one of Africa's major e-waste generators. The global E-waste Monitor (2024) indicates that, in Africa, the largest volumes of e-waste currently come from Egypt (690 kT), followed by South Africa (530 kT), Nigeria (500 kT), and Algeria (330 kT). In South Africa, e-waste generation is increasing at three times the rate of solid municipal waste, leading to an impending landfill crisis in major metropolitan areas. The country faces several problems, including inadequate infrastructure, a lack of public education on the topic, and behavioral barriers (Saha et al., 2021).

Therefore, this study aims to explore the business model innovation and technological advancement for sustainable e-waste management using EWaste Africa in South Africa. The rationale for selecting this enterprise is that EWaste Africa prides itself on commissioning Africa's first lighting waste recycling plant and is a market leader in electronic, solar and lighting waste recycling (EWaste Africa.net). This study seeks to achieve its overall aim by pursuing the following objectives:

- To analyze the business model innovations employed by EWaste Africa to ensure sustainability and growth in e-waste management.
- To identify and evaluate the innovative solutions and technological advancements implemented by EWaste Africa for effective e-waste recycling and management.
- To examine the key challenges faced by EWaste Africa in managing e-waste and the strategies used to address these challenges.
- To evaluate the effectiveness of EWaste Africa's waste management practices and their impact on sustainability.

This study is significant as the dynamics of EWaste Africa's operations provide a foundation for e-waste management that can lead to the enhancement of sustainable e-waste management in South Africa and developing countries in general. This study contributes to the continuing discourse concerning e-waste management in South Africa and provides novel perspectives for researchers, policymakers, and other diverse stakeholders. The remainder of this study is divided into five sections. The literature review is followed by the methodology, presentation of findings, discussion, and conclusions.

9.2 Literature Review

9.2.1 Business Model Innovation for Sustainable E-Waste Management

Electrical and electronic gadgets enhance the living standards of the people globally (Saha et al., 2021). They provide opportunities for education, and make communication and access to information easier, among other things (Cheshmeh et al., 2023). Nonetheless, e-waste is an economic, environmental, and social problem (Gaur et al., 2024). Thus, several countries are attempting to cut e-waste generation and manage it properly. Nevertheless, these countries are facing a myriad of challenges.

Several countries face numerous challenges in e-waste management. These challenges include the absence or indistinct policies, regulations, and rules (Gaur et al., 2024). For instance, literature has indicated that Extended Producer Responsibility (EPR) regulations are not being adequately implemented or there are delays in their implementation (Leclerc & Badami, 2020; Park et al., 2018). Further, developed countries send their e-waste to developing countries due to the lack of implementation of regulations (Awasthi et al., 2018; Zacho et al., 2018). Similarly, Saha et al. (2021) indicated that developed countries dump electronic and electrical devices in Africa. As a result, e-waste has been piling in developing countries that cannot effectively and efficiently recycle it and that do not have regulations that promote sustainable e-waste management practices (Mishra & Mishra, 2023). Ahirwar and Tripathi (2021) postulated that the majority of Oceanian and African economies lack e-waste management laws or frameworks to address the problem of e-waste and informal recycling.

In many countries, particularly developing countries, there is no awareness among consumers when it comes to the significance of responsible e-waste disposal. Gaur et al. (2024) propounded that consumers, in place of dumbing products to the manufacturers after using them so that they can be recycled, discard them or dump them into a landfill. Nowadays, we have seen an increase in green products, which do not lead to environmental degradation. However, customers continue to by non-eco-friendly products due to their ignorance of the benefits of green products (Mehta & Chahal, 2021).

The majority of e-waste is being recycled in the informal sector. In 2019, roughly 17.4% (9.3 Mt) of the universal e-waste produced was gathered and recycled in the formal sector, whilst the remaining 44.3 Mt (82.6%) of e-waste was not accounted for

(Forti et al., 2020). Perkins et al. (2014) opined that 25% of e-waste is recycled in formal recycling hubs with enough employee protection. Existing literature indicate that, in Africa, e-waste management is under the control of the informal sector. Ahirwar and Tripathi (2021) opined that the majority of the e-waste, principally in Asian and African countries, is handled by the informal sector. The authors further proffered that in Africa this is caused by the absence of particular e-waste regulations or their application.

9.2.2 Technology Advancement for Sustainable E-Waste Management

Existing literature has indicated that the rate and volume at which e-waste is generated are more than the rate and volume at which it is recycled. Given that e-waste is not biodegradable and comprises of substantial amounts of hazardous and environmentally sensitive chemicals, it ends up piling in landfills and ultimately poses dangers to the environment and its inhabitants. e-Waste is a critical problem that results in environmental degradation and severe human health consequences if it is not managed safely (Magalini, 2016). The e-waste problem is worsened by the fact that these devices are diverse when it comes to size, function, weight, material components, and environmental effect at end-of-life (see Magalini, 2016; Shittu et al., 2021). Moreover, the gadget complexity of e-waste makes it challenging to sort and recycle e-waste.

Even though the rate and volume at which e-waste is generated are exponentially increasing, the lack of infrastructure is a challenge (Cheng et al., 2019; Gunarathne et al., 2020; Saha et al., 2021). Several countries and regions lack recycling units designed to handle the intricacies of e-waste. This often leads to improper disposal approaches such as incineration, uncontrolled dismantling, and landfilling. Existing literature indicates that the infrastructure challenge is exacerbated by the increasing sophistication and daily generation of electronic and electrical gadgets (Kavil et al., 2025; Liu et al., 2023). While some stages of e-waste recycling, including certain mechanical dismantling processes, can be resource intensive, it is important to note that manual dismantling, when performed safely and with appropriate protective measures, can be highly efficient and cost-effective for recovering materials from e-waste, particularly in regions with lower labor costs. This underscores the need for proper training and safety protocols in all recycling processes.

Formal recyclers are affected by technological challenges. For the formal sectors, e-waste recycling is challenging since they lack skilled labor with sufficient knowledge concerning recycling tactics (Gaur et al., 2024). Also, Azadnia et al. (2021) signified that the formal sectors are implementing out-of-date technologies and approaches. Under the same line of reasoning, existing literature has indicated that recyclers are finding it difficult to adopt new practices like green recycling and biological treatment of e-waste (Bocken & Konietzko, 2023; Chakrabarty & Nandi, 2021; Jin et al., 2020). Challenges in the application of efficient e-waste management also include community cultural challenges (e.g., deficiency of convenient recycling choices, nonexistence of consciousness concerning the environmental effect of e-waste, and culture of disposability where individuals gladly substitute old gadgets with new ones) and economic challenges (e.g., lack of funds to acquire new technologies) (see Bouzon et al., 2016; Chaudhary et al., 2017; Gaur et al., 2024; Kumar and Dixit, 2018; Mhatre et al., 2023). Jangre et al. (2022) articulated that, in a developing country, managing e-waste is challenging due to the following diverse barricades hampering the process: "lack of customer awareness about return;" "less policies addressing e-waste problem;" "lack of long-term planning" and "insensitiveness of public towards environmental issues."

9.3 Methodology

This section discusses the methodology used for the empirical research conducted in this study, as illustrated in Figure 9.1.

In Figure 9.1, the study used a revelatory case study, which is a type of in-depth single case study design that aims to reveal something unique, previously unknown, or poorly understood about the topic (Yin, 2018). In view of this, the study explores a sustainable electronic waste recycling, using EWaste Africa as the

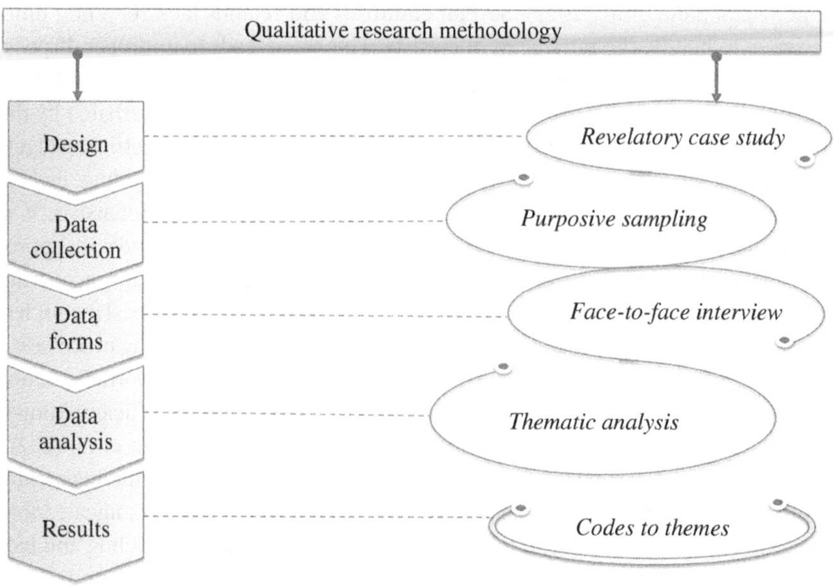

Figure 9.1 Qualitative research methodological model. *Source:* Asiedu (2024) / Telecommunications Association.

case. This case was selected because it is a case of interest (Stake 2005). The choice of a revelatory case study design is particularly useful when the case represents a unique single case (Yin, 2014), and data collection can also be a single long-form interview. Therefore, the study used this design to effectively gain in-depth insights into strategic and operational aspects from the owner's perspectives.

In selecting the data source, the study purposively selected the Founder and Chief Executive Officer of EWaste Africa's as the sole interviewee. The purposive sampling method was adopted since it better matches the aim of the study, thus promoting the study's robustness and reliability (Campbell et al., 2020). To gather data, the study used face-to-face interviews, as it allows participants to respond meticulously and honestly and disclose sensitive information (Schober, 2018). In addition, Schober (2018) indicated that face-to-face interviews can build rapport easily and permit the interviewer to collect richer qualitative data by asking follow-up questions. The study used semi-structured interviews because they are exploratory in nature (Patton, 2002), and responses were recorded with the interviewee's consent.

For data analysis, the study used qualitative thematic analysis with Atlas.ti software. The study followed the qualitative data analysis process using thematic analysis, first preparing the data for analysis by transcribing the recordings and cleaning the transcripts. Codes were then created and grouped into clusters called categories, from which themes and aggregated themes were formed (Braun et al., 2014; Miles et al., 2020). The findings were then displayed using graphical display methods (Miles et al., 2020). The intent of using graphical display methods is to provide ready access to information and convey a message, a discovery, or a particular perspective on specific data or topics (Iliinsky, 2010).

9.4 Findings and Discussions

This section presents and discusses the business model innovation and technological advancement for sustainable e-waste management using EWaste Africa in South Africa. Four aggregated themes emerged from analyzing the Founder and Chief Executive Officer of EWaste Africa's narrative: business model innovation; innovative solutions and technological advancement; addressing challenges; and evaluating impacts and effectiveness, as presented later.

9.4.1 Business Model Innovation

The first aggregated theme that emerged was business model innovation. This is characterized by two main themes: business models and growth and stakeholder engagement, as discussed thereof.

9.4.1.1 Business Models and Growth

This theme is explained by two subthemes, namely business models EWaste Africa employs to ensure sustainability and growth, and future trajectories and potential growth areas for EWaste Africa.

EWaste Africa's Sustainable Business Models This outlines the business models EWaste Africa employs to ensure sustainability and growth. The analysis from the Founder and Chief Executive Officer of EWaste Africa's narrative has been categorized under five main areas (see Figure 9.2).

First, EWaste Africa adopts a fee-for-service model as its primary business approach, wherein corporations are charged for compliant e-waste management services. The Founder and Chief Executive Officer of EWaste Africa explained:

> "Currently, clients pay us for compliance. Ensuring their compliance adds value and provides a strong value proposition for them."

Additionally, EWaste Africa employs the EPR model, whereby producers assume responsibility for the end-of-life management of their products and compensate EWaste Africa to fulfill their recycling targets. The Founder and Chief Executive Officer of EWaste Africa stated, *"Going forward, producers are now paying us if they have recycling targets in place."*

Furthermore, EWaste Africa emphasizes the long-term economic value of recovered raw materials as a future revenue stream. Although currently reliant on service fees, the company's long-term vision involves transitioning to a model driven by the intrinsic value of extracted materials from e-waste. This approach aims to reduce dependency on compliance fees, as the Founder and Chief Executive Officer of EWaste Africa noted, "*We believe that the first two models will eventually phase out, and the natural economics of the value inherent in these raw materials will take precedence.*"

Moreover, corporate activation days are part of EWaste Africa's business models. These activation days, while free in the short term for participating

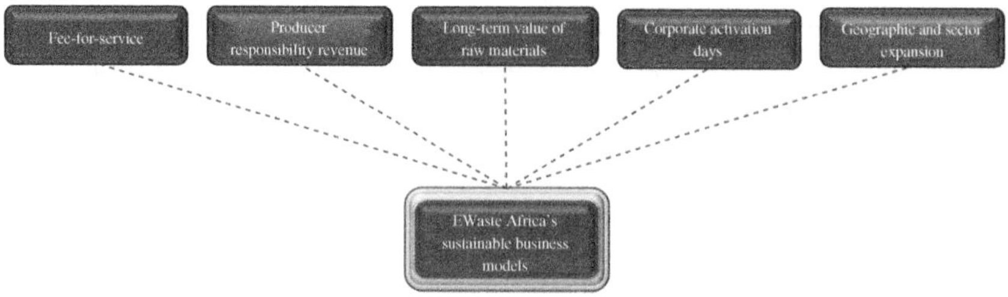

Figure 9.2 EWaste Africa's sustainable business model. *Source:* Authors' own idea from the field data (2025).

corporations, serve as a marketing and brand-building tool. They enhance visibility and potentially lead to future client acquisition and partnerships as noted by the Founder and Chief Executive Officer of EWaste Africa:

> "One unique approach we've introduced from an advertising perspective is that we actively go out, set up, and install the necessary facilities. We also engage in some media outreach, encouraging people to bring in their old waste, assuring them that it will be recycled ethically and in full compliance with regulations."

Lastly, geographic and sector expansion is a key business model for EWaste Africa. Expanding into new geographic areas and waste streams, such as solar panels and batteries, is a strategic move to enhance market reach and diversify revenue. The Founder and Chief Executive Officer of EWaste Africa stated:

> "In 2016, we expanded into Gauteng and began the process of licensing a comprehensive electronic waste recycling facility. To strengthen compliance and add credibility, we also obtained our ISO 9001, 14001, and 45001 accreditations. Additionally, we are exploring the development of our facility within a special economic zone, which is a strategic move for us."

Future Trajectories and Potential Growth Areas for EWaste Africa The future trajectories and potential growth areas for EWaste Africa are illustrated in Figure 9.3 and presented therein.

Solar panel recycling expansion is identified as the most significant future growth area for EWaste Africa. The Founder and Chief Executive Officer of EWaste Africa

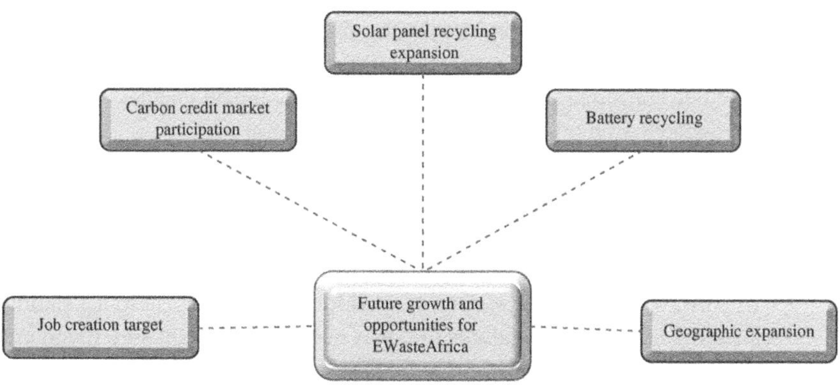

Figure 9.3 Future growth and opportunities for EWaste Africa. *Source:* Authors' own idea from the field data (2025).

highlighted that solar panel recycling is expected to become the largest part of their business by 2025. The Founder and Chief Executive Officer of EWaste Africa noted:

> "*The coolest thing we've done recently, and the most exciting, is the recycling of solar panels. This is probably going to be the biggest part of our business by 2025.*"

Moreover, EWaste Africa aims to create a significant number of jobs, reflecting ambitious growth plans. The Founder and Chief Executive Officer of EWaste Africa indicated that they aim to reach 200 employees by 2029, potentially sooner. He noted:

> "Our target is ambitious. I personally had to retrench over 200 people in my corporate career, and I think I will break even at 200. We are on track to hit that by 2029, and I think we might achieve it a year or even a year and a half earlier."

9.4.1.2 Stakeholder Engagement

This is the second theme that emerged under the aggregated theme business model innovation, and it is characterized by the key stakeholders involved in EWaste Africa's operations and their roles (Figure 9.4).

Corporate clients (waste generators and compliance seekers) are key stakeholders of EWaste Africa. They generate significant e-waste volumes and need compliant disposal solutions. They pay EWaste Africa for their services. The Founder and Chief Executive Officer of EWaste Africa mentioned:

> "We initially had a number of good corporate social responsibility clients, however a lot of corporate South Africa came on board just before the implementation of the ban in 2016 and these clients paid us to ensure that they are kept compliant. By ensuring that they are compliant means that we have a value add and a value proposition that they are willing to continue to pay for."

Consumers, as waste generators within the framework of EPR, are becoming increasingly significant stakeholders, particularly as EPR expands producer accountability to include consumer-generated waste. This development enables consumers to utilize designated drop-off points for responsible disposal.

Government legislation (norms and standards, EPR) drives the e-waste management sector and EWaste Africa's growth. However, enforcement is a potential gap. The Founder and Chief Executive Officer of EWaste Africa noted:

> "In 2013, when I left the corporate world, there was a specific legislation called the norms and standards that was published. I'm not sure if you are aware of the National Environmental Waste Management Act in detail, but it's the Act that governs waste management in South Africa. This evolution of this legislation is what ultimately drove the growth of our business to where we are right now."

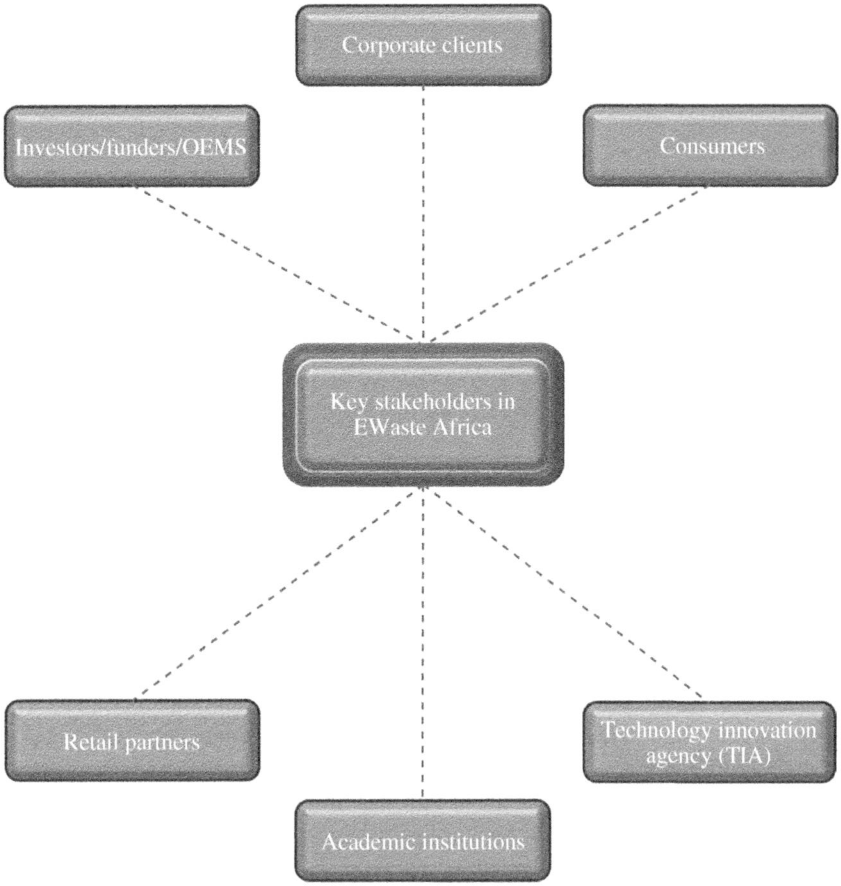

Figure 9.4 Key stakeholders in EWaste Africa. *Source:* Authors' own idea from the field data (2025).

Technology Innovation Agency (TIA) is a key funding partner, particularly for innovation projects such as the LPX™ plant, facilitating the commercialization of university research. The Founder and Chief Executive Officer of EWaste Africa stated, *"It's a partnership that we have established with TIA, who coincidentally also funds the Innovation hub, an organization that supports the incubation of green business's in South Africa."*

Investors and funders providing growth capital are critical in supporting EWaste Africa's expansion and development. The Founder and Chief Executive Officer of EWaste Africa highlighted:

> "We are currently in our first Series-A funding round to scale up our solar panel recycling process and increase our electronic waste processing capacity."

Academic institutions, such as the University of KwaZulu-Natal (UKZN), serve as technology and R&D partners and are essential stakeholders. For instance, UKZN contributes research and technological expertise, which EWaste Africa commercializes, exemplified by the LPX™ plant. The Founder and Chief Executive Officer of EWaste Africa emphasized:

> "We have collaborated with academic institutions, particularly UKZN, to find ways to commercialize a project they had been working on for years. Our business incorporates various elements that can effectively bridge academia and commercialization."

Retail partners, including major chain stores such as Pick n Pay, play a vital role in the e-waste collection network by offering essential infrastructure for consumer waste disposal via drop-off points. The Founder and Chief Executive Officer of EWaste Africa stated:

> "We are currently rolling out drop-off points at all our Pick n Pay outlets across South Africa, aiming to reach 250 stores by June next year."

Original equipment manufacturers (OEMs), including automotive manufacturers such as Toyota, provide specific waste streams, such as electric vehicle batteries, which EWaste Africa manages through specialized handling processes. The Founder and Chief Executive Officer of EWaste Africa explained:

> "Toyota's electric vehicle batteries are now being processed through our facilities, and we ship them to Europe for further recycling."

9.4.2 Innovative Solutions and Technologies

This is the second aggregated theme that emerged after analyzing the Founder and Chief Executive Officer of EWaste Africa's narratives. This is explained by two main themes namely innovative solutions being used by EWaste Africa to address the challenges, and strategies that EWaste Africa uses to address these challenges.

9.4.2.1 Innovative Solutions by EWaste Africa

Specific innovative solutions being used by EWaste Africa to address the challenges are presented here. From the narratives of the Founder and Chief Executive Officer of EWaste Africa, four key subthemes emerged, culminating in the main theme of innovative solutions by EWaste Africa, as illustrated in Table 9.1.

The first subtheme focuses on the development of the LPX™ plant, a pioneering innovation in rare earth metal extraction. The findings indicate that the LPX™ plant represents a significant technological advancement by EWaste Africa, developed in collaboration with the UKZN and TIA. This initiative aims to address

Table 9.1 Thematic analysis of innovative solutions and technologies used by EWaste Africa

Aggregated themes	Themes	Subthemes
Innovative solutions and technologies	Innovative solutions by EWaste Africa	*LPX™ plant development (rare earth extraction innovation).*
		Innovative Glass-Recycled Pavers
		Solar panel recycling facility (pioneering in Africa
		Corporate activation days (awareness and collection).
	Strategies employed by EWaste Africa	*Establishing recycling infrastructure (facilities).*
		Charging for compliance
		Leveraging extended producer responsibility (EPR) and market dynamics
		Technology and innovation
		Enhancing consumer accessibility (drop-off points)

the challenge of phosphor powder waste while enabling the extraction of valuable rare earth metals. The Founder and Chief Executive Officer of EWaste Africa remarked:

> "The LPX™ plant—we own the rights to the commercialization of the patent as EWaste Africa. The plant is currently being commissioned at our Pietermaritzburg facility. It represents a world-first innovation for the extraction of rare earth metals from lumina phosphor powder."

The second subtheme focuses on innovative glass-recycled pavers as part of glass recycling advancements. The findings highlight the development of pavers and low-cost housing blocks made from recycled glass derived from e-waste as a pioneering solution to the challenge of glass waste, particularly given the lack of recycling infrastructure in South Africa. This innovation is further emphasized in the Founder and Chief Executive Officer of EWaste Africa's statement:

> "Over the years, we developed a process that integrates recycled glass into the production of pavers and low-cost housing blocks, such as concrete blocks with hollow centers. Currently, through our novel technology, we have achieved a 98% diversion of byproducts from landfill. The pavers on the right-hand side consist of 80% recycled material sourced from our electronic waste."

The third subtheme centers on solar panel recycling facility (pioneering in Africa). The analysis highlights that establishing a solar panel recycling facility is an innovative move, positioning EWaste Africa at the forefront of a rapidly growing waste stream and a nascent industry in Africa. The Founder and Chief Executive Officer of EWaste Africa propounded that:

> "The coolest thing we've done most recently, and the most exciting, and it's probably going to be the biggest part of our business come 2025, is the recycling of solar panels. Just to give you an idea, we commissioned the plant probably 18 months ago. We've done over 20 000 panels to date."

The fourth subtheme emphasizes corporate activation days (awareness and collection). From the Founder and Chief Executive Officer of EWaste Africa's narrative, EWaste Africa embarks on corporate activation days, which are an innovative approach to raising awareness, promoting e-waste recycling, and facilitating collection, particularly targeting corporate employees and the public. As the Founder and Chief Executive Officer of EWaste Africa stated:

> "*Something different we've started from an advertising perspective is free corporate activation days. We go out, set up, and place e-waste bins within their facilities. We do a bit of media to say, "Bring in your old waste, and we'll ensure it is recycled ethically and compliantly.*"

9.4.2.2 Strategies Employed by EWaste Africa

The strategies implemented by EWaste Africa in its operations are outlined in this section. An analysis of the Founder and Chief Executive Officer of EWaste Africa's narrative reveals five key subthemes under the overarching theme of strategic solutions for sustainable e-waste management, as presented in Table 9.1.

The first subtheme focuses on the establishment of recycling infrastructure. According to the Founder and Chief Executive Officer of EWaste Africa's narrative, developing specialized facilities is a fundamental solution. EWaste Africa is actively constructing the necessary physical infrastructure for e-waste processing, beginning with lighting waste and subsequently expanding to broader e-waste streams, including solar panels. The Founder and Chief Executive Officer of EWaste Africa remarked, *"In 2014, we established what was then Africa's first lighting recycling facility."*

The second subtheme addresses charging for compliance. The narrative highlights that EWaste Africa's business model involves charging corporations for the compliant disposal of e-waste. This approach mitigates the "cost-negative" challenge by generating a revenue stream through responsible disposal services. As the Founder and Chief Executive Officer of EWaste Africa explained, *"We charge corporations for the service, ensuring that it is managed in a legally compliant manner."*

The third subtheme explores leveraging EPR and market dynamics. The narrative underscores the significance of EPR legislation in shifting the responsibility for e-waste management to producers, fostering a more sustainable and expansive market for e-waste recycling services. The Founder and Chief Executive Officer of EWaste Africa stated, *"Now that EPR exists, it contributes to driving what I call the consumer market."*

The fourth subtheme highlights technology, innovation, patents, and industry development. The findings indicate that innovation is central to EWaste Africa's strategy. The development of new technologies, such as the LPX™ plant and advancements in solar panel recycling, is pivotal solutions for managing complex waste streams and extracting valuable materials. The Founder and Chief Executive Officer of EWaste Africa emphasized,

> "This plant ... we own the rights to the commercialization of the patent as EWaste Africa. The plant is currently being commissioned at our Pietermaritzburg facility. It represents a world-first innovation for the extraction of rare earth metals from luminophosphor powder.

The fifth subtheme focuses on enhancing consumer accessibility through the establishment of drop-off points, underscoring expansion efforts. The Founder and Chief Executive Officer of EWaste Africa's narrative highlights that effective consumer e-waste management requires accessible collection facilities. By strategically situating drop-off points in retail stores, the public is encouraged to engage in responsible disposal practices. The Founder and Chief Executive Officer of EWaste Africa stated:

> "We are currently rolling out drop-off points across all our Pick n Pay outlets in South Africa, with the goal of reaching 250 stores by June next year."

9.4.3 Addressing Challenges

This is the third aggregated theme that emerged after analyzing the Founder and Chief Executive Officer of EWaste Africa's narratives. It is characterized by two main themes: key challenges that are addressed by EWaste Africa's operations, and key challenges in managing e-waste are presented in Table 9.2 and as discussed later.

9.4.3.1 Key Challenges That Are Addressed by EWaste Africa's Operations

This subsection outlines the key challenges addressed by EWaste Africa's operations. An analysis of the narratives provided by the Founder and Chief Executive Officer of EWaste Africa reveals four central subthemes, as illustrated in Table 9.2.

The first subtheme focuses on environmental hazard diversion (landfill ban). This aspect highlights the critical environmental risk posed by hazardous e-waste

Table 9.2 Thematic analysis of challenges addressed by EWaste Africa's operations

Aggregated themes	Themes	Subthemes
Addressing challenges	Key challenges that are addressed by EWaste Africa's operations	*Environmental hazard diversion (Landfill ban)*
		Finite resource depletion
		Lack of e-waste management infrastructure/solutions
		Cost-negative waste streams
	Key challenges in managing e-waste	*Lengthy licensing processes*
		Client onboarding time (corporate procurement)
		Building credibility as a new entrant
		Economic impact of COVID-19 (business disruption)
		Cash flow management (high growth and research and development (R & D) costs)

contaminating landfills and surrounding ecosystems. The Founder and Chief Executive Officer of EWaste Africa emphasized their commitment to mitigating this issue, stating, *"Ultimately, we are diverting hazardous chemicals away from our land and water, preventing seepage into landfill sites."*

The second subtheme addresses finite resource depletion, underscoring the issue of resource scarcity. E-waste contains valuable materials that are lost when improperly disposed of. EWaste Africa's work contributes to resource recovery, as the Founder and Chief Executive Officer of EWaste Africa asserted, *"There's a finite amount of these resources that must be taken into consideration."*

The third subtheme highlights the insufficiency of e-waste management infrastructure and solutions. Before legislative intervention, despite the evident need for e-waste management, there was limited initiative or implementation of effective solutions. In this context, legislation acted as a pivotal catalyst, creating the necessary market conditions and momentum for companies such as EWaste Africa. The Founder and Chief Executive Officer of EWaste Africa underscored this point by stating:

> "Legislation is necessary; without it, there is no way to convert or make this process work, highlighting the role of regulatory frameworks in establishing a viable e-waste management industry."

The fourth subtheme examines cost-negative waste streams, emphasizing the economic challenge of managing e-waste that lacks inherent market value in its raw materials. EWaste Africa is addressing this issue by implementing sustainable waste management solutions, despite the absence of immediate profitability. The Founder and Chief Executive Officer of EWaste Africa noted:

> "Solar panels, batteries, and lighting, which constitute the majority of the products we handle, are all cost-negative—there's not enough financial return on the back end to justify this commercial process."

9.4.4 Key Challenges in Managing E-waste

This subsection outlines the key challenges EWaste Africa encounters in e-waste management. An analysis of the Founder and Chief Executive Officer of EWaste Africa's narratives identifies six central subthemes under the overarching theme of challenges to business viability and sustainable growth, as presented in Table 9.2.

First, the findings highlight lengthy licensing processes as a major challenge in e-waste management. According to the Founder and Chief Executive Officer of EWaste Africa, bureaucratic hurdles—particularly those related to environmental licensing—resulted in substantial delays and financial strain from the outset. He asserted:

> "It took much longer than I expected to go live. At one point, we were installing the machinery, but the building still lacked a roof. Delaying the machinery would have meant keeping the installers for another week, which I could not afford."
>
> "Business is tough, and this is the rosy picture. It does not mean that we are not under massive stress's at the moment. High growth brings significant financial strain, particularly from a cash flow perspective."

Second, client onboarding time (corporate procurement) was identified as another key challenge. The findings indicate that the lengthy sales cycle for corporate clients—especially navigating procurement processes—hinders revenue generation and business growth.

The Founder and Chief Executive Officer of EWaste Africa elaborated:

> "Corporate clients typically take anywhere between nine to twelve months, from initial contact to the closure of an eventual sale, this is after we have spent months establishing who the potential paying clients will be. Even if we provide a unique offering that meets their needs, it sometimes can take a further three to six months to be onboarded by them."

Third, establishing credibility as a new entrant posed a considerable challenge. The findings highlight that gaining trust and reputation in an unfamiliar industry—particularly when competing with established players—was a significant initial hurdle. The Founder and Chief Executive Officer of EWaste Africa articulated:

> "One of the biggest challenges at the outset was entering the industry as an unknown entity. I had no prior experience in this sector, having previously worked in the fast-moving consumer goods industry. Establishing credibility was difficult, as it sometimes meant stepping on others' toes. Initially, we lacked industry recognition, which made gaining traction even more challenging."

Fourth, the economic impact of COVID-19 (business disruption) severely affected operations. The pandemic significantly reduced revenue and threatened the company's survival. The Founder and Chief Executive Officer of EWaste Africa explained:

> "The hardest challenge we faced was COVID-19. The 2016 ban had taken for lighting waste, 2021 had arrived, and the electronic waste ban had been implemented. We had anticipated this. We had been preparing the business for years by investing in our expansion and bootstrapping the capex requirements, but then COVID-19 hit. At that point, waste management was not a priority for most clients, as the shutting down of corporate South Africa, also meant that there was no waste generated by the corporate market, which resulted in our turnover being halved from 2020 to 2021."

Fifth, cash flow to fund the extensive R&D expenditure needed in a rapidly developing industry remains an ongoing challenge. The findings underscore that financial stability is a persistent struggle, particularly when balancing rapid expansion, substantial R&D investments, and the need to maintain a highly skilled workforce. This issue is exacerbated by the necessity of allocating significant funds to R&D initiatives to keep pace with technological advancements. The Founder and Chief Executive Officer of EWaste Africa emphasized:

> "High growth brings considerable financial strain, particularly from a cash flow perspective. While having a strong team is essential, skilled personnel come at a high cost, making it crucial to ensure the right level of expertise within the organization."
>
> "While managing daily business operations, we are simultaneously driving innovation and developing a solar process, all while covering the expenses associated with R&D."

9.4.5 Evaluating Impacts and Effectiveness

This is the fourth aggregated theme that emerged after analyzing the Founder and Chief Executive Officer of EWaste Africa's narratives. It is explained by two main themes: environmental, economic, and social impacts of EWaste Africa's innovations, and EWaste Africa's waste management effectiveness and impacts as discussed later.

9.4.5.1 The Environmental, Economic, and Social Impacts of EWaste Africa's Innovations

The environmental, economic, and social impacts of EWaste Africa's innovations are described in this subsection. The findings gave rise to three subthemes under the main theme of the triple bottom line impacts of EWaste Africa's innovations, as illustrated in Figure 9.5.

EWaste Africa's recycling processes save energy compared to producing new materials, contributing to energy conservation. In support of this, the Founder and Chief Executive Officer of EWaste Africa stated that *"That's 32 megawatt hours of savings."*. The social impacts of EWaste Africa's innovations comprise job creation, specifically "uncannibalized jobs," meaning new jobs that do not displace existing ones. The Founder and Chief Executive Officer of EWaste Africa highlighted that *"We currently have forty jobs. I call it uncannibalized jobs, meaning that if we didn't exist, these jobs wouldn't exist ... We created a completely new industry."*

Skills development and team growth are some of the social impacts of EWaste Africa's innovations. EWaste Africa provides opportunities for skill development and career growth, even for individuals without formal education, contributing to social upliftment. The Founder and Chief Executive Officer of EWaste Africa stated:

> *"Our first employee, has grown with no formal education, starting with one day a week at EWaste Africa and now a member of our national leadership team. This by itself is a beautiful story."*

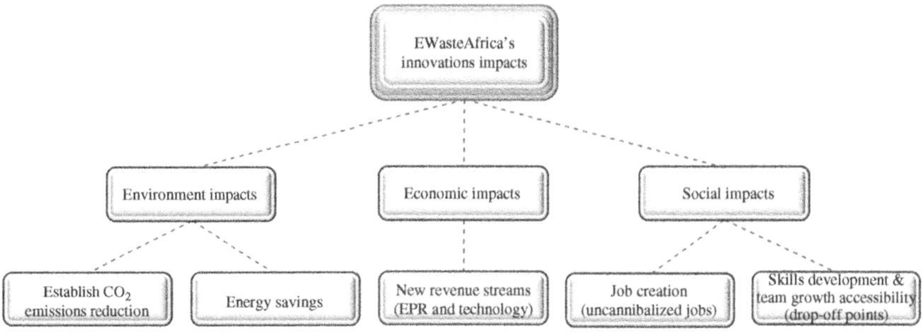

Figure 9.5 EWaste Africa innovations impacts. *Source*: Authors' own idea from the field data (2025).

Figure 9.6 Measuring EWaste Africa's success. *Source:* Authors' own idea from the field data (2025).

The economic effects of EWaste Africa's innovations include new revenue streams from EPR legislation and technological innovation. These create new business opportunities in the e-waste sector, making it more economically viable and sustainable as indicated by the Founder and Chief Executive Officer of EWaste Africa:

> "*EPR moves responsibility to the producer for the end-of-life of their products. This regulation is one of the factors that drives our growth. We've seen almost a 100% growth rate in the last 14 months.*"

9.4.5.2 Assessing EWaste Africa's Waste Management Effectiveness and Impacts

This section examines EWaste Africa's waste management effectiveness and its broader impacts. Through an analysis of the Founder and Chief Executive Officer of EWaste Africa's narrative, six subthemes emerged under the overarching theme of evaluating EWaste Africa's success, as illustrated in Figure 9.6.

To begin with, EWaste Africa measures the effectiveness and impact of their waste management through quantifiable job creation. The most direct metric mentioned is the number of jobs created, which is currently 40. This shows the direct employment generated by their operations, as indicated by the Founder and Chief Executive Officer of EWaste Africa "*We are currently sitting at forty jobs.*"

The concept of "uncannibalized jobs" is essential, as it extends beyond merely quantifying employment figures. It specifically measures the net new jobs generated as a direct result of EWaste Africa's operations, underscoring its distinct impact on the labor market by creating employment opportunities that would not otherwise exist. The Founder and Chief Executive Officer of EWaste Africa explained:

> "I call it uncannibalized jobs, meaning that if we didn't exist, these jobs wouldn't exist. It's not like opening another landfill site down the road and competing for landfill waste; these are newly created jobs that wouldn't have existed otherwise."

EWaste Africa measures its impact through industry/sector creation. The Founder and Chief Executive Officer of EWaste Africa stated that "*We created a completely new industry.*" This indicates a broader, systemic impact. EWaste Africa contributes to the development of a new sector within the economy, specifically in e-waste management and related areas.

Innovation and market development are key indicators through which EWasteAfrica evaluates its effectiveness. The Founder and Chief Executive Officer of EWaste Africa emphasized that the company's pioneering efforts in advancing e-waste management technologies represent a significant technological breakthrough and contribute to the expansion of new markets within the broader waste management and environmental sustainability sectors.

Additionally, EWaste Africa assesses its impact through benchmarking against international industry standards and fostering global industry awareness. The Founder and Chief Executive Officer of EWaste Africa remarked:

> "We recognize that solar panel recycling has gained traction in some countries, and we are actively engaging in discussions with several of these companies overseas."

This shows EWaste Africa's awareness of international best practices and a desire to benchmark itself against global standards. Finally, technological and operational capability is used by EWaste Africa to measure its effectiveness. The Founder and Chief Executive Officer of EWaste Africa stated that "*We are confident that we've secured the technology to be able to do it internally.*" This showcases that EWaste Africa's ability to implement and internalize advanced technologies is a key indicator of their operational success and ability to deliver on their goals.

9.5 Discussions

This section discusses the findings of the business model innovation and technological advancement for sustainable e-waste management using EWaste Africa in South Africa.

The first research objective sought to analyze the business model innovations employed by EWaste Africa to ensure sustainability and growth in e-waste management. The findings revealed that EWaste Africa employs innovative and evolving business models to ensure sustainability and growth in the e-waste sector. These models include a core fee-for-service model for corporate compliance, leveraging producer responsibility revenue streams through EPR legislation, a strategic long-term shift towards raw material value extraction to enhance economic sustainability beyond compliance fees, utilizing corporate activation days as a marketing and brand-building tool, and a geographic and sector expansion strategy to diversify revenue streams and increase market reach. These business models demonstrate a dynamic and adaptive approach to building a sustainable and scalable e-waste management enterprise.

The findings further revealed that EWaste Africa employs some specific strategic solutions for sustainable e-waste management, demonstrating business model innovation and technological advancement. These strategies include establishing specialized recycling infrastructure, implementing a fee-for-service model charging corporations for compliant recycling, leveraging EPR legislation to drive market demand, developing innovative technologies (like the LPX™ plant for enhanced resource recovery and pioneering solar panel recycling), and enhancing consumer accessibility through convenient drop-off points at retail locations. Under the same line of reasoning, Garg (2021) and Arya and Kumar (2020) propounded that e-waste management can be enhanced through strategies such as eco-product development, efficient and effective application of e-waste policies, polluter pays' rule, orders, and regulations like EPR, advance recycling fee, adopting technological and green innovations, and strategic partnerships among supply chain associates. These strategies showcase a comprehensive and innovative approach to addressing the challenges in the e-waste sector. They are essential because e-waste generation in South Africa is increasing, while the collection and recycling rates are low.

Interestingly, the findings indicate that EWaste Africa is strategically positioned for substantial future growth and expansion, leveraging emerging opportunities within the e-waste and green economy sectors. Key growth areas include the expansion of solar panel recycling to establish it as a dominant business segment, entry into battery recycling—particularly for lithium-ion batteries—strategic geographic expansion to access solar energy markets across various regions, and ambitious job creation targets reflecting significant employment growth. These prospective developments highlight EWaste Africa's potential to further scale its influence and strengthen its position as a leader in sustainable e-waste management across Africa.

Lastly, the findings revealed that EWaste Africa operates within a complex stakeholder ecosystem, engaging diverse actors crucial to its business model and the broader e-waste management sector. Key stakeholders include corporate clients who generate waste and seek compliant recycling solutions, consumers engaged through EPR and accessible collection points, government and regulatory bodies that establish legislation and influence market dynamics, TIA and academic institutions that provide funding and R&D partnerships, retail partners that facilitate consumer collection networks, OEMs that generate specific waste streams, and investors or funders who provide capital for growth. Effective stakeholder engagement and collaboration are essential for EWaste Africa's success and the advancement of sustainable e-waste management in South Africa. This is supported by Cao et al. (2024) and Alblooshi et al. (2022) who indicated that stakeholder involvement is vital for sustainable e-waste management.

The second research objective aimed to identify and evaluate the innovative solutions and technological advancements implemented by EWaste Africa to enhance e-waste recycling and management. The findings reveal that EWaste Africa demonstrates significant technological progress through the development and implementation of various innovative solutions to promote sustainable e-waste management. First, the company established the LPX™ plant in collaboration with UKZN and TIA to facilitate the extraction of rare earth metals from phosphor powder waste. Second, it developed novel pavers, using recycled glass from e-waste to manufacture pavers and low-cost housing blocks, achieving a 98% diversion from landfill. Third, they established a solar panel recycling facility, positioning themselves as pioneers in this emerging industry in Africa. Lastly, they conduct corporate activation days to raise awareness, promote e-waste recycling, and facilitate collection.

In terms of strategies, EWaste Africa focuses on establishing recycling infrastructure, starting with lighting and expanding to broader e-waste and solar panels. They charge corporations for compliant e-waste disposal, creating a revenue stream. They leverage EPR legislation to shift responsibility to producers and create a sustainable market for e-waste management services. Innovation is central, with the development of new technologies like the LPX™ plant and solar panel recycling. Lastly, they enhance consumer accessibility by establishing drop-off points in convenient locations, aiming to reach 250 stores by June next year.

The third research objective aimed at examining the key challenges faced by EWaste Africa in managing e-waste and the strategies used to address these challenges. The findings suggest that EWaste Africa addresses critical challenges hindering sustainable e-waste management in South Africa. These challenges include preventing environmental hazards from the landfill disposal of e-waste, mitigating resource depletion through material recovery, overcoming the lack of adequate e-waste management infrastructure, and managing cost-negative waste streams to ensure economic viability. This is not surprising since, even though the purpose of South Africa's legislation on e-waste management is changing, the present waste infrastructure is still weak (Moyo et al., 2022).

Also, several regulations related to e-waste management were introduced in South Africa, but a number of corporates are not sure where to start from. Our findings agree with several studies, which have indicated that there are numerous challenges that hinder sustainable e-waste management (Gaur et al., 2024; Osibanjo & Nnorom, 2007; Shittu et al., 2021). These challenges highlight the urgent need for innovative and sustainable e-waste management solutions, which EWaste Africa aims to provide.

Despite its innovative approach, EWaste Africa faces significant business challenges that impact its operations to achieving viability and sustainable growth. These challenges include lengthy bureaucratic processes for obtaining environmental licenses, extended client onboarding times due to complex corporate procurement procedures, the initial hurdle of building credibility as a new entrant, severe business disruption and revenue loss due to the economic impact of COVID-19, persistent cash flow management pressures exacerbated by rapid growth and R&D investments, and difficulties in shortening of the sales management process. Researchers like Moyo et al. (2022) and Ichikowitz and Hattingh (2020) also indicated that e-waste management companies face several challenges that impact their operations. Overcoming these challenges is crucial for EWaste Africa to achieve long-term sustainability and scale its impact.

The fourth research objective sought to evaluate the effectiveness of EWaste Africa's waste management practices and their impact on sustainability. The findings revealed that EWaste Africa's innovations generate positive triple bottom line impacts, demonstrating its contribution to sustainable development. Environmentally, their operations significantly reduce CO_2 emissions and conserve energy compared to traditional disposal methods. Socially, EWaste Africa creates "uncannibalized jobs," providing new employment opportunities and fostering skills development within the local community. Economically, the company establishes new revenue streams through EPR mechanisms and technological advancements, enhancing the economic viability of e-waste management and fostering growth in the green economy. These impacts underscore the holistic sustainability benefits of EWaste Africa's business model and technological innovations. In line with these findings, Shaikh et al. (2020) and Magalini (2016) outlined the benefits of the downstream effects of e-waste recycling.

In addition, EWaste Africa employs a multi-faceted approach to measure its waste management effectiveness and overall success, going beyond simple metrics. These measures include quantifiable job creation as a direct indicator of economic and social impact, "additionality" of job creation by highlighting "uncannibalized jobs," creation of a new industry sector in e-waste management, pioneering of innovation and market development, benchmarking against international industry standards to ensure operational excellence and industry awareness, and leveraging technological and operational capability. Basically, these are key indicators of EWaste Africa's ability to deliver effective and advanced recycling solutions. These comprehensive measurement metrics demonstrate EWaste Africa's commitment to accountability and continuous improvement in its pursuit of sustainable e-waste management.

9.6 Conclusion

In this study, we analyzed the operations of EWaste Africa, a South African e-waste management organization. This study highlights the urgent need for sustainable e-waste management and provides insights into the innovative solutions employed by EWaste Africa.

EWaste Africa exemplifies a successful and multi-faceted approach to sustainable e-waste management in South Africa. The company employs a dynamic and evolving business model that integrates fee-for-service operations, EPR compliance, material recovery initiatives, and strategic partnerships. Key technological advancements—such as the LPX™ plant, novel paver technology, and solar panel recycling—further establish EWaste Africa as an industry leader in innovation. The company's operations clearly contribute to positive environmental outcomes (reduced CO_2 emissions, resource conservation), social outcomes (job creation, skills development), and economic outcomes (new revenue streams, green economy growth). However, EWaste Africa faces challenges like bureaucratic hurdles, client onboarding, cash flow management, and the need to build trust. Despite these issues, effective stakeholder engagement and a focus on triple-bottom-line impact position the company for continued growth and leadership in the expanding African e-waste sector. Therefore, the study recommends that EWaste Africa should advocate and collaborate with government agencies to streamline licensing and permitting processes for e-waste management companies.

The practical implication of this study is that the dynamics of EWaste Africa's operations provide a foundation for e-waste management that can enhance sustainable e-waste management in South Africa and developing countries in general. Theoretically, this study contributes to the ongoing discourse concerning e-waste management in Africa. It provides novel perspectives for researchers, policymakers, and other diverse stakeholders. Future studies can investigate the feasibility of replicating EWaste Africa's business model and technological solutions in other African countries, considering variations in regulatory frameworks and infrastructure.

References

Ahirwar, R., & Tripathi, A.K. (2021). E-waste management: A review of recycling process, environmental and occupational health hazards, and potential solutions. *Environmental Nanotechnology, Monitoring & Management, 15*, 100409.

Alblooshi, B.G.K.M., Ahmad, S.Z., Hussain, M., & Singh, S.K. (2022). Sustainable management of electronic waste: Empirical evidences from a stakeholders' perspective. *Business Strategy and the Environment, 31*(4), 1856–1874.

Althaf, S., Babbitt, C.W., & Chen, R. (2021). The evolution of consumer electronic waste in the United States. *Journal of Industrial Ecology*, *25*(3), 693–706.

Arya, S., & Kumar, S. (2020). E-waste in India at a glance: Current trends, regulations, challenges and management strategies. *Journal of Cleaner Production*, *271*, 122707.

Asiedu, E.M. (2024). Incidents and impacts on operator revenue in the telecommunications sector: A multiple case-study approach. *Journal of Telecommunications and the Digital Economy*, *12*(1), 65–93. https://doi.org/10.18080/jtde.v12n1.852

Awasthi, A.K., Wang, M., Awasthi, M.K., Wang, Z., & Li, J. (2018). Environmental pollution and human body burden from improper recycling of e-waste in China: A short-review. *Environmental Pollution*, *243*, 1310–1316.

Azadnia, A.H., Onofrei, G., & Ghadimi, P. (2021). Electric vehicles lithium-ion batteries reverse logistics implementation barriers analysis: A TISM-MICMAC approach. *Resources, Conservation and Recycling*, *174*, 105751.

Babbitt, C.W., Kahhat, R., Williams, E., & Babbitt, G.A. (2009). Evolution of product lifespan and implications for environmental assessment and management: A case study of personal computers in higher education. *Environmental Science & Technology*, *43*(13), 5106–5112.

Baldé, C. P., Kuehr, R., Yamamoto, T., McDonald, R., D'Angelo, E., Althaf, S., Bel, G., Deubzer, O., Fernandez-Cubillo, E., Forti, V. & Gray, V. (2024). Global e-waste monitor 2024. *United Nations University (UNU), International Telecommunication Union (ITU) & International Solid Waste Association (ISWA)* (pp. 1–147). Geneva/Bonn.

Bocken, N., & Konietzko, J. (2023). Experimentation capability for a circular economy: A practical guide. *Journal of Business Strategy*, *44*(6), 406–414.

Bouzon, M., Govindan, K., Rodriguez, C.M.T., & Campos, L.M. (2016). Identification and analysis of reverse logistics barriers using fuzzy Delphi method and AHP. *Resources, Conservation and Recycling*, *108*, 182–197.

Braun, V., Clarke, V., & Rance, N. (2014). How to use thematic analysis with interview data. In *The counselling & psychotherapy research handbook* (Vol. 3, pp. 183–197). Sage Publication.

Campbell, S., Greenwood, M., Prior, S., Shearer, T., Walkem, K., Young, S., Bywaters, D., & Walker, K. (2020). Purposive sampling: complex or simple? Research case examples. *Journal of Research in Nursing*, *25*(8), 652–661. https://doi.org/10.1177/1744987120927206

Cao, D., Puntaier, E., Gillani, F., Chapman, D., & Dewitt, S. (2024). Towards integrative multi-stakeholder responsibility for net zero in e-waste: A systematic literature review. *Business Strategy and the Environment*, *33*(8), 8994–9014.

Chakrabarty, A., & Nandi, S. (2021). Electronic waste vulnerability: Circular economy as a strategic solution. *Clean Technologies and Environmental Policy*, *23*, 429–443.

Chaudhary, K., Mathiyazhagan, K., & Vrat, P. (2017). Analysis of barriers hindering the implementation of reverse supply chain of electronic waste in India. *International Journal of Advanced Operations Management*, *9*(3), 143–168.

Cheng, C.P., Lin, C.H., Wen, L.C., & Chang, T.C. (2019). Determining environmental costs: A challenge in a governmental e-waste recycling scheme. *Sustainability*, *11*(19), 5156.

Cheshmeh, Z.A., Bigverdi, Z., Eqbalpour, M., Kowsari, E., Ramakrishna, S., & Gheibi, M. (2023). A comprehensive review of used electrical and electronic equipment management with a focus on the circular economy-based policy-making. *Journal of Cleaner Production*, *389*, 136132.

Forti, V., Balde, C. P., Kuehr, R., & Bel, G. (2020). The global e-waste monitor 2020, United Nations University (UNU), International Telecommunication Union (ITU) & International Solid Waste Association (ISWA) (p. 120). Bonn/Geneva/Rotterdam.

Garg, C.P. (2021). Modeling the e-waste mitigation strategies using grey-theory and DEMATEL framework. *Journal of Cleaner Production*, *281*, 124035.

Gaur, T.S., Yadav, V., Mittal, S., & Sharma, M.K. (2024). A systematic review on sustainable e-waste management: Challenges, circular economy practices, and a conceptual framework. *Management of Environmental Quality: An International Journal*, *35*(4), 858–884.

Geneva Environment Network. (2024). *The growing environmental risks of e-waste.* https://www.genevaenvironmentnetwork.org/resources/updates/the-growing-environmental-risks-of-e-waste/

Gunarathne, N., de Alwis, A., & Alahakoon, Y. (2020). Challenges facing sustainable urban mining in the e-waste recycling industry in Sri Lanka. *Journal of Cleaner Production*, *251*, 119641.

Henzler, M., Eisinger, F., Gaurav, J., Hemkhaus, M., Sinha, S., Mahesh, P., & Mehra, G. (2017). Building the link: Leveraging formal-informal partnerships in the Indian ewaste sector. In Daniel, H. & Ellen, G. (Eds.), *Deutsche Gesellschaft für Internationale Zusammenarbeit (GIZ)* (p. 10). GmbH.

Ichikowitz, R., & Hattingh, T.S. (2020). Consumer e-waste recycling in South Africa. *South African Journal of Industrial Engineering*, *31*(3), 44–57.

Iliinsky, N. (2010). On beauty. In Steele, J. & Iliinsky, N. (Eds.), *Beautiful visualization: Looking at data through the eyes of experts* (pp. 1–13). Sebastopol, CA: O'Reilly Media.

International Telecommunication Union. (2024). *The global e-waste monitor 2024.* https://www.itu.int/en/ITU-D/Environment/Pages/Publications/The-Global-E-waste-Monitor-2024.aspx

Jangre, J., Prasad, K., & Patel, D. (2022). Analysis of barriers in e-waste management in developing economy: An integrated multiple-criteria decision-making approach. *Environmental Science and Pollution Research*, *29*(48), 72294–72308.

Jin, H., Frost, K., Sousa, I., Ghaderi, H., Bevan, A., Zakotnik, M., & Handwerker, C. (2020). Life cycle assessment of emerging technologies on value recovery from hard disk drives. *Resources, Conservation and Recycling, 157*, 104781.

Kavil, Y.N., Alshemmari, H., Alkasbi, M.M., Alelyani, S.S., Orif, M.I., & Al-Farawati, R.K. (2025). Electronic waste and environmental diplomacy: How GCC e-waste management interfaces with the Stockholm Convention. *Journal of Hazardous Materials Advances, 17*, 100610.

Kumar, A., & Dixit, G. (2018). Evaluating critical barriers to implementation of WEEE management using DEMATEL approach. *Resources, Conservation and Recycling, 131*, 101–121.

Leclerc, S.H., & Badami, M.G. (2020). Extended producer responsibility for e-waste management: Policy drivers and challenges. *Journal of Cleaner Production, 251*, 119657.

Liu, K., Tan, Q., Yu, J., & Wang, M. (2023). A global perspective on e-waste recycling. *Circular Economy, 2*(1), 100028.

Magalini, F. (2016). Global challenges for e-waste management: The societal implications. *Reviews on Environmental Health, 31*(1), 137–140.

Mehta, P., & Chahal, H.S. (2021). Consumer attitude towards green products: Revisiting the profile of green consumers using segmentation approach. *Management of Environmental Quality: An International Journal, 32*(5), 902–928.

Mhatre, P., Gedam, V.V., Unnikrishnan, S., & Raut, R.D. (2023). Circular economy adoption barriers in built environment – A case of emerging economy. *Journal of Cleaner Production, 392*, 136201.

Miles, M.B., Huberman, M.A., & Saldaña, J. (2020). *Qualitative data analysis: A method sourcebook* (4th ed.). Thousand Oaks, CA: Sage Publications.

Mishra, Y., & Mishra, S.P. (2023). E-waste management and legal concepts in India. In Grima, S., Sood, K., & Ozen, E. (Eds.), *Contemporary studies of risks in emerging technology, Part B* (pp. 299–312). Emerald Publishing Limited.

Moyo, T., Sadan, Z., Lotter, A., & Petersen, J. (2022). Barriers to recycling e-waste within a changing legal environment in South Africa. *South African Journal of Science, 118*, 1–8.

Needhidasan, S., Samuel, M., & Chidambaram, R. (2014). Electronic waste – An emerging threat to the environment of urban India. *Journal of Environmental Health Science and Engineering, 12*, 1–9.

Osibanjo, O., & Nnorom, I.C. (2007). The challenge of electronic waste (e-waste) management in developing countries. *Waste Management & Research, 25*(6), 489–501.

Park, J., Díaz-Posada, N., & Mejía-Dugand, S. (2018). Challenges in implementing the extended producer responsibility in an emerging economy: The end-of-life tire management in Colombia. *Journal of Cleaner Production, 189*, 754–762.

Patton, M.Q. (2002). *Qualitative research & evaluation methods* (3rd ed.). Thousand Oaks, CA: Sage.

Perkins, D.N., Drisse, M.N.B., Nxele, T., & Sly, P.D. (2014). E-waste: A global hazard. *Annals of Global Health*, *80*(4), 286–295.

Saha, L., Kumar, V., Tiwari, J., Rawat, S., Singh, J., & Bauddh, K. (2021). Electronic waste and their leachates impact on human health and environment: Global ecological threat and management. *Environmental Technology & Innovation*, *24*, 102049.

Schober, M.F. (2018). The future of face-to-face interviewing. *Quality Assurance in Education*, *26*(2), 290–302.

Shaikh, S., Thomas, K., Zuhair, S., & Magalini, F. (2020). A cost-benefit analysis of the downstream impacts of e-waste recycling in Pakistan. *Waste Management*, *118*, 302–312.

Shittu, O.S., Williams, I.D., & Shaw, P.J. (2021). Global e-waste management: Can WEEE make a difference? A review of e-waste trends, legislation, contemporary issues and future challenges. *Waste Management*, *120*, 549–563.

Stake, R.E. (2005). Qualitative case studies. In Denzin, N.K. & Lincoln, Y.S. (Eds.), *The SAGE handbook of qualitative research* (3rd ed., pp. 443–466). London, Thousand Oaks: Sage Publications.

The global E-waste Monitor. (2024). *ITU and UNITAR*.

United Nations Environment Management Group (n.d.). *Inter-agency issue management group on tackling e-waste*. Geneva, Switzerland: UN Environment Management Group. UN Environment Management Group. Retrieved from https://unemg.org/our-work/emerging-issues/innter-agency-issue-management-group-on-tackling-e-waste/

Yin, R.K. (2014). *Case study research: Design and methods*. Sage Publications.

Yin, R.K. (2018). *Case study research and applications: Design and methods* (6th ed.). Thousand Oaks, CA: Sage.

Zacho, K.O., Bundgaard, A.M., & Mosgaard, M.A. (2018). Constraints and opportunities for integrating preparation for reuse in the Danish WEEE management system. *Resources, Conservation and Recycling*, *138*, 13–23.

10

A Sustainable Electronic Waste Recycling: A Case Study of Basadi E-Waste

Marcia Mkansi[1], *Ernest Marfo Asiedu*[2], *and Frank Ranganai Matenda*[2]

[1] *Office of Graduate Studies and Research, University of South Africa, Pretoria, South Africa*
[2] *Department of Operations Management, University of South Africa, Pretoria, South Africa*

10.1 Introduction

Electronic waste (e-waste) refers to electronic or electrical gadgets that have been discarded. These products are not working, unwanted, and nearing or at their end-of-life. They include equipment like microwaves, electric cookers, heaters, laptops, smartwatches, and smartphones. The list is endless. In today's world, e-waste is a global challenge, with roughly 57 million tonnes produced in 2021, of which only approximately 9 million tonnes were appropriately recycled (Grandhi et al., 2024). The authors (Grandhi et al., 2024) further articulated that, according to the Global E-waste Monitor 2024, e-waste generation has passed 65 billion tonnes due to an increase in the adoption and reliance on technologies. Basically, e-waste is one of the fastest increasing wastes of the waste stream (Doan et al., 2019).

Rapid urbanization, population growth, advanced industrialization, human longing for new electronic and electrical devices, and advancement of information and communication technologies have amplified a swift upsurge in diverse kinds of wastes (Faibil et al., 2023; Sharma et al., 2023). Xu et al. (2021) and Doan et al. (2019) opined that e-waste generation has increased substantially mainly due to unceasing technological advancements and reduced life cycles of electronic and electrical devices. Interestingly, technological advances are moving at an

Green Innovations in Supply Chain Management: Case Studies and Applications, First Edition. Edited by Marcia Mkansi, Godfrey Mugurusi, Aaron L. Nsakanda, Temidayo Akenroye, and Frank Tietze.

Published 2026 by John Wiley & Sons, Inc.

astonishing speed and, as a result, numerous electrical and electronic gadgets that are thrown away are still working.

Even though electronic and electrical devices have improved the standard of living of individuals, the exponential increase of e-waste has negative effects on human health, economy, and environment due to toxic and hazardous chemicals like cadmium, beryllium, lead, and mercury, which cause severe dangers to water, soil, air, and wildlife. Chen et al. (2020) propounded that e-waste contains more than 1000 elements that need efficient and effective processes for their gathering, separation, transportation, recycling, storage, and burning. Existing literature has indicated that unsuitable treatment and mismanagement in dealing with e-waste result in severe economic (e.g., e-waste treatment and hidden externalized costs for community), health (e.g., damages in deoxyribonucleic acid (DNA), inconsistent thyroid functions), and environmental (e.g., ecosystems contamination and leaching in groundwater) challenges. On the other hand, e-waste comprises several valuable materials such as metals (e.g., gold, silver, copper, nickel, and palladium), glass, and plastics, which can be reused.

E-waste should be properly recycled and disposed of since it contains both toxic and hazardous chemicals and valuable materials. The e-waste problem is worsened by inappropriate landfill disposal, illegal recycling, and illegal exports, among other things (Grandhi et al., 2024). For instance, second-hand computers are illegally exported to South Africa and Nigeria from the United States of America (Andeobu et al., 2021), where e-waste is recycled in the informal sector and disposed in landfills. Interestingly, the e-waste challenge has attracted significant attention from both developed and developing countries. Developed countries are associated with some of the best practices for addressing the challenge of e-waste. Nonetheless, 80% of global e-waste is exported to developing countries since e-waste recycling is expensive (Xu et al., 2021). Worryingly, developing economies are characterized by insufficient recycling infrastructure, poor enforcement of regulations, and inadequate management resources for appropriate recycling (Wang et al., 2016).

In Africa, the e-waste problem is amplified by the nonexistence of responsiveness by the people when it comes to e-waste issues and insufficient e-waste frameworks and regulations (Grandhi et al., 2024; Moyen Massa & Archodoulaki, 2023). Grandhi et al. (2024) indicated that, according to the data from the Global E-waste Monitor, Africa produces approximately 30 million tonnes of e-waste but recycles 3.6 million tonnes only. Therefore, to reduce e-waste, well-thought-out attempts and efficient and effective policy interventions are required. Green innovations, such as artificial intelligence-powered separation, robotic sorting systems, bioleaching, and hydrometallurgy, play a crucial role in e-waste reduction and management. They effectively and efficiently extract treasured metals from e-waste and minimize e-waste's impact on the environment.

In this study, we analyze the operations of Basadi E-waste, a South African black female-owned corporate that focuses on e-waste, data destruction, data sanitization, and refinery. Particularly, in the context of Basadi E-waste, we examine motivations and core elements of green innovations, and practical implementation of green innovation in e-waste recycling and the challenges involved. South Africa is one of the fastest growing countries globally. Although South Africa's e-waste management industry is in its infancy, the country is one of the major generators of e-waste in Africa. Forti et al. (2020) indicated that, in 2019, South Africa produced 461 kilotons of e-waste. South Africa has in place some legislations and industry initiatives to manage e-waste. Nevertheless, the country's e-waste management practices face a myriad of challenges in effectively and efficiently regulating and managing the increasing volume of e-waste and its effect on the environment, economy, and human health.

The rest of the chapter is structured as follows. Section 10.2 provides the literature review, and Section 10.3 outlines the methodology. Findings of the study are articulated in Sections 10.4 and 10.5. Section 10.6 concludes the study and provides the limitations of the study and avenues for future research.

10.2 Literature Review

10.2.1 The Growing Nature of E-Waste

E-waste generation reached unparalleled levels. It is widely documented that e-waste is one of the fastest increasing solid waste streams globally. This is powered mainly by, among other things, rapid technological advancements, industrialization, urbanization, higher levels of disposable income, consumers' hunt for new electronic and electrical devices, short life cycles, and limited repair alternatives for electronic and electrical gadgets (Baldé et al., 2024). In 2019, global e-waste generation has reached a record of 53.6 million metric tonnes (Faibil et al., 2023), which is an average of 7.3 kg per capita (Forti et al., 2020). Forti et al. (2020) further articulated that, since 2014, the universal generation of e-waste increased by 9.2 million metric tonnes and is anticipated to increase to 74.7 million metric tonnes by 2030. Given that e-waste is associated with substantial environmental and health risks, its rapid increase is a matter of concern. The e-waste challenge is worsened by the fact that the recycling attempts are not matching the universal increase of e-waste generation. Forti et al. (2020) indicated that, in 2019, formal gathering and recycling were 9.3 million metric tonnes, which is 17.4% of e-waste produced. Baldé et al. (2024) postulated that the increase in e-waste generation is outstripping the increase in formal recycling by a factor of approximately 5 and has therefore surpassed the increase in formal and environmentally robust gathering and recycling.

10.2.2 Impediments to Sustainable E-Waste Management

Sustainable e-waste management is hindered by a multiplicity of challenges. In practice, the bigger portion of e-waste is recycled in the informal sector, which is dangerous to people and the environment. Baldé et al. (2024) propounded that, in 2022, 62 billion kg of e-waste were generated globally, which is an average of 7.8 kg per capita per year; 22.3% of this e-waste was formally gathered and recycled in an environmentally rigorous way. Gollakota et al. (2020) and Awasthi et al. (2018) indicated that 20% of e-waste generated globally is managed via formal practices, while the remaining 80% is managed by implementing conventional practices. Rautela et al. (2021) proffered that roughly 82.6% of e-waste is recycled informally in unorganized sectors globally. Since most e-waste is handled outside the formal gathering and recycling arrangements, due to noncompliant e-waste management, 58,000 kg of mercury and 45 million kg of plastics comprising brominated flame retardants are released into the environment each year, adversely impacting the environment and human health (Baldé et al., 2024). Pregnant women and children are predominantly vulnerable as a result of their exposure pathways and developmental position (World Health Organization, 2025). Informal recycling procedures usually comprise harmful and unsafe techniques of dealing with e-waste like open burning, acid baths, incineration, and acid leaching to extract treasured materials.

The sustainable management of e-waste needs sound infrastructure from gathering to recycling and disposal. Nonetheless, a lack of adequate infrastructure is one of the major challenges in e-waste management. Generally, there is a shortage of disposal facilities, collection points, and processing and recycling facilities. The lack of infrastructure hampers the appropriate management of e-waste, resulting in environmental and health risks. Without proper infrastructure, these risks continue. Gaur et al. (2024), Mir and Chang (2024), and Lakshmi and Raj (2017) indicated that e-waste management can be deterred by insufficient infrastructure.

The lack of awareness among diverse stakeholders like consumers and corporates is another challenge that impedes sustainable e-waste management. Some individuals are not aware of suitable disposal and recycling alternatives for electrical and electronic gadgets. This results in unfitting e-waste management. People may throw away e-waste if they are not aware of the best practices of e-waste management (Oliveira et al., 2020; Santos & Ogunseitan, 2022). In support of this, Santos and Ogunseitan (2022) stated that e-waste reverse logistics system can only function when stakeholders, who must know their roles to attain the policy, are integrated and engaged.

Lack of data and research on e-waste and lack of technical expertise in e-waste recycling are some of the challenges that hinder sustainable e-waste management (Santos & Ogunseitan, 2022). Research on e-waste management promote sustainable e-waste management by revealing sustainable solutions to the challenge of e-waste disposal and recycling. Further, the sophistication of e-waste needs advanced technologies that can be used, especially in recycling. Herat (2021) opined that formal e-waste recycling processes need investment in cutting-edge technologies to promote the effective and efficient recovery of treasured metals while hazardous chemicals are managed and treated properly. Rautela et al. (2021) opined that challenges in handling e-waste emanate from poor infrastructure, lack of technical skills, inadequate financial support, and inactive community engagement.

10.2.3 Solutions and Future Directions in Sustainable E-Waste Management

Nowadays, both developing and developed economies are paying attention to sustainable e-waste management. However, the challenges of e-waste in developing countries have been amplified by the transboundary movement of e-waste from developed to developing countries. Baldé et al. (2024) stated that "5.1 billion kg of e-waste were shipped across borders in 2022. Of this, an estimated 3.3 billion kg were shipped from high-income to middle- and low-income countries through uncontrolled and undocumented transboundary movements, accounting for 65% of the total transboundary movement of e-waste globally." This leads to a huge amount of e-waste in developing countries. Nevertheless, Mishra and Mishra (2023) propounded that developing economies are not able to manage or recycle this huge amount of e-waste efficiently and effectively, and they do not possess rules and policies that deal with sustainable e-waste recycling activities.

The above-discussed challenges reveal that addressing the problem of e-waste efficiently and effectively requires collaborative attempts from several stakeholders such as consumers, governments, manufacturers, and recyclers, in conjunction with the designing and adoption of novel and inventive techniques to appropriately gather, manage, and recycle e-waste. Technological advancements possess a lot of promise in e-waste management transformation since they are efficient and effective in handling e-waste (Liang et al., 2024; Sharma & Dutt, 2024). The incorporation of advanced green innovations can revolutionize e-waste management. Existing literature has indicated that the future of e-waste management is contingent on the adoption of advanced, eco-friendly technologies that give precedence to the recovery of resources and minimize environmental impact[1]. Some examples of advanced green technologies leading the way are pyrolysis for plastic recovery, blockchain for e-waste tracking,

1 https://www.linkedin.com/pulse/revolutionizing-global-e-waste-management-advanced-green-ali-mustafa-teguf/

and renewable energy-powered facilities[1]. Under the same line of reasoning, Liang et al. (2024) opined that novel green technologies like supercritical fluid extraction, bioleaching, lanmodulin-derived peptides, and ionic liquid extraction are environmentally friendly since they curb the implementation of toxic chemicals and minimize energy use.

10.3 Methodology

This section discusses the methodology used for the empirical research conducted in this study, as illustrated in Figure 10.1.

The study used an exploratory case study research design to gain better understanding of the operations of Basadi E-waste (Figure 10.1). An exploratory case study research design is considered as an investigation into an undefined problem in order to obtain an in-depth understanding of the phenomenon (Elman et al., 2020; Saunders et al., 2016). Since the overall aim of our study is to explore a sustainable electronic waste recycling using Basadi E-waste, it is therefore suitable for the study. This research design seeks to address new problem with little or no prior research as it focuses on collecting

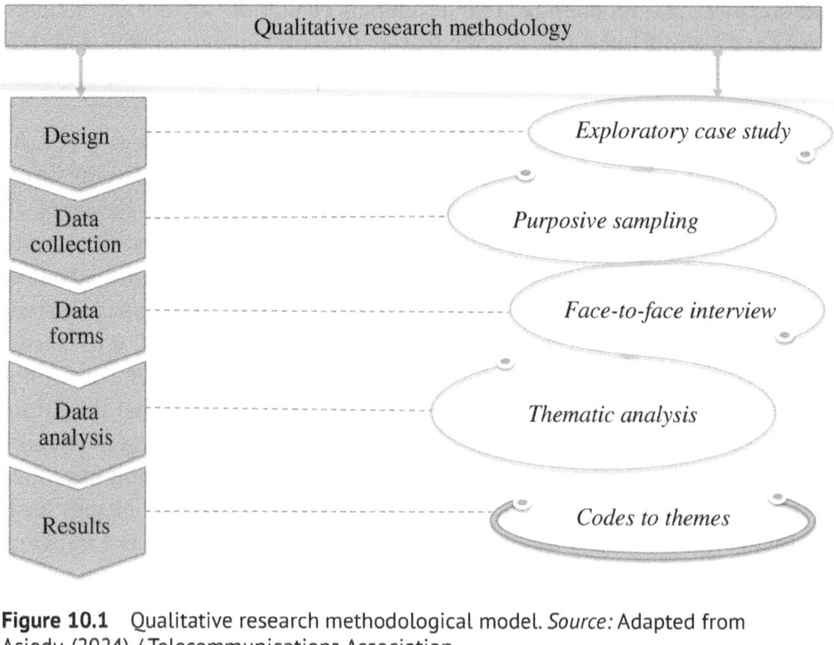

Figure 10.1 Qualitative research methodological model. *Source:* Adapted from Asiedu (2024) / Telecommunications Association.

secondary or primary data in an unstructured format and interpreting it using formal procedures (Bernd, 2017).

In selecting data source, the study purposively selected the founder and managing director as the sole interviewee. An exploratory case study research design enables the researcher to be as creative as possible in order to gain the most understanding of a subject (Tegan, 2021; Voxco, 2021). Therefore, the study used this design to effectively gain in-depth insights into strategic and operational aspects from the owner's perspectives.

To gather data, an in-depth, semi-structured face-to-face interview was conducted with the founder and managing director of Basadi E-waste (Figure 10.1). Face-to-face interviews permit controlled interactions, sophisticated questioning, and probe mechanisms (Jennings, 2005). Kunz et al. (2024) propounded that face-to-face interviews give interviewers a distinct chance to gather additional data, known as interviewer-observed paradata, concerning participants and nonparticipants, their living environment, and the interview status quo itself. Further, face-to-face interviews capture verbal and nonverbal ques, emotions, and behaviors. The data that was collected through an in-person interview was subjected to content analysis.

For data analysis, the study used the qualitative thematic analysis with Atlas. ti software (Figure 10.1). The data analysis consists of examining, categorizing, tabulating, testing, or otherwise recombining evidence to draw empirically based conclusions (Yin, 2009). The study used explanation-building pattern-matching method, which aims to analyze the case study data by building an explanation about the case (Yin, 2009).

10.4 Findings

This section presents the findings of the study.

10.4.1 Motivations and Core Elements of Green Innovation

Basadi E-waste motivations and core elements of green innovation are outlined in this subsection. From the founder's narratives, two main themes emerged as presented in Table 10.1.

10.4.1.1 Green Innovation Motivations

Green innovation motivations is explained by three subthemes. The first subtheme highlights that Basadi E-waste was socially driven to access green innovation. The corporate aims to empower women within a male-dominated

Table 10.1 Basadi E-waste motivations and core elements of green innovation

Codes	Subthemes	Themes
Gender inclusion in e-waste industry	Socially driven green innovation	Green innovation motivations
Bridging the gap in IT and e-waste sectors		
Equitable access to green innovation opportunities		
Addressing industry disparities		
Environmental consciousness	Environmental problem as catalyst for green business	
Identifying gaps in e-waste management		
Commitment to true recycling		
Problem-solving mindset	Solution-oriented mindset for green innovation	
Driving innovation through solutions		
Passion for environmental impact		
Byproduct innovation from e-waste	Byproduct innovation from e-waste plastic	Core elements of green innovation
Research-driven processing		
E-waste plastic recycling		
Material repurposing		
3D printing filaments from e-waste plastics	Diverse range of innovative green products	
Assistive technology innovation		
Sustainable construction materials		
Upcycling e-waste into functional products		
Zero-waste-to-landfill approach	Zero-waste value chain innovation (comprehensive recycling)	
Comprehensive e-waste processing		
Full repurposing of e-waste materials		

industry, ensuring equitable access to green innovation opportunities, thereby addressing industry disparities. Also, this green business closes the gap between information technology (IT) and e-waste industries. The quote from the founder to support this is as follows:

> *"So, the ultimate goal for me to start the company was because I know that e-waste is almost like an IT industry since we deal a lot with electronic stuff. The purpose of this company was to give females a chance to tap into this male-dominated and white-dominated industry. That's how the company came about."*

The second subtheme centers on environmental problems as a catalyst for green business. This emphasizes the recognition of the environmental problem of e-waste and the lack of proper recycling as a key driver for starting Basadi E-waste. This showcases a problem-solving approach by adopting green innovation. The founder commented:

> *"But then I realized along the way that they're not actually recycling; they're refurbishing and reselling. So, I saw a gap and thought, we're having an environmental issue in our country with e-waste because electronic waste is hazardous. I then decided that we could try to eradicate or at least have an impact on saving the planet and the environment by recycling."*

Lastly, the third subtheme indicates that the adoption of green innovation is motivated by the desire to find solutions, i.e., having solution-oriented mindset. It underscores the founder's drive as a "solutionist" and her desire to address environmental problems, highlighting a proactive and purposeful approach to green innovation. The founder asserted:

> *"I am passionate about being a solutionist. You know, I'm someone who wants to come up with solutions. So, I saw that there's an environmental problem that I can solve. I think that's what drove me a lot."*

10.4.1.2 Core Elements of Green Innovation

This theme is characterized by the following subthemes: by-product innovation from e-waste plastic, diverse range of innovative green products, and zero-waste value chain innovation (comprehensive recycling) (Table 10.1).

To begin with, the first core element of green innovation involves creating byproducts from e-waste plastic (WEEP), a material often discarded. This demonstrates a value-added approach to recycling and resource recovery. The founder stated:

> *"So, we then decided, why not create byproducts from electronic waste? To cut a long story short, we started doing our research and processing. Currently, we are processing electronic waste plastic, which we call WEEP. We have created byproducts from the plastic in e-waste."*

A diverse range of innovative green products is another core element of green innovation. This element provides concrete examples of innovative green products being developed (such as 3D filaments, digital crutches, roof tiles), showcasing the breadth of Basadi E-waste's green product innovation and its application to different sectors. The founder noted:

> *"So firstly, we're making 3D filaments. A 3D filament is like raw material for 3D printing from the plastic in e-waste. We then said we can also create digital crutches for people with impaired mobility. Additionally, we can create roof tiles for low-cost housing, like RDP (Reconstruction and Development Programme) houses, using this plastic."*

The last core element of green innovation entails zero-waste value chain innovation (comprehensive recycling). This explains the innovative approach of creating a "zero-waste" value chain, processing and repurposing every component of e-waste, showcasing a holistic and comprehensive green innovation in recycling methodology. The founder highlighted:

> *"Like I said earlier, what I believe sets us apart from many recycling companies is that we have stretched our value chain to achieve zero waste to landfills. We process every little e-waste component and repurpose every e-waste gadget."*

10.4.2 Practical Implementation of Green Innovation in E-Waste Recycling and Challenges Involved

In this subsection, the findings related to the practical implementation of green innovation in e-waste recycling and the challenges involved are outlined (Table 10.2). Four main themes emerged from the founder's narratives.

10.4.2.1 Green Innovation Sustainable Practices

Green innovation sustainable practices are explained through three subthemes. The first subtheme highlights a commitment to genuine recycling rather than just refurbishment, demonstrating a focus on environmental integrity and deeper sustainability within their operations. The quote to support this is as follows:

> *"Along the way, because I used to work for an e-waste company that called themselves an e-waste recycling company, I realised that they're not actually recycling; they're refurbishing and reselling."*

Table 10.2 Practical implementation of green innovation in e-waste recycling and challenges involved

Codes	Subthemes	Themes
Distinction between recycling and refurbishing	Environmental integrity	Green innovation sustainable practices
Commitment to true recycling		
Local job creation over exporting e-waste	Local value retention	
Empowering the township economy		
Environmental impact monitoring & reporting	Accountability	
Collaboration with research institutions for sustainability		
High capital costs for machinery & infrastructure	Financial barriers to green tech adoption (high costs)	Scaling green innovation challenges
Funding challenges for equipment procurement		
Inadequate feeder stock for processing	Securing consistent e-waste feedstock (supply chain challenge)	
Dependency on external e-waste supply		
Lack of awareness about e-waste hazards	Low public awareness of e-waste issues (market development hurdle)	
Public education on e-waste		
Competition and funding support	Government and institutional funding support	Green innovation support ecosystem
Collaboration with academic institutions		
Collaboration with research institutions	Academic and research partnerships	
Collaboration with academic institutions		
Site provision by TUT	Infrastructure and talent support from Technical Institutions	
Partnership for student internships		
Free e-waste collection with value-added services	Value-added services (free collection with certification and audit)	Green innovation sustainable business models

(*Continued*)

Table 10.2 (Continued)

Codes	Subthemes	Themes
Compliance as a value proposition for corporate clients		
Credibility through certification		
ISO compliance as a strategic advantage	Compliance & ISO Certification as competitive advantage	
Credibility and trust with corporate clients		
Flat fees for the clients	Affordable e-waste disposal cost by clients	
Minimal payment for e-waste from client		

The second subtheme exhibits a commitment to economic sustainability within the local township economy by prioritizing local job creation and value retention rather than solely focusing on exporting materials for profit, demonstrating a socially conscious sustainable practice. The founder commented:

> *"Because many companies or recycling companies sell components or some of their materials to international companies to make money, which is great, we thought, why don't we create jobs within our country? Especially since I'm more of a township girl, I wanted everything I do to benefit the township economy as well."*

The third subtheme demonstrates that accountability through environmental impact monitoring and reporting, and collaboration with research institutions for sustainability is one of the green innovation sustainable practices. The founder noted:

> *"And in terms of the environmental impact, we send out our reports and work to the CSIR (Council for Scientific and Industrial Research) under NCPC-SA (National Cleaner Production Centre of South Africa). We report to them on a monthly basis. For everything we process, we send them the tonnages when reporting."*

10.4.2.2 Scaling Green Innovation Challenges

The theme of scaling green innovation challenges is explained through three subthemes. Financial barriers to green technology adoption serve as a key challenge for scaling green innovation. These financial challenges include high capital costs associated with acquiring necessary machinery and infrastructure, and funding challenges for equipment procurement. Lack of finance is a common barrier for green startups. The founder asserted:

> *"But in terms of acquiring the machines in our name, we haven't achieved that because they are quite pricey. One machine or plant to process, meld, create, and mold these things costs around 5 to 7 million. So, that's a lot. Currently, we're still trying to seek funding for the procurement of these machines."*

Securing consistent e-waste feedstock, i.e., a supply chain problem, is another challenge. The main reasons being inadequate feeder stock for processing and dependency on external e-waste supply. Securing a reliable and consistent supply of e-waste feedstock is crucial for maintaining operations and scaling up byproduct production. This issue highlights a supply chain sustainability challenge. The founder noted:

> *"So, for me, that's the biggest challenge because, although we have a facility and some feeder stock, I'd rate it a 4 out of 10. That's another challenge because if we don't get feeder stock like e-waste from companies or individuals, it obviously affects the processing of byproducts."*

Lastly, low public awareness of e-waste issues is another challenge. This explains the challenge of low public awareness regarding e-waste hazards and management, which hinders market development for e-waste recycling services and feedstock sourcing, highlighting a market-related challenge to green innovation adoption. The quote that supports this is as follows:

> *"So, we've come across challenges where we are still behind in understanding what e-waste is and the dangers of e-waste in South Africa. Creating awareness to let people know about e-waste and its hazardous chemicals is a significant challenge."*

10.4.2.3 Green Innovation Support Ecosystem

Green innovation support ecosystem, the third theme, is explained through subthemes.

The first subtheme highlights the crucial role of government and institutional funding in enabling the corporate to start and develop its green innovations. This underlines the importance of financial support for green startups. The response of the founder is as follows:

> *"Fortunately for us, the competition we won at the Innovation Hub provided us with funding. We also received funding from the NYDA (National Youth*

Development Agency). Additionally, we were given the opportunity to work on TUT's (Tshwane University of Technology) main campus in Soshanguve, where they deal with Information Technology. They have a workshop with some machinery that we can use."

The second subtheme underlines the vital role of partnerships with academic and research institutions such as the University of Johannesburg (UJ) and CSIR. These partnerships provide access to expertise, testing facilities, Research & Development (R&D) support, and even client networks, demonstrating the power of collaboration in fostering green innovation. This was evidenced by the founder's responses in the following quote:

"And with UJ, we have signed an MOU (Memorandum of Understanding) with them as well. They help us a lot with everything from testing to analysis to prototyping. Additionally, we have a MOA (Memorandum of Agreement) with the CSIR. They assist us in getting clients for e-waste and help with a lot of testing."

The third subtheme underscores the support from TUT in providing physical space and potential access to talent (internships), illustrating diverse forms of institutional support beyond just funding and R&D. This contributes to a supportive ecosystem for green innovation. These are the founder's words:

"Honestly speaking, our relationship with TUT isn't very strong. They just provided us with a site in the townships where we can place our container. The advantage for us is that it's a black-owned company. We also told TUT that since it's an IT university campus, if they need IT or electronics graduates to do practicals, we would gladly allow them to work with us."

10.4.2.4 Green Innovation Sustainable Business Models

Green innovation sustainable business models, explained through three subthemes, are the last theme that emerged from the founder's narrative. One of the business models for green innovation sustainability centers on value-added services, which manifests through free e-waste collection, compliance as a value proposition for corporate clients, and credibility through certification. This makes it attractive to companies seeking compliance and responsible disposal. The quote to support this is as follows:

"No, we collect from most of our clients for free. Usually, companies want an offer, so we need to pay them. But most of our clients don't require payment because we provide them with a recycling certification and an audit report on all the devices, we collected from them. Compliance is quite important,

> *especially when dealing with corporate clients. Fortunately, we are ISO (International Organization for Standardization) compliant with ISO 9001 and 14001, which cover quality and environmental standards. This compliance gives us an advantage, making us appear serious and knowledgeable to corporate companies. As a result, many companies just give us their waste."*

Another business model underscores ISO certification as a key element of their business model, providing credibility and a competitive advantage when dealing with corporate clients who prioritize compliance and environmental standards. The founder highlighted:

> *"Fortunately for us, we are ISO compliant with ISO 9001 and 14001, which cover quality and environmental standards. This compliance gives us an advantage. When speaking to corporate companies, it shows that we are serious and knowledgeable."*

Lastly, zero-waste value chain innovation (comprehensive recycling) is one of the business models. This model manifests through flat fees for the clients and minimal payment for e-waste from client. Affordable e-waste disposal cost by clients shows flexibility, suggesting negotiated pricing strategies depending on the client and waste stream, but generally relying on value-added services for revenue. The founder asserted:

> *"Only one of our clients, a mine in Burgersfort, requires payment for their waste. However, it's very minimal; we pay around 0.50c per kilogram of their electronic waste."*

10.5 Discussions

The findings reveal that there are multifaceted motivations that drive green innovation in e-waste management. In view of this, Basadi E-waste's green innovation is not only driven by environmental concerns but also powered by social and economic motivations. The social drivers of green motivation include inclusion of women in a male-dominated e-waste industry (i.e., fostering economic opportunities within the township economy), bridging the gap between IT and e-waste sectors, equitable access to green innovation opportunities, and addressing industry disparities. Environmental drivers of green innovation comprise environmental consciousness, identifying gaps in e-waste management and commitment to true recycling, while economic drivers include problem-solving mindset, driving innovation through solutions, and passion for environmental impact. These multifaceted motivations strengthen the company's commitment and broaden its impact. Under the same line of reasoning, existing literature has indicated that there is a rising quantity of literature on the drivers of green innovation (Bothongo & Kinyar, 2025; Shahzad et al., 2022).

There are several core elements of green innovation, as discovered in this study: by-product innovation from e-waste plastic, diverse range of innovative green products, and zero-waste value chain innovation (comprehensive recycling). Byproduct innovation from e-waste plastic contains byproduct innovation from e-waste, research-driven processing, e-waste plastic recycling, and material repurposing. A diverse range of innovative green products comprise 3D printing filaments from e-waste plastics, assistive technology innovation, sustainable construction materials, and upcycled e-waste into functional products. Zero-waste value chain innovation includes zero-waste-to-landfill approach, comprehensive e-waste processing, and full repurposing of e-waste materials. These discoveries demonstrate that Basadi E-waste has adopted a strong circular economy model. Basadi E-waste's value-added recycling distinguishes it from other companies primarily focused on refurbishment or exporting raw materials. Our findings are supported by several studies such as Hossain and Sahajwalla (2024) and Nassani et al. (2023).

Interestingly, green innovation sustainable practices are rooted in environmental integrity, local value retention, and accountability. Environmental integrity manifests through distinction between recycling and refurbishing, and commitment to true recycling. Local value retention is associated with local job creation over exporting e-waste and empowering the township economy. Accountability manifests through environmental impact monitoring and reporting, and collaboration with research institutions for sustainability. This discussion indicates that Basadi E-waste has a holistic approach toward sustainability that manifests itself through the integration of sustainable practices into its operations.

Even though green innovations introduce several benefits in e-waste management, scaling green innovation faces significant challenges (Ghulam & Abushammala, 2023; Sakhuja et al., 2022). These challenges are financial barriers to green technology adoption, securing consistent e-waste feedstock, and low public awareness of e-waste issues. Financial barriers include high capital costs for machinery and infrastructure, and funding challenges for equipment procurement. The problem of securing consistent e-waste feedstock includes inadequate feeder stock for processing and dependency on external e-waste supply. Low public awareness of e-waste issues includes the lack of awareness about e-waste hazards and public education on e-waste. These challenges act as significant hurdles to growth and broader market penetration, and they highlight the systemic barriers that green startups often encounter.

Green innovation support ecosystem is crucial for early-stage green ventures. This indicates that strategic partnerships with diverse stakeholders are vital in sustainable e-waste management (Alblooshi et al., 2022; Cao et al., 2024). Basadi E-waste's progress is significantly enabled by a supportive ecosystem comprising government and institutional funding support, partnerships with academic and research institutions and access to infrastructure and talent support from technical institutions. The support network is essential for overcoming resource constraints, accessing expertise, and establishing credibility, particularly for early-stage green ventures.

Basadi E-waste employs green innovation sustainable business models that focus on providing value-added services to corporate clients. Green innovative business models enhance sustainability and attract corporate clients. These green innovative business models include value-added services (free collection with certification and audit), compliance, and ISO certification (which enhances credibility and provides a competitive advantage), and affordable e-waste disposal costs for clients. This makes Basadi E-waste's services attractive.

10.6 Conclusion

The study adopted a qualitative descriptive research design to explore the operations of Basadi E-waste, a South African black female-owned corporate that focuses on e-waste, data destruction, data sanitization, and refinery. Specifically, in the context of Basadi E-waste, we examine motivations and core elements of green innovations, and practical implementation of green innovation in e-waste recycling and the challenges involved. The findings of this study can assist e-waste management companies, consumers, and policymakers in their attempts to shift toward sustainable e-waste management. The findings of the study provide key lessons for sustainable e-waste management (Figure 10.2), and the expectation is that they will revolutionize the manner in which e-waste is being processed.

I) Integrate Social and Economic Goals with Environmental Motivation
Green businesses are more effective when they integrate social, economic, and environmental goals. By aligning their models with social empowerment and local economic development, they can gain stronger support and provide broader benefits to society.

II) Focus on Value Creation from Waste Streams
True green innovation involves creating value from waste, moving beyond basic recycling to produce valuable products. This approach not only generates new revenue streams but also significantly reduces environmental impact.

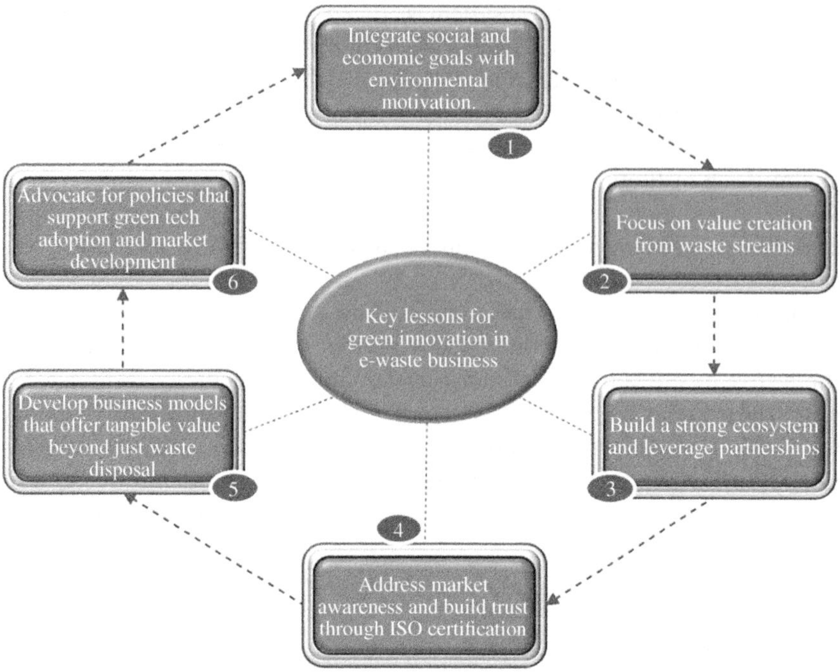

Figure 10.2 Key lessons for green innovation in e-waste business. *Source:* Authors' compilation (2025).

III) Build a Strong Ecosystem and Leverage Partnerships
Building a strong ecosystem through partnerships is crucial for early-stage green ventures. Collaborating with government institutions, academic bodies, and other businesses can provide essential funding, expertise, and market opportunities, helping these ventures overcome resource constraints and scale their innovations

IV) Address Market Awareness and Build Trust Through ISO Certification
Raising market awareness and building trust through certifications like ISO is vital. Educating the public and corporations about e-waste and responsible recycling can enhance market access and demonstrate a commitment to quality and environmental standards.

V) Develop Business Models That Offer Tangible Value Beyond Just Waste Disposal
In competitive markets, sustainable businesses need to offer more than just waste disposal services. Providing value-added services such as certifications and compliance reports can attract clients who prioritize sustainability.

VI) Advocate for Policies That Support Green Tech Adoption and Market Development

Advocating for policies that support green technology adoption and market development is essential. Green startups often face challenges like high capital costs and low awareness. Supporting government policies and incentives for green technology can create a more enabling environment for these businesses to thrive and scale.

In summary, these findings and lessons give important insights into green innovation in e-waste recycling, using Basadi E-waste as an example. They provide practical insights for entrepreneurs, policymakers, and researchers who want to improve sustainable waste management and support the green economy.

Even though the study has provided valuable insights, it has some limitations. The study examined the operations of a single company, which may restrict the generalizability of the findings to other companies. Therefore, the study can be enhanced by examining the operations of several e-waste management companies. Another limitation is related to the data collection method. A face-to-face interview with the founder was adopted. Using face-to-face interviews and quantitative surveys via a mixed-method approach can enhance the findings of this study.

References

Alblooshi, B.G.K.M., Ahmad, S.Z., Hussain, M., & Singh, S.K. (2022). Sustainable management of electronic waste: Empirical evidences from a stakeholders' perspective. *Business Strategy and the Environment, 31*(4), 1856–1874.

Andeobu, L., Wibowo, S., & Grandhi, S. (2021). An assessment of e-waste generation and environmental management of selected countries in Africa, Europe and North America: A systematic review. *Science of the Total Environment, 792*, 148078.

Asiedu, E.M. (2024). Incidents and impacts on operator revenue in the telecommunications sector: A multiple case-study approach. *Journal of Telecommunications and the Digital Economy, 12*(1), 65–93. https://doi.org/10.18080/jtde.v12n1.852

Awasthi, A.K., Wang, M., Wang, Z., Awasthi, M.K., & Li, J. (2018). E-waste management in India: A mini-review. *Waste Management & Research, 36*(5), 408–414.

Baldé, C.P., Kuehr, R., Yamamoto, T., McDonald, R., D'Angelo, E., Althaf, S., Bel, G., Deubzer, O., Fernandez-Cubillo, E., Forti, V., Gray, V., Herat, S., Honda, S., Iattoni, G., Khetriwal, D. S., di Cortemiglia, V. L., Lobuntsova, Y., Nnorom, I., Pralat, N., & Wagner, M. (2024). *The global e-waste monitor 2024.* International Telecommunication Union and United Nations Institute for Training and Resources, Geneva/Bonn.

Bernd, R. (2017). Theory and methodology of exploratory social science research. *International Journal of Science and Research Methodology*, *5*(4), 129–150.

Bothongo, K., & Kinyar, A. (2025). Factors associated with the adoption of eco-innovation and its effect on environmental performance in the US household cleaning sector. *Technological Forecasting and Social Change*, *210*, 123898.

Cao, D., Puntaier, E., Gillani, F., Chapman, D., & Dewitt, S. (2024). Towards integrative multi-stakeholder responsibility for net zero in e-waste: A systematic literature review. *Business Strategy and the Environment*, *33*(8), 8994–9014.

Chen, D., Faibil, D., & Agyemang, M. (2020). Evaluating critical barriers and pathways to implementation of e-waste formalization management systems in Ghana: A hybrid BWM and fuzzy TOPSIS approach. *Environmental Science and Pollution Research*, *27*(35), 44561–44584.

Doan, L.T.T., Amer, Y., Lee, S.H., & Phuc, P.N.K. (2019). Strategies for E-waste management: A literature review. *International Journal of Energy and Environmental Engineering*, *13*(3), 157–162.

Elman, C., Gerring, J., & Mahoney, J. (Eds.) (2020). *The production of knowledge: Enhancing progress in social science*. Cambridge, England: Cambridge University Press. https://doi.org/10.1017/9781108762519

Faibil, D., Asante, R., Agyemang, M., Addaney, M., & Baah, C. (2023). Extended producer responsibility in developing economies: Assessment of promoting factors through retail electronic firms for sustainable e-waste management. *Waste Management & Research*, *41*(1), 117–142.

Forti, V., Balde, C.P., Kuehr, R., & Bel, G. (2020). *The global e-waste monitor 2020: Quantities, flows and the circular economy potential. United Nations University, United Nations Institute for Training and Research – Co-hosted SCYCLE Programme* (p. 120). Bonn/Geneva and Rotterdam: International Telecommunication Union, and International Solid Waste Association.

Gaur, T.S., Yadav, V., Mittal, S., & Sharma, M.K. (2024). A systematic review on sustainable E-waste management: Challenges, circular economy practices, and a conceptual framework. *Management of Environmental Quality: An International Journal*, *35*(4), 858–884.

Ghulam, S.T., & Abushammala, H. (2023). Challenges and opportunities in the management of electronic waste and its impact on human health and environment. *Sustainability*, *15*(3), 1837.

Gollakota, A.R., Gautam, S., & Shu, C.M. (2020). Inconsistencies of e-waste management in developing nations – Facts and plausible solutions. *Journal of Environmental Management*, *261*, 110234.

Grandhi, S.P., Dagwar, P.P., & Dutta, D. (2024). Policy pathways to sustainable e-waste management: A global review. *Journal of Hazardous Materials Advances*, *16*, 100473.

Herat, S. (2021). E-waste management in Asia Pacific region: Review of issues, challenges and solutions. *Nature, Environment and Pollution Technology*, *20*(1), 45–53.

Hossain, R., & Sahajwalla, V. (2024). Current recycling innovations to utilize e-waste in sustainable green metal manufacturing. *Philosophical Transactions A*, *382*(2284), 20230239.

Jennings, G.R. (2005). Business, social science methods used in. In Kempf-Leonard, K. (Ed.), *Encyclopedia of social measurement* (pp. 219–230). Elsevier.

Kunz, T., Daikeler, J., & Ackermann-Piek, D. (2024). Interviewer-observed paradata in mixed-mode and innovative data collection. *International Journal of Market Research*, *66*(1), 14–26.

Lakshmi, S., & Raj, A. (2017). A review study of e-waste management in India. *Asian Journal of Applied Science and Technology*, *1*, 33–36.

Liang, B., Gu, J., Zeng, X., Yuan, W., Rao, M., Xiao, B., & Hu, H. (2024). A review of the occurrence and recovery of rare earth elements from electronic waste. *Molecules*, *29*(19), 4624.

Mir, M.A., & Chang, S.K. (2024). Saudi Arabia e-waste management strategies, challenges and opportunities, effect on health and environment: A strategic review. *Emerging Contaminants*, *10*, 100357.

Mishra, Y., & Mishra, S.P. (2023). E-waste management and legal concepts in India. In *Contemporary studies of risks in emerging technology, Part B* (pp. 299–312). Emerald Publishing Limited.

Moyen Massa, G., & Archodoulaki, V.M. (2023). Electrical and electronic waste management problems in Africa: Deficits and solution approach. *Environments*, *10*(3), 44.

Nassani, A.A., Hussain, H., Condrea, E., Grigorescu, A., Yousaf, Z., & Haffar, M. (2023). Zero waste management: Investigation of green technology, the green supply chain, and the moderating role of CSR intentions. *Sustainability*, *15*(5), 4169.

Oliveira, J.D., Oliveira Neto, J.F., Silva, M.M., & Santos, S.M. (2020). E-waste mistakenly disposed of as recyclable waste: A case study from Brazil. *Clean: Soil, Air, Water*, *48*. https://doi.org/10.1002/clen.202000115

Rautela, R., Arya, S., Vishwakarma, S., Lee, J., Kim, K.H., & Kumar, S. (2021). E-waste management and its effects on the environment and human health. *Science of the Total Environment*, *773*, 145623.

Sakhuja, D., Ghai, H., Bhatia, R.K., & Bhatt, A.K. (2022). Management of e-waste: Technological challenges and opportunities. In *Handbook of solid waste management: Sustainability through circular economy* (pp. 1523–1557). Singapore: Springer Nature.

Santos, S.M., & Ogunseitan, O.A. (2022). E-waste management in Brazil: Challenges and opportunities of a reverse logistics model. *Environmental Technology & Innovation*, *28*, 102671.

Saunders, M., Lewis, P., & Thornhill, A. (2016). *Research methods for business students* (7th ed.). Harlow, Essex: Pearson Education Limited.

Shahzad, M., Qu, Y., Rehman, S.U., & Zafar, A.U. (2022). Adoption of green innovation technology to accelerate sustainable development among manufacturing industry. *Journal of Innovation & Knowledge*, *7*(4), 100231.

Sharma, S., & Dutt, V. (2024). Contribution of e-waste management in green computing. In *Sustainable management of electronic waste* (pp. 397–412). Retrieved from https://onlinelibrary.wiley.com/

Sharma, H., Kumar, H., & Mangla, S.K. (2023). Enablers to computer vision technology for sustainable E-waste management. *Journal of Cleaner Production*, *412*, 137396.

Tegan, G. (2021). *A guide to exploratory research*. https://www.scriber.com/methodology/exploratory-research

Voxco. (2021). *Exploratory research: Pros and cons*. https://www.voxco.com/blog/exploratory-research-pros-and-cons/32

Wang, Z., Zhang, B., & Guan, D. (2016). Take responsibility for electronic-waste disposal. *Nature*, *536*(7614), 23–25. https://doi.org/10.1038/536023a

World Health Organisation. (2025). *Electronic waste (e-waste)*. https://www.who.int/news-room/fact-sheets/detail/electronic-waste-(e-waste)#:~:text=E%2Dwaste%20is%20one%20of,%2C%20dumped%2C%20and%20illegally%20exported

Xu, Y., Yeh, C.H., Liu, C., Ramzan, S., & Zhang, L. (2021). Evaluating and managing interactive barriers for sustainable e-waste management in China. *Journal of the Operational Research Society*, *72*(9), 2018–2031.

Yin, R. (2009). *Case study research: Design and methods* (4th ed.). SAGE Publications.

11

Toward Sustainable Consumption: A Case of PLAN3T Application

Marcia Mkansi and Frank Ranganai Matenda

Department of Operations Management, College of Economic and Management Sciences, University of South Africa, Pretoria, South Africa

11.1 Introduction

Consumerism has ushered in the need to protect consumers. However, nowadays, consumers have become a danger to the society due to the environmental footprint of their choices. The world is presently consuming more resources than ever, surpassing the planet's ability for regeneration (United Nations Environment Program, 2025a). In the same vein, Wackernagel et al. (2021) propounded that environmental degradation like the build-up of greenhouse gases, acidification of oceans, and reduction of groundwater is speeding up since the demand of natural resources by people is progressively surpassing the biological rate of regeneration by the earth. Economies need 1.75 Earths to maintain the present resource consumption rate in the world (McCarthy & Sánchez, 2019). However, the world cannot use 1.75 Earths without having devastating effects. Natural resources are getting depleted, global economic activities are increasing consumption, pollution, and waste, and the gap between the poor and rich is expanding, adversely affecting education, health, empowerment, and equity (see United Nations Environment Program, 2025). The regeneration capacity of the earth has become a significant bottleneck for the human economy (Wackernagel et al., 2021). Economic development theory and practice highlight the significance of natural resources (Wackernagel et al., 2021).

Globally, rising incomes and consumption are the drivers of resource use and environmental degradation (Gulati & Naudé, 2017). Carbon footprint amount is

Green Innovations in Supply Chain Management: Case Studies and Applications,
First Edition. Edited by Marcia Mkansi, Godfrey Mugurusi, Aaron L. Nsakanda,
Temidayo Akenroye, and Frank Tietze.

Published 2026 by John Wiley & Sons, Inc.

linked to consumption and production patterns. Gulati and Naudé (2017) articulated that consumer lifestyle and behavior substantially influence the use of energy and the consequential emissions, and a significant share of global greenhouse gas emissions emanating from production are connected to consumption habits. Unconscious life patterns and consumption habits of consumers amplify the carbon footprint amount generated (Kuduz, 2022). This is worrisome considering that, for instance, consumer spending is accountable for roughly 60% of the gross domestic product in Organisation for Economic Co-operation and Development countries. The unsustainability of consumption habits and of the environmental dangers presented by consumerism is shown by the insertion of sustainable consumption under Sustainable Development Goal (SDG) 12 (Gulati & Naudé, 2017).

Present consumer behavior changes have led to lifestyle polarization (Santos-Corrada et al., 2024). Some consumers have a consumerist lifestyle (Santos-Corrada et al., 2024) while others have a highly consumerist lifestyle, i.e., anthropocentric behavior (Perera et al., 2022). Anthropocentric behavior refers to the behavior that prioritizes humans over other species. It results in the sidelining and suppression of nonhuman entities. Consumers with a highly consumerist lifestyle are not worried about the environment. Kotler et al. (2021) proffered that minimalist or ecocentric consumers are bothered about the environment and sustainable consumption. In practice, all life in an ecosystem should be respected and preserved, and human beings are not superior beings (Santos-Corrada et al., 2024).

Since humankind has only one planet to depend on for natural resources, sustainable consumption plays an important role in mother earth. Economies should sustainably consume resources to prevent depleting the planet. Sustainable consumption is one of the most significant requirements for future sustainable development. Sustainable consumption was defined at the Oslo Symposium in 1994 (Norwegian Ministry of Environment, 1994) as, "the use of services and related products, which respond to basic needs and bring a better quality of life while minimizing the use of natural resources and toxic materials as well as the emissions of waste and pollutants over the life cycle of the service or product so as not to jeopardize the needs of further generations." Glavič (2021) opined that sustainable consumption refers to the use of services and products that have a minimal effect on the environment and allow future generations to attain their needs. Sustainable consumption and sustainable production enable the attainment of 2030 SDGs as indicated in SDG 12 (Adebayo et al., 2021). SDG 12 calls out for the application of the United Nations 10-year framework of programs on sustainable consumption and sustainable production configurations (Fischer et al., 2023).

The world now has a task of responding to environmental challenges caused by unsustainable consumption patterns. Nevertheless, countries are finding it difficult to discover solutions since restricting consumption is politically sensitive.

Santos-Corrada et al. (2024) opined that the existing literature has overlooked consumers' role in sustainable consumption practices. Sustainable consumption is substantially influenced by goals and values embedded in a community. Therefore, sustainable lifestyles are essential for future sustainable consumption (Glavič, 2021). For instance, Adebayo et al. (2021) postulated that consumption habits are drivers of emissions, known as consumption-premised emissions. Modifying lifestyle and consumption habits is essential in enhancing sustainable consumption behavior. Akenji and Chen (2016) proffered that "a sustainable lifestyle is a cluster of habits and patterns of behavior embedded in a society and facilitated by institutions, norms and infrastructures that frame individual choice, in order to minimise the use of natural resources and generation of wastes, while supporting fairness and prosperity for all." United Nations Environment Program (2025b) articulated that "sustainable lifestyles are considered as ways of living, social behaviours and choices, that minimise environmental degradation (use of natural resources, carbon dioxide (CO_2) emissions, waste and pollution) while supporting equitable socio-economic development and better quality of life for all."

Since people's lifestyle decisions are placing the planet at risk, a targeted action is required to foster consumption patterns that would decrease environmental degradation and promote the attainment of the needs of humanity. Sustainable consumption practices deal with diverse patterns of consumer behavior and should be enhanced. Sustainable consumption encourages consumers to select brands that seem to be ecofriendly (Canales, 2018). In practice, selecting these eco-friendly brands is a challenge. However, of late, an extensive range of technologies has been designed to bring consumption to more sustainable levels by helping consumers in selecting ecofriendly products and services. Technology has brought sustainable behavior pleas for behavioral change toward sustainable consumption habits, which can assist in achieving sustainability goals. Therefore, environmentally sustainable consumption has become a vital research area of late.

This study presents PLAN3T GmbH's free application (also known as PLAN3T), a reward system for sustainable consumption that helps users in their pursuit for a more environmentally conscious lifestyle and analyzes its potential in promoting sustainable consumption. This application is a digital platform that can be securely connected to customers' bank accounts within minutes and automatically rewards consumers who are environmentally conscious for sustainable purchasing decisions and climate protection in daily life. With PLAN3T, users can track their ecological footprint, i.e., they can track and measure their carbon footprint, obtain personal recommendations on how to decrease it, and get cashback for sustainable purchases. PLAN3T GmbH was founded in 2020, and its mission is to make sustainable consumption affordable and easy for everybody.

11.2 Challenges

Globally, mass consumption is putting pressure on resources and the environment, resulting in ecological degradation and prospective dangers to the well-being of people. There has been consensus at an international level that the main driver of the non-stop deterioration of the global environment is the unsustainable consumption and production patterns since the United Nations Conference on Environment and Development in Rio de Janeiro in 1992. The consumption of services and products affects the environment in a myriad of ways. Unsustainable consumption contributes to global warming, biodiversity loss, increasing waste, food insecurity, pollution, and resource depletion. These issues emanate from the use of natural resources, which is far beyond the capacity of the earth to replenish them. Maintaining the present consumption habits is not an alternative.

In 2022, the Johannesburg Plan of Implementation was agreed at the World Summit on Sustainable Development. At this summit, sustainable consumption and production were recognized as vital requirements for sustainable development. It was concluded that communities should make significant changes in the way they consume and produce if they want to attain global sustainable development. The Johannesburg Plan of Implementation also indicates that all countries should foster sustainable consumption and production patterns. Interestingly, it has been documented that roughly 24% of individual CO_2 footprint emanates from eating habits, and 26% of global emissions are because of diet. This implies that the unsustainable patterns of consumption and production should be changed.

Even though consumers play a vital role in cutting the carbon footprint amount emanating from individual consumption, techniques for alerting consumers about these issues and for lowering the carbon footprint emanating from individual consumption are restricted. Also, most customers want to buy sustainable products, but they are not able to do it in their daily life due to various reasons. For instance, it has been stated that over 60% of European consumers covet to buy sustainable products, but approximately 26% can buy them in their daily life, mainly because environmentally friendly products and services are more expensive than conventional ones. Furthermore, globally, millions of people are prepared to modify their lifestyles, but they do not know how to change them. Literature has indicated that if the unconscious consumption model stays, the carbon footprint amount from consumption will continue to rise. If our climate is remarkably changing, then why consumers cannot change their habits to reduce their carbon footprint? This is a billion-dollar question that needs answers.

Thus, it is important to develop frameworks, techniques, and approaches that improve sustainable consumption at the individual level, which mainly depends on the behavior of consumers. The challenges of attaining sustainable consumption are linked to, among other things, existing technologies, implementation of resources, and consumer demands. Consumers have a savings potential when it comes to sustainable consumption. The overarching challenge in attaining sustainable consumption is to promote sustainable lifestyles by encouraging consumers to make better choices and to demand better services and products. This indicates that consumers' decision-making should be better harnessed for sustainability.

11.3 Overcoming the Challenges

PLAN3T addresses the challenge of unsustainable consumption in the following ways, which inspire consumers to shift toward sustainable consumption by making it more accessible, affordable, and rewarding:

- **Promotion of greener consumption choices**: The application provides options that inspire consumers to buy sustainable goods and services.
- **Open banking integration**: The application can be connected to one or more bank accounts within a few minutes. It works smoothly with numerous banks. This integration enables consumers to get cashback from PLAN3T GmbH's sustainable partners and track their carbon footprint, enabling them to reduce and offset their carbon footprints. Interestingly, bank connection renewal is done every three months, and bank connection can be terminated at any time.
- **Awareness and behavioral change**: The application alerts consumers regarding the environmental influences of their selections when purchasing goods and services. This is critical since consumers need to know products and services that increase or reduce their carbon footprint amount. This awareness enhances behavioral changes as far as sustainable consumption is concerned. It has been documented that modifications in lifestyle and consumer behavior can lead to a decline of up to 70% of global greenhouse gas emissions by 2050.
- **Cashbacks, bonuses, and discounts**: PLAN3T GmbH gives unique cashbacks, bonuses, and discounts when consumers buy from their sustainable partner corporates (Figures 11.1 and 11.2). Specifically, consumers can get "Planet Coins" via reading personalized sustainability tips, CO_2 offsets, inviting friends, and answering quizzes. "Planet Coins" refers to the virtual currency in which consumers get their cashback paid out. When customers shop with sustainable partner corporates, they get up to 5% of their turnover in "Planet Coins." These coins are then used to redeem discounts and to pay for discounted services or products or can be donated to climate protection projects. This incentivizes consumers to make environmentally friendly selections by making sustainable services and products more financially accessible and affordable for everyone. In this manner, consumers are encouraged to opt for

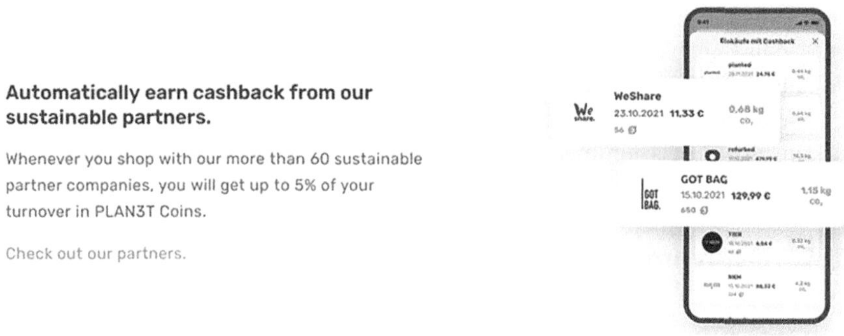

Figure 11.1 Cashback from sustainable partners. *Source:* Image from PLAN3T GmbH.

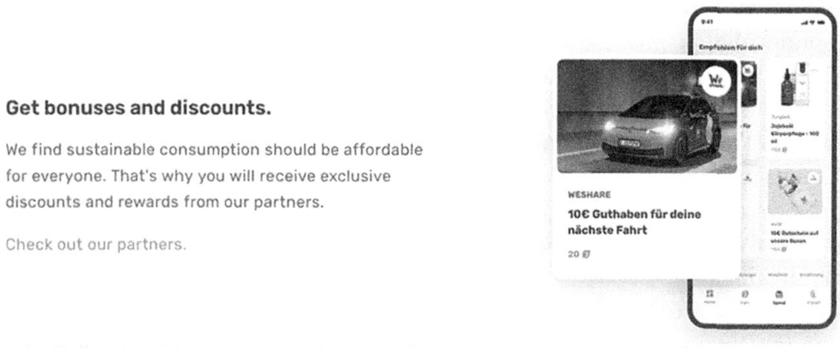

Figure 11.2 Bonuses and discounts. *Source:* Image from PLAN3T GmbH.

environmentally friendly services and products. This approach enhances sustainable consumption patterns by making them more economically attractive and viable.

- **Social and climate protection projects**: "Planet Coins" can be used at over 60 sustainable partner brands and corporates (e.g., Inyova, VAYS, release, Pukka, tink, SHARE NOW, racoon, and GEPA) or can be donated to effective climate protection projects (see Figure 11.3). In these projects, trees are planted, and solar plants are built, among other things.
- **Partnering with sustainable brands**: PLAN3T GmbH partners with corporates and brands that are committed to sustainability, i.e., it collaborates with brands and corporates that support their mission. The startup's partner network incorporates brands like Enpal, tibber, SodaStream, refurbed, planted, WeShare, and TIER. This not only supports sustainable services and products but also warrants that consumers have access to ethical alternatives. PLAN3T GmbH encourages responsible consumer behavior and warrants that the brands and

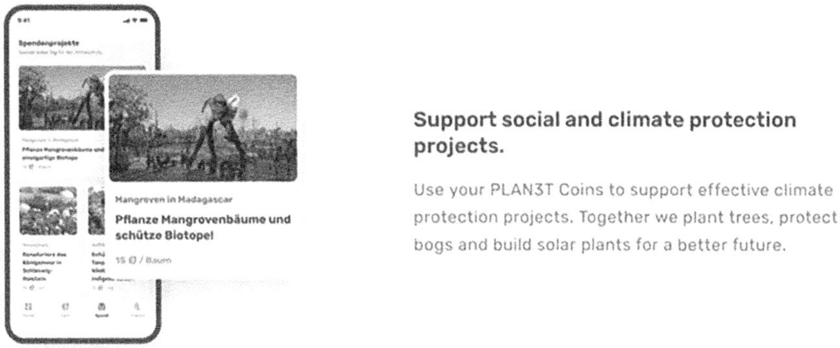

Figure 11.3 Social and climate protection projects. *Source:* Image from PLAN3T GmbH.

corporates they team up with meet the high standards required for a substantial positive environmental influence.

- **Cost reduction**: The application has been designed to offer a solution to the financial challenge when it comes to sustainable purchasing. When consumers make sustainable purchasing decisions, they are given rewards. Consumers can stay motivated and protect their wallets. Thus, this solution makes sustainable consumption easy, affordable, and accessible to all kinds of consumers.
- **Data analytics**: With PLAN3T, consumers can track their ecological footprint. Once the customer's bank account is connected to the application, the customer can get cashback from sustainable partners and can track their carbon footprint. Customers can see how much CO_2 makes their consumption decisions in daily life (see Figure 11.4). This would help consumers reduce their ecological footprint. Basically, the application allows consumers to compute their carbon emissions, comprehend their influence on the environment, and take practical steps to cut and offset these emissions. The calculation of

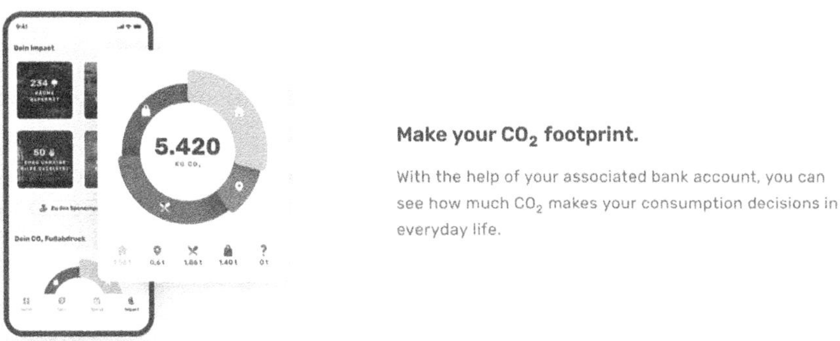

Figure 11.4 Carbon footprint. *Source:* Image from PLAN3T GmbH.

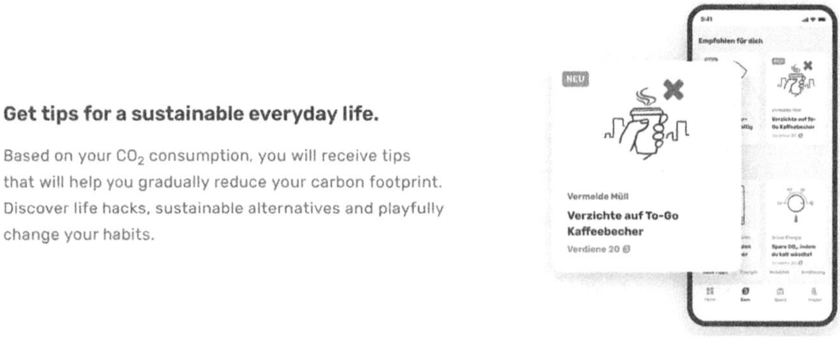

Figure 11.5 Tips for sustainable daily life. *Source:* Image from PLAN3T GmbH.

carbon emissions is executed with the assistance of PLAN3T GmbH partner eco.income, whose model is Technischer Überwachungsverein (which is German for Technical Inspection Association) certified.

- **Tips and gamification**: Consumers get tips that assist them to progressively cut their CO_2 footprint based on their CO_2 consumption (see Figure 11.5). They can discover life hacks and sustainable options and modify their behavior in a fun and easy way. Also, the application has numerous other exciting features.
- **Real-time feedback and visualization**: The application provides immediate feedback on the possible environmental effects of consumer purchases and visualizes the environmental effects of these consumer purchases. This real-time feedback helps consumers comprehend the direct influences of their purchases, allowing them to get a synopsis of the emissions produced by their consumption. Premised on this, environmentally friendly options are proposed to users.

11.4 Method

Web scraping, also known as web harvesting or web data extraction, was adopted in this study to extract data from the website of PLAN3T GmbH. Web scraping is a process by which data is collected from websites and sorted and saved for further research. We implemented web scraping to gather textual data. Web scraping is regarded as one of the most powerful and effective methods of collecting qualitative data (see Dongo et al., 2021; Hillen, 2019). Dongo et al. (2021) indicated that web scraping is fast and flexible in terms of collecting data. Hillen (2019) postulated that, in web scraping, costs are low, data is gathered in real-time, customized data sets can be designed, and there is transparency in how the data was gathered. Thus, web scraping has been widely used in management sciences.

11.5 Impact of the Innovation

PLAN3T GmbH, the Hamburg-based startup, has launched its application, which is the first reward system for sustainable consumption that helps its users in their pursuit of a more environmentally conscious lifestyle. In spring 2021, PLAN3T GmbH launched the beta version of its application with the support of cc.systems. The application has been downloaded approximately 10,000 times and can be found in App Store and Google Play.

PLAN3T GmbH has partnered up with several eco-conscious corporates and brands to enlarge its network. By partnering with eco-friendly corporates and brands, the startup can benefit from amplified visibility. Of late, the startup has swiftly enlarged its partner network, which now incorporates well-familiar brands like NAYA, The Fair Cottage, tibber, GREENFORCE, Enpal, Anifree Shoes, refurbed, SodaStream, WeShare, planted, and TIER. This argument is supported by the following quotes from the partners and/or brands:

Kilian Kaminski, Founder of refurbed:

> "With refurbed we joined PLAN3T early on because their app offers a unique solution to reward consumers for sustainable actions."

Philip von Have, Co-founder and Chief Executive Officer of Blue Farm:

> "PLAN3T offers a great platform to connect environmental-conscious consumers with sustainable brands like us at Blue Farm. We look forward to scale our impact together."

The user base of the PLAN3T GmbH's application has been increasing. In Spring 2021, PLAN3T GmbH launched the beta version of its application in German application stores. Currently, 6000 registered users have averted and rewarded for over 250 tonnes of CO_2. Consumers also get "Planet Coins" that can be used at over 60 sustainable partner corporates and brands or donated to effective climate protection projects. Users value the application's effectiveness in promoting environmentally friendly purchases through cashback rewards. The startup has been receiving positive feedback from the users. Positive quotes from the users and supporters include the following:

Mark Griffiths, Global Leader Climate Business at WWF:

> "The responsibility to reduce one's personal or company carbon footprint is an absolute priority, to achieve a net zero world aligned with the United Nations Climate Agreement. I am thus delighted to see PLAN3T providing innovative solutions to incent actions that deliver immediate carbon reductions."

Maria de Luna i Visa, Head of Partnerships bei Organic Basics:

> "PLAN3T makes the transition to sustainable living engaging and rewarding, but most importantly, it's easy, educational, and fun!"

The startup has enticed investors who support sustainable initiatives. Recently, the corporate stated that it has raised a pre-seed funding of €750,000. Contributors in the pre-seed financing round include well-known Berlin-based early-stage investor APX, Hamburgische Investitions- und Förderbank (IFB), and distinguished business angels from the German digital and advertising sectors, including Guido Syré, Marc Sasserath, and Mutabor founders Heinrich Paravicini and Johannes Plass.

PLAN3T GmbH has been recognized and praised within the eco-friendly community for its innovative approach in promoting sustainable consumption. It has received positive attention from the eco-friendly community for its work. Also, the PLAN3T GmbH's application has been expanded by an open banking integration, fortifying the position of the startup in the fintech space.

PLAN3T GmbH is striving to become certified as a B Corporation. Its status now is "B Corporation Pending." Certified B Corporations are pioneers in addressing the most significant societal problems. Several startups desire to send a sign to their stakeholders that they are striving to become certified as a B Corporation. Thus, "B Corporation Pending" status assists these startups transmitting precisely that signal and indicating their obligation. The "B Corporation Pending" status provides startups time to get ready for the thorough process of full B Corporation certification.

11.6 Business Model of the Organization

The PLAN3T GmbH's application is free to all users. Thus, any individual can download it, making it a go-to application for sustainable consumption. This enhances the visibility of the startup and makes sustainable consumption affordable and easy and assists consumers to save money and CO_2. Consequently, the startup has a network of more than 60 partner brands and corporates. Each partner offers a sustainable option to services and products on the market. Interestingly, PLAN3T GmbH designed its own impact rating, which it uses to rate its prospective partner brands and corporates. Therefore, the startup can make sure that the partner brands and corporates are green as they claim to be.

The business model of PLAN3T GmbH rotates around offering a subscription-based service. The application users pay a monthly fee to access resources and tools that assist them in tracking their carbon footprint, supporting efficient carbon offset projects, and adopting sustainable habits. Carbon offset projects are developed to attain rapid and lasting CO_2 cuts whilst providing social benefits and generating positive effects in developing economies.

PLAN3T GmbH gets revenue from subscription fees and from partnerships with corporates and brands, which provide carbon offsetting as a component of their corporate social responsibility enterprises. Further, the startup retains users by giving them rewards for sustainable behaviors, inspiring them to ask family and friends to join them in using the application. This nurtures a society of environmentally responsible people.

11.7 Learning Outcomes

Generally, technology is a vital component of the climate change solution. With the advent of artificial intelligence, technology has become more indispensable in addressing the climate change issue. Specifically, technology from startups plays a crucial role in bringing consumption to more sustainable levels. Sustainable consumption contributes to the reduction of carbon emissions.

Sustainable consumption needs strategic partnerships among diverse stakeholders across diverse sectors in different countries. These stakeholders include consumers, government, manufacturers, and retailers, among others. Thus, for startups to thrive, they need strategic partnerships with these stakeholders. For instance, since consumers get "Planet Coins" that can be used at sustainable partner corporates and brands or donated to effective climate protection projects, partnerships with these partner corporates and brands and climate protection projects are vital.

Partnerships with banks are essential because PLAN3T GmbH's application is linked to banks for consumers to get cashback from PLAN3T GmbH's sustainable partners and track their carbon footprint. Since the analysis of bank data is executed, there is a dire need to collaborate with partners that are certified and allowed to examine bank data securely and in conformity with data protection regulations.

Most importantly, for startups like PLAN3T GmbH to thrive, there is dire need for strategic association with investors or providers of finance that support sustainable initiatives. Investors need to trust the owners of the startup and to have confidence in the business model of the startup. With support and funding from investors, the startup can make consumption patterns more sustainable and affordable.

Data protection is vital since huge amounts of sensitive information are at stake. The information that is at stake include bank data and consumption patterns. This implies that the highest security standards should be adhered to, and the latest technology should be adopted to protect consumers' data from unauthorized and third-party access.

11.8 Plans to Further Develop the Initiative

PLAN3T GmbH is a Germany-based startup that has a plan to expand into other European economies. This will enable the startup to access new markets and reach a bigger customer base, increase growth, and fill gaps in foreign markets where other players are lacking, thereby fortifying their competitive edge. To position the startup in the fintech space, the owners are planning to progressively automate their application with the assistance of an open banking integration. Also, PLAN3T GmbH plans to assess the novel topographies of the beta version of their application and then make them accessible in Austria and Germany. Further, PLAN3T GmbH plans considerably enhance the CO_2 tracking feature to suggest sustainable alternatives that measurably reduce CO_2 emissions.

11.9 Conclusion

PLAN3T GmbH is a Hamburg-based fintech startup that promotes climate action and sustainability by offering tools to cut and offset carbon emissions. The startup uses an application that acts as a reward system for sustainable consumption that helps its users in their pursuit of a more environmentally conscious way of life. PLAN3T GmbH assists eco-conscious corporates and people and generates money via subscription fees and partnerships. It helps consumers to save money and CO_2.

References

Adebayo, T.S., Udemba, E.N., Ahmed, Z., & Kirikkaleli, D. (2021). Determinants of consumption-based carbon emissions in Chile: An application of non-linear ARDL. *Environmental Science and Pollution Research*, *28*(32), 43908–43922.

Akenji, L., & Chen, H. (2016). *A framework for shaping sustainable lifestyles: Determinants and strategies*. United Nations Environment Program (UNEP): Nairobi, Kenya.

Canales, H.J.S. (2018). El greenwashing y su repercusión en la ética empresarial. *Neumann Business Review*, *4*(1), 28–43.

Dongo, I., Cardinale, Y., Aguilera, A., Martinez, F., Quintero, Y., Robayo, G., & Cabeza, D. (2021). A qualitative and quantitative comparison between Web scraping and API methods for Twitter credibility analysis. *International Journal of Web Information Systems*, *17*(6), 580–606.

Fischer, M., Foord, D., Frecè, J., Hillebrand, K., Kissling-Näf, I., Meili, R., Peskova, M., Risi, D., Schmidpeter, R., & Stucki, T. (2023). Sustainable consumption.

In *Sustainable business: managing the challenges of the 21st century* (pp. 105–116). Cham: Springer International Publishing.

Glavič, P. (2021). Evolution and current challenges of sustainable consumption and production. *Sustainability*, *13*(16), 9379.

Gulati, M., & Naudé, L. (2017). *Chewing over consumption-based carbon emissions accounting*. chrome-extension://efaidnbmnnnibpcajpcglclefindmkaj/https://wwfafrica.awsassets.panda.org/downloads/WWF_2017_IKI_Futures_Consumption-based_Carbon_Emissions_Accounting.pdf?24561/Chewing-over-consumption-based-carbon-emissions-accounting.

Hillen, J. (2019). Web scraping for food price research. *British Food Journal*, *121*(12), 3350–3361.

Kotler, P., Kartajaya, H., & Setiawan, I. (2021). *Marketing 5.0: Tecnología para la humanidad*. LID Editorial.

Kuduz, N. (2022). Decreasing carbon footprint for sustainable consumption. In Grima, S., Özen, E., & Romānova, I. (Eds.), *Managing risk and decision making in times of economic distress, part A (contemporary studies in economic and financial analysis, Vol. 108A)* (pp. 215–230). Leeds: Emerald Publishing Limited. https://doi.org/10.1108/S1569-37592022000108A042

McCarthy, J., & Sánchez, E. (2019). *The world is using natural resources faster than ever before*. https://www.globalcitizen.org/en/content/overshoot-day-natural-resources/.

Norwegian Ministry of Environment (1994). *Report of the symposium on sustainable consumption*. Oslo, Norway: Ministry of Environment.

Perera, C.R., Kalantari Daronkola, H., & Johnson, L.W. (2022). Anthropocentric, biospheric and egobiocentric environmental values and green product purchase intention: The mediating effect of environmental identity. *Journal of Consumer Behaviour*, *21*, 1–17.

Santos-Corrada, M.D.L.M., Méndez-Tejeda, R., Flecha-Ortiz, J.A., & Lopez, E. (2024). An analysis of sustainable consumption practices through the role of the consumer behavior in the circular economy. *Journal of Consumer Behaviour*, *23*(1), 229–242.

United Nations Environment Program. (2025b). *Sustainable lifestyles*. https://www.unep.org/explore-topics/resource-efficiency/what-we-do/sustainable-lifestyles.

United Nations Environment Program (UNEP). (2025a). *Sustainable consumption and production policies*. https://www.unep.org/explore-topics/resource-efficiency/what-we-do/sustainable-consumption-and-production-policies#:~:text=Sustainable%20Consumption%20and%20Production%20(known,efficiency%20and%20promoting%20sustainable%20lifestyles.

Wackernagel, M., Hanscom, L., Jayasinghe, P., Lin, D., Murthy, A., Neill, E., & Raven, P. (2021). The importance of resource security for poverty eradication. *Nature Sustainability*, *4*(8), 731–738.

12

Green Initiatives of South Africa's Leading Retailers: A Comparative Analysis of Sustainability Strategies

Elizma S. Bok and Leonie B. Louw

Department of Operations Management, University of South Africa, Pretoria, South Africa

12.1 Introduction

The South African retail landscape is marked by a complex interplay of formal and informal sectors, diverse formats, and significant economic contributions. As the largest retail market in Sub-Saharan Africa and ranked 20th globally, it encompasses sectors such as clothing, groceries, and furniture, reflecting local and international influences (Ridge et al., 2015). Since the end of apartheid in 1994, the sector has evolved with greater access to modern retail formats and a focus on customer experience (Heidig et al., 2017; Strydom, 2013).

A key feature of South African retail is the coexistence of formal and informal sectors. Informal retail, particularly spaza shops in townships, plays a crucial role in local economies by providing essential goods and services and contributing to employment (Eyk et al., 2022; Hare & Walwyn, 2019). However, these shops face competition from local and migrant businesses, complicating their operations (Malgas & Zondi, 2019, 2020).

The formal retail sector has grown through modern formats and big data analytics to enhance efficiency and customer engagement (Ridge et al., 2015; Sharma & Gautam, 2017). The rise of e-retail has transformed shopping behaviors, offering convenience and variety (Alexander & Mason, 2017). Despite advancements, challenges such as attracting skilled labor and adapting to changing consumer preferences persist (Heidig et al., 2017; Idowu & Ndidiamaka, 2018).

Green Innovations in Supply Chain Management: Case Studies and Applications, First Edition. Edited by Marcia Mkansi, Godfrey Mugurusi, Aaron L. Nsakanda, Temidayo Akenroye, and Frank Tietze.

Published 2026 by John Wiley & Sons, Inc.

Global retail giants, such as Walmart, have reshaped competitive dynamics, with Walmart's acquisition of Massmart driving local retailers to innovate and expand into African markets (Dakora & Slabbert, 2014; Engelbrecht & Rensburg, 2018; Kenny, 2014). The COVID-19 pandemic further disrupted the sector, particularly informal retailers, prompting government interventions to support food security and economic stability, though long-term sustainability remains a concern (Abe & Mugobo, 2021; Wegerif, 2020).

South Africa's retail sector presents both opportunities and challenges. Its diverse consumer base and retail environments drive growth, yet competition from informal traders, labor shortages, and evolving consumer expectations pose significant hurdles. Formal retailers must differentiate their offerings and enhance service quality to compete effectively (Sharma & Gautam, 2017).

Despite establishing retail management programs in universities, the sector struggles to attract young talent, leading to operational inefficiencies and high turnover rates that impact service delivery (Heidig et al., 2017; Idowu & Ndidiamaka, 2018). Additionally, the adoption of e-commerce, while offering growth potential, presents challenges for small and medium enterprises (SMEs) due to high costs and limited technical expertise (Ismail et al., 2011; Ridge et al., 2015).

Sustainability is increasingly influencing South African retail, driven by consumer demand, regulatory frameworks, and competitive pressures. Retailers must integrate sustainability into their operations and supply chains to remain viable. However, complexities in supply chain transparency and compliance costs pose challenges, especially for SMEs (Berning & Venter, 2015; Haywood et al., 2021; Niehaus et al., 2018).

The digital divide further complicates sustainability efforts, with varying levels of internet access impacting smaller retailers (Jibril et al., 2020). The gap in comparative research on African retailers' green initiatives highlights the need for further study to better understand sustainability practices in this context (Ruiz–Real et al., 2018; Wang et al., 2013). Addressing these research gaps can help inform strategies to enhance sustainability efforts among South African retailers.

12.1.1 Objectives, Significance, and Scope of the Innovation

The primary objective of this chapter is to analyze and compare the sustainability strategies adopted by the top five retailers in South Africa. To achieve this objective, the following sub-objectives have been set:

- Identify key sustainability initiatives implemented by these retailers.
- Evaluate the effectiveness of their sustainability strategies in addressing environmental, social, and governance (ESG) challenges.
- Compare the approaches adopted by different retailers to determine best practices.
- Assess the impact of sustainability efforts on customer perceptions and financial performance.

- Provide recommendations for enhancing sustainability practices in the South African retail sector.

Understanding the sustainability strategies of leading South African retailers is critical in promoting responsible business practices and aligning with global sustainability trends. This study contributes to the literature by:

- Providing insights into the sustainability challenges and opportunities within the South African retail sector.
- Offering a benchmark for other retailers seeking to improve their sustainability initiatives.
- Informing policymakers and industry stakeholders about the effectiveness of current sustainability efforts and potential areas for improvement.
- Enhancing consumer awareness of sustainable retail practices and their broader societal impact.

The scope of this chapter is limited to the top five retailers in South Africa selected based on market share, revenue, and sustainability disclosures. The analysis focuses on their sustainability strategies across key areas such as supply chain management, energy efficiency, waste reduction, and corporate social responsibility (CSR). The study covers the 2021/2022 financial reports of the five identified retailers to capture recent trends and developments. It does not include smaller retailers or those operating solely in the informal sector but provides insights applicable across the broader retail landscape.

12.1.2 Chapter Outline

This chapter explores the sustainability strategies of South Africa's leading retailers, focusing their efforts to integrate green initiatives into their operations. The chapter begins with an overview of the South African retail landscape, highlighting the sector's complexities and the growing importance of sustainability. It then examines key sustainability initiatives implemented by top retailers, followed by a comparative analysis of their approaches. The chapter further explores the impact of these initiatives on customer perceptions and financial performance. Finally, recommendations are provided for enhancing sustainability practices in the sector, considering emerging trends and regulatory requirements.

12.1.3 Definition of Key Terms

For this chapter, the definitions of key terms are as follows:

Consumer behavior: This term describes individuals' decision-making processes and actions regarding purchasing goods and services, particularly concerning sustainability and environmental consciousness. Understanding consumer behavior is critical for organizations promoting green products effectively (Guerreiro & Pacheco, 2021; Zhang et al., 2023).

Corporate social responsibility (CSR): CSR encompasses the efforts made by organizations to operate ethically, considering their impact on society and the environment. This includes initiatives promoting social good, environmental stewardship, and ethical governance (Ateş, 2022; Pinheiro et al., 2023).

Environmental, social, and governance (ESG) metrics: ESG metrics are standards used to evaluate an organization's operations regarding environmental impact, social responsibility, and governance practices. These metrics help investors assess their investments' sustainability and ethical impact (Choi & Hong, 2022).

Green bonds: These are fixed-income financial instruments specifically earmarked to raise funds for projects with positive environmental impacts. The issuance of green bonds is a part of a broader strategy to finance sustainable development initiatives (Yeow & Ng, 2021).

Greenwashing: This term refers to organizations misleading consumers regarding the environmental benefits of their products or practices. It often involves exaggerating or fabricating claims about sustainability to enhance corporate image without implementing substantive environmental improvements (Del Río et al., 2023; Xu et al., 2023).

Stakeholder engagement: This concept involves all parties affected by an organization's operations, including employees, consumers, suppliers, and the community, in decision-making processes. Effective stakeholder engagement fosters trust and transparency in corporate practices (Nandkeolyar & Chen, 2022).

Sustainability: Sustainability refers to the capacity to maintain specific processes or states indefinitely. In an organization context, it involves practices that do not deplete resources or harm ecological systems, ensuring that future generations can meet their needs (Blaha et al., 2021; Morseletto, 2022).

Transparency: In corporate sustainability, transparency refers to the openness with which organizations communicate their practices and policies regarding environmental and social impacts. High levels of transparency can enhance trust among consumers and stakeholders (Nandkeolyar & Chen, 2022).

12.2 The State of the Art

Sustainability in retail has become a key focus globally, driven by consumer demand, regulatory requirements, and corporate responsibility. South African retailers are adopting green initiatives such as energy efficiency, waste reduction, and ethical sourcing to align with global sustainability trends while addressing local socio-economic challenges. The circular economy, stakeholder engagement, and technological advancements like blockchain for supply chain transparency are shaping sustainability efforts. Despite progress, challenges such as resource constraints and operational inefficiencies persist. This chapter provides a comparative analysis of leading South African retailers to identify best practices and opportunities for improvement in sustainability strategies.

12.2.1 The Triple Bottom Line (TBL) Framework

The Triple Bottom Line (TBL) framework, introduced by John Elkington in 1997, provides a comprehensive model for evaluating sustainability across three critical dimensions: economic, social, and environmental performance. This model emphasizes balancing financial outcomes with social equity and environmental stewardship, making it particularly relevant for analyzing green initiatives in various sectors, including retail.

The economic dimension of the TBL focuses on the financial viability of sustainable practices. Retailers increasingly recognize that adopting green initiatives can lead to cost savings through improved efficiencies, waste reduction, and energy conservation (Ahamat, 2017; Kurnia, 2023). Sustainable supply chain practices can enhance profitability by reducing operational costs and increasing consumer demand for eco-friendly products (Hahn et al., 2014). However, it is crucial to ensure that these economic benefits do not compromise social equity or environmental integrity.

The social dimension addresses the impact of business practices on stakeholders, including employees, customers, and communities. Retailers are under pressure to adopt socially responsible practices that promote fair labor conditions, community development, and customer well-being (Marzouk, 2023; Munjal & Sharma, 2021). Consumers increasingly support businesses that demonstrate a commitment to social responsibility (Arviana & Wibisono, 2023). For example, initiatives such as local sourcing and fair trade contribute to social performance and local economies.

The environmental dimension evaluates the ecological impact of business operations. Retailers are adopting green initiatives such as sustainable sourcing, waste reduction, and energy-efficient practices to minimize their environmental footprint (Sari, 2023; Wahid, 2012). Growing consumer awareness and regulatory pressures drive businesses to integrate environmental considerations into their strategies (Alhaddi, 2015).

The interplay among these dimensions underscores the need for a holistic sustainability approach. Munjal and Sharma (2021) note that the TBL framework enables a comprehensive assessment and reporting of sustainability practices, facilitating an integrated understanding of performance across economic, social, and environmental aspects. Addressing potential tensions among these dimensions is essential (Hahn et al., 2014).

The practical application of TBL is evident in industries such as hospitality and manufacturing. Boley and Uysal (2013) illustrate how hotels achieve competitive synergy by integrating sustainability practices, enhancing brand reputation, and reducing operational costs. In manufacturing, Malik et al. (2015) demonstrate how the lignocellulosic biofuel industry utilizes TBL principles to assess economic viability, social impacts, and environmental benefits, aligning with sustainable development goals.

12.2.2 The Circular Economy Model

The circular economy model is a transformative approach for analyzing green initiatives, particularly in retail and production. This model emphasizes recycling, reusing, and sustainable production to minimize waste and maximize resource efficiency by promoting a closed-loop system that continuously cycles resources back into production.

A core tenet of the circular economy is designing products for easy disassembly and repurposing at the end of their life cycle (Shen et al., 2019). Retailers can significantly reduce food waste through innovative practices such as local e-commerce platforms and agile logistics (Shen et al., 2019). Such strategies align with consumer preferences for sustainable products.

The circular economy also prioritizes sustainable production by using renewable resources and minimizing environmental degradation. Upadhyay and Akter (2021) emphasize that adopting circular business models enhances customer satisfaction and fosters brand loyalty. The transition to circularity requires systemic shifts, with Urbinati et al. (2017) proposing a taxonomy of business models that establish closed production systems.

Collaboration among stakeholders is crucial to the circular economy. Heyes et al. (2018) highlight the role of collaborative business models in resource sharing and waste reduction. Ghisellini et al. (2016) argue that the circular economy offers a balanced interplay between environmental and economic systems, fostering sustainable development.

The fashion and electronics industries provide notable examples of circular economy adoption. Patagonia encourages customers to repair and recycle their products, fostering a culture of sustainability. Apple's take-back programs recover valuable materials and minimize electronic waste, exemplifying the closed-loop system's effectiveness.

12.2.3 Stakeholder Theory

The stakeholder theory provides a valuable framework for analyzing green initiatives by emphasizing collaboration among businesses, consumers, NGOs, and other stakeholders. It underscores the importance of addressing diverse stakeholder interests to achieve sustainability goals.

Stakeholder engagement enhances decision-making and corporate social responsibility (Manetti, 2011). Organizations that actively involve stakeholders can better understand their social and environmental impacts and align strategies with stakeholder expectations. Wang et al. (2014) argue that collaboration helps address sustainability concerns and create funding opportunities for sustainable projects.

Transparency and accountability are essential elements of stakeholder engagement. Kaur and Lodhia (2014) highlight that incorporating stakeholder feedback into sustainability reporting enhances credibility and trust. However, navigating complex relationships and power dynamics requires an inclusive decision-making approach (Barrett et al., 2021). Conflicts may arise due to competing interests. Tuffour (2023) notes that inadequate regulatory frameworks can hinder collaboration, limiting the effectiveness of green initiatives. Addressing such challenges through open dialogue fosters alignment toward sustainability goals.

Examples of stakeholder collaboration include the shipping industry, where multi-stakeholder initiatives promote environmental and social responsibility (Parviainen et al., 2017). In SMEs, stakeholder engagement improves the TBL performance and strengthens community relationships (Kurnia, 2023).

12.2.4 Conclusion

The TBL, circular economy model, and stakeholder theory offer complementary frameworks for implementing green initiatives across industries. Integrating these approaches enables businesses to achieve sustainable practices that balance economic viability, social equity, and environmental stewardship. Practical applications in sectors such as hospitality, manufacturing, fashion, electronics, and shipping demonstrate how these frameworks drive sustainability and foster stakeholder collaboration.

12.3 The Case: Top JSE Listed Retailers in South Africa (Green Initiatives)

12.3.1 Methodology

To examine the green initiatives of the selected retailers, a mixed-method approach was employed, incorporating qualitative and quantitative techniques. A qualitative approach was used to explore the sustainability strategies and policies adopted by the top five retailers in South Africa. This consisted of a document analysis to review integrated/financial reports, sustainability report, CSR disclosures, and corporate policies from the specific retailers. A quantitative approach was used to assess the performance of retailers based on sustainability indicators. This included an analysis of recycling rates, circular economy initiatives, and waste diversion efforts.

Data evaluation involved triangulation, ensuring consistency and reliability by cross-verifying information from multiple sources. Qualitative data was analyzed using thematic analysis to identify patterns and trends, while quantitative data was subjected to statistical analysis to assess progress in sustainability performance.

Additional factors influencing the selection of the top five retailers, namely Shoprite, Woolworths, Pick n Pay, Spar, and MassMart, were based on their market size, share price, and revenue. This methodological approach provides a robust framework for understanding the green initiatives undertaken by the leading retailers in South African and their impact on the broader industry.

12.3.2 Overview of Top Five South African Retailers

The top five biggest retailers listed on the Johannesburg Stock Exchange (JSE) in South Africa based on financial annual reports (from 2021/22) are Shoprite, Woolworths, Pick n Pay, Spar, and MassMart (Business Tech, 2022, 2024). Moreover, the chapter delves into the role of the leading retailer's green initiatives supporting the United Nations' 17 Sustainable Development Goals (SDGs), with a particular focus on SDG 12 (responsible consumption and production).

Figure 12.1 displays the top five leading retailers annual financial statement information.

From Figure 12.1, it is evident that Shoprite is the largest retailer in South African, focusing on grocery and household goods. Woolworths is ranked as second, Pick n Pay as third, Spar in fourth place, and MassMart as number five based on investor indicators and group finances. Various key brands operate under the top five leading retailers, which provide a diverse range of offerings to customers. For example, Woolworths offers high-end retail offering premium food, clothing, and homeware with key brands such as Country Road and Trenery. In contrast, MassMart presents a large format warehouse store that offers bulk groceries, furniture, and more. These top five leading retailers dominate the South African market across various segments and play a crucial role in South Africa's economy, providing employment and meeting the needs of a diverse customer base.

12.3.3 Green Initiatives

An extensive literature review was conducted to investigate the green initiatives of the leading retailers in South Africa. Table 12.1 displays the key themes and description of green initiatives for the top five retailers in South Africa. Massmart

Figure 12.1 Leading retailers group finances. *Source:* Adapted from Business Tech (2022).

is also identified as the leading producer of renewable energy in South Africa's retail industry (Massmart, 2024b).

According to literature, the key themes of green initiatives identified are based on the SDG 12, which focuses on ensuring sustainable consumption and production patterns. The top five leading retailers in South Africa adopt sustainable practices that include promoting responsible consumption, reducing waste, encouraging sustainable production practices, consumer education and awareness, supporting circular economy initiatives, collaborating with stakeholders and measuring impact on sustainability. The green initiatives consist of sustainability reports and environmental and social reports from the top five leading retailers in South Africa ranging between 2022 and 2024.

12.3.4 Comparative Analysis of Green Initiatives

A desktop study of the top five leading South African retailers provides a comparative analysis of green initiatives, outlining the specific sustainability features embedded in their business practices. This evaluation offers insights into the measurable impact of green initiatives on retailers' success along with their focus on and contribution to SDG 12. Word Cloud was used to extract key words identified from Table 12.1.

Table 12.1 Green initiatives

Key themes	Description	Green initiatives	Retailer				
			Shoprite	Pick n Pay	Spar	Woolworths	Mass Mart
Promoting responsible consumption	Encourages sustainable purchasing by offering rewards for ecofriendly products and practices.	R1.00 back when you reuse planet bag (reusable and recyclable) and swipe your Xtra savings rewards card.	✓				
Reducing waste	Incentivizes recycling and the use of sustainable packaging to reduce waste.	Recyclable coffee bags				✓	
		Recyclable sauce pouches				✓	
		Customers hand in their old reusable shopping bags at stores.	✓	✓	✓	✓	✓
		Hanger recycling—hangers rescued and recycled/paper and cardboard used for footwear hangers/plastic hangers from 100% recycled materials/plastic bags from 100% recycled material.	✓	✓	✓	✓	✓
		Paper and paperboard used in delis, bakeries, and fresh fish departments are certified by the Forest Stewardship Council or Sustainable Forestry Initiative/FSC certified.	✓	✓	✓	✓	✓

		Recycling labels on private label packaging.	✓	✓	✓	✓	✓
		Closed loop system on packaging that is reusable, recyclable, or compostable—sustainable packaging such as carrier bags and fresh food packaging.	✓	✓	✓	✓	✓
		Bio-digester bins to process wet and organic waste efficiently at distribution centers/e-waste recycling/recycling bins in stores.	✓	✓		✓	
Encouraging sustainable production practices	Rewards consumers for buying from companies with sustainable supply chains, promoting industrial practices.	Support local economies by creating diversified sustainable and resilient/inclusive supply chain partnerships—develop local and small suppliers.	✓	✓	✓	✓	✓
Consumer education and awareness	Educates consumers on environmental impact and sustainable consumption, fostering informed choices.	Discovery Vitality HealthyFood program incentivizes and rewards customers for making healthy food-purchasing decisions—customers link their loyalty card to Vitality to receive cashbacks on healthy food items.	✓			✓	
Supporting circular economy initiatives	Promotes circular economy by rewarding recycling or purchasing refurbished products.	Reverse logistics—taking products from their end-of-life point back into the supply chain to get more value or to dispose of them responsibly.	✓	✓	✓	✓	✓
		Coffee grounds collected and sent to organic waste recyclers (e.g., mushroom farming)				✓	

(Continued)

Table 12.1 (Continued)

			Retailer				
Key themes	**Description**	**Green initiatives**	**Shoprite**	**Pick n Pay**	**Spar**	**Woolworths**	**Mass Mart**
		Coffee grounds are available upon request for customers to use in their home composting.			✓	✓	
		Local recyclers recycle old reusable bags returned by customers.				✓	
		Returning disposable coffee cups to distribution centers.				✓	
		Collected by a third-party recycler and processed.		✓	✓		
		Selling reverse osmosis treated municipal water—customers can refill their bottles at stores.		✓			
		Recycled plastic crates, trolley baskets, and plastic shelving/hopping trollies.	✓				
		Reverse vending recycle machine in stores, e.g., deposit return system of redeemable coupons, airtime and data etc.	✓		✓	✓	
		Condensate from refrigeration units is captured and used for non-potable water needs.					✓

Collaboration with stakeholders	Collaborates with NGOs and the government to create loyalty programs supporting community sustainability.	Partnership with NGOs—MySchool MyVillage MyPlanet loyalty programme.				✓	
		The Packa-Ching Polyco recycling program—rewards individuals who gather recyclables with immediate cashless payments for plastics, glass, cans, and paper.	✓				
		Surplus food donated to community beneficiaries to reduce the environmental impact of food waste.	✓	✓	✓	✓	✓
		Community-based gardens ensure food security as well as enhance resilience to climate change by saving water and planting trees.	✓	✓		✓	
		Collaboration with the South African Sustainable Seafood Initiative (SASSI) to ensure products are sustainably sourced and on SASSI's green list.	✓	✓	✓	✓	✓
		Designs of reusable shopping bags to create awareness and funding for local NGOs as proceeds of bags sold are donated to charities.	✓	✓		✓	
		SPAR Rural Hub programme develops and supports farmers through grant funding, loans, and training in sustainable farming,			✓		

(Continued)

Table 12.1 (Continued)

Key themes	Description	Green initiatives	Retailer				
			Shoprite	Pick n Pay	Spar	Woolworths	Mass Mart
		SPAR Zwartkops river restoration project—removing plastics and other rubbish from the river through awareness campaigns and donations.			✓		
		Massmart partnership with Retail Credit Solutions (RCS) introduced a solar energy funding solution for customers.					✓
Measuring impact on sustainability	Develops metrics to assess waste reduction and sustainable purchases, ensuring accountability.	Renewable energy tariff program with Eskom to source wind energy/renewable electricity procurement	✓	✓	✓	✓	✓
		Electrical delivery vehicles for online deliveries/heavy-duty electric truck for local store deliveries/electric forklifts.	✓		✓	✓	✓
		Closed-door refrigeration technology in-store to reduce energy usage and reduction in food waste/improvements in refrigeration systems/ climate-friendly natural refrigeration.	✓	✓		✓	✓
		Renewable energy consumption—rooftop solar systems to reduce energy usage.	✓	✓	✓	✓	✓

	Energy-efficient LED lighting to reduce carbon emission	✓	✓		✓	✓
	Transportation—enhanced fleet energy efficiency with solar photovoltaic (PV) panels on trailers/electric vans.	✓			✓	
	Solar electric drive refrigeration for trucks/battery-electric truck refrigeration systems to improve supply chain logistics.	✓	✓	✓	✓	
	Sourcing sustainable cotton/equitable amount of clothing sourced locally.		✓		✓	

Source: Filho et al. (2019), Stauropoulou et al. (2022), Seridaran and Noor (2021), Lorgnier (2023), Rubio et al. (2022), Perepelkin and Wilson (2018), Massmart (2024a), Pick n Pay (2024), Shoprite holdings ltd (2024), Shoprite holdings ltd (2023), The Spar Group Ltd (2023), and Woolworths holdings limited (2022).

The Word Cloud was considered an effective and efficient tool for identifying key issues based on the description of each green initiative. Figure 12.2 presents the generated Word Cloud, highlighting the key issues that emerged such as recycling, packaging, recycling, energy, solar, reusable, refrigeration, bags, and sustainable.

Based on the key aspects identified, retailers are placing greater emphasis on sustainable initiatives, focusing on environmentally responsible practices to drive long-term impact. Moreover, retailers are actively implementing sustainable initiatives and appear to favor partnerships and collaborations to achieve their environmental goals.

In Table 12.2 a summary is provided regarding the similarities, differences, strengths, and weaknesses of green initiatives identified in Table 12.1.

From Table 12.2, it can be seen that retailers are demonstrating an industry-wide commitment to green initiatives. The three critical dimensions (economic, social, and environmental) of the TBL framework are also supported by retailers such as sustainable operations lowering long-term expenses, community programs and waste reduction contributing to environmental preservation. At the same time, retailers focus on recycling and sustainable packaging, which demonstrate alignment with the circular economy models, improving resource efficiency.

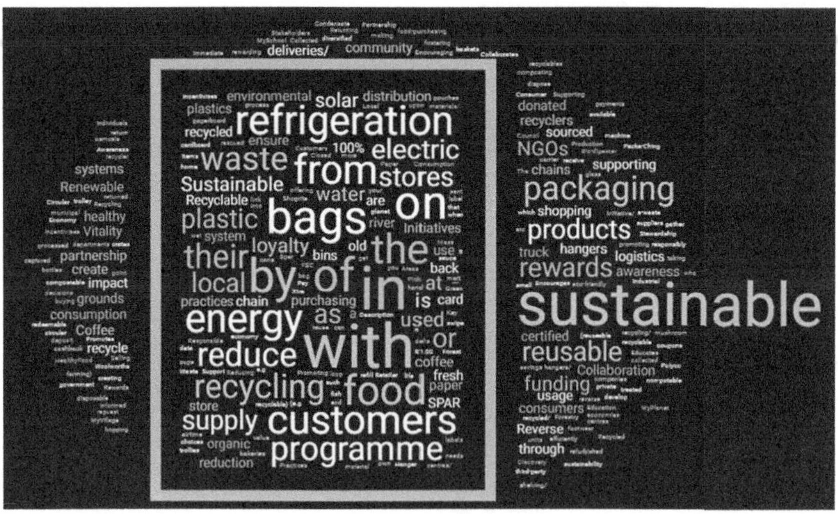

Figure 12.2 Word Cloud. *Source:* Authors' own compilation.

Table 12.2 Green initiative analysis

Green initiatives			
Similarities	**Differences**	**Strengths**	**Weaknesses**
• *Plastic reduction*—implementation of reusable bags	• Not all retailers offer premium sustainability efforts	• *Alignment with circular economy practices*—reduction in landfill waste	• *Scalability challenges*—some sustainability initiatives are not universally implemented as all stores
• *Renewable energy investments*—installation of solar panels	• A lack of sustainable purchasing rewards for eco-friendly products	• *Long-term cost savings*—investments in solar energy to reduce expenses	• Cost barriers—the high cost of sustainable packaging might limit expansion
• *Sustainable Sourcing*—committed to sourcing sustainable products from suppliers	• More investments are required in large-scale renewable energy projects	• *Consumer awareness*—campaigns promoting responsible consumption	
• *Waste Management and Recycling*—introducing and implementing several recycling initiatives		• *Water conservation measures*—implementation of water-efficient fixtures	

12.4 Impact of Innovation

When evaluating the key findings of effective green initiatives, it is important to measure effectiveness consistently. In this case, the effectiveness of green initiatives can be evaluated based on the extent to which SDG 12 is addressed. Here are some ways in which SDG12 applies to the discussion:

a) The TBL framework expands traditional business success metrics beyond financial profit to include environmental and social dimensions, making it highly relevant to SDG 12. Retailers implementing take-back programs for plastic packaging align with TBL by reducing waste (environmental), creating jobs in recycling (social), and cutting costs on raw materials (economic).

b) The Circular Economy model shifts away from the traditional linear "take-make-dispose" economy to a regenerative system where resources are reused, recycled, and repurposed. South African retailers adopting zero-waste packaging and returnable containers help close the production loop, directly contributing to SDG 12 targets on reducing per capita waste.
c) The stakeholder theory argues that businesses must consider the interests of all stakeholders—beyond shareholders—when making decisions. Massmart and Woolworths partnership with suppliers to ensure ethical and sustainable sourcing demonstrates stakeholder engagement in promoting responsible consumption.

12.5 Learning Outcomes

South African retailers are increasingly focusing on green initiatives to enhance sustainability and reduce environmental impact. The primary objective of this chapter is to analyze and compare the sustainability strategies adopted by the top five retailers in South Africa. Through the examination of green initiatives in South Africa, readers can take away several important insights namely (i) operational efficiency, which focuses on the implementation of green technologies like solar power and efficient logistics that can reduce emissions. (ii) adopting circular economy practices can lead to innovative waste management solutions. (iii) retailers can benefit from collaborating on sustainability initiatives and transparent reporting. (iv) ensuring suppliers adhere to environmental standards is crucial for sustainable green initiatives. These initiatives demonstrate how South African retailers are actively contributing to environmental sustainability and global trends.

12.6 Plans to Develop Further Initiatives

As retailers increasingly acknowledge the rising significance of sustainability, they should continue to expand and enhance their green initiatives. This section outlines strategic plans for advancing these initiatives, integrating innovative technologies, and strengthening stakeholder engagement. Retailers should enhance circular economy practices by incorporating refill stations and returnable packaging, which will significantly contribute to reduce waste.

The aim of retailers should be to expand partnerships with local and ethical suppliers, ensuring that raw materials are sustainably sourced. Certifications such as Fair Trade and Rainforest Alliance will play a greater role in procurement decisions to enhance responsible consumption. Enhancing consumer awareness through educational campaigns will encourage sustainable shopping habits among consumers. Initiatives such as eco-labeling, digital carbon footprint trackers, and green loyalty programs will empower customers to make environmentally responsible choices. The integration of AI, big data, and the internet of things (IoT) will enhance resource management, optimize logistics, and reduce waste. Smart energy management systems and AI-driven supply chain analytics will play a key role in improving sustainability performance in the retail industry.

Strategic recommendations for retailers to develop sustainability initiatives involve enhanced sustainability practices and improved corporate governance. Retailers should focus on clear and honest communication regarding their sustainability efforts, implementing robust reporting mechanisms that provide stakeholders with accurate information about environmental practices. Retailers should also plan to align more closely with national and international sustainability regulations, advocating for stronger policies that support green business practices. Increased transparency through sustainability reporting and ESG disclosures will further enhance accountability.

12.7 Conclusion

This chapter highlighted the critical role that green initiatives play in promoting sustainability and corporate responsibility within the South African context. Retailers are continuously evolving their sustainability strategies by integrating renewable energy, circular economy practices and ethical sourcing. By expanding their green initiatives, retailers aim to foster long-term sustainability while enhancing business resilience and consumer trust.

The triple bottom line, circular economy, and stakeholder theory collectively provide a structured framework for understanding and implementing green initiatives aligned with SDG 12. Moving forward, retailers must prioritize authenticity and stakeholder collaboration to ensure sustainable green initiatives to create long-term value for the environment and society.

References

Abe, I., & Mugobo, V. (2021). Post-covid-19 retail landscapes. *Harvard Deusto Business Research*, *10*(2), 337–345. https://doi.org/10.48132/hdbr.364

Ahamat, A. (2017). Is Islamic banking and finance doing enough? Shaping the sustainable and socially responsible investment community. *Asian Social Science*, *13*(3), 170. https://doi.org/10.5539/ass.v13n3p170

Alexander, B., & Mason, R. (2017). The South African e-retail change agenda: A curriculum development perspective. *The Electronic Journal of Information Systems in Developing Countries*, *82*(1), 1–21. https://doi.org/10.1002/j.1681-4835.2017.tb00604.x

Alhaddi, H. (2015). Triple bottom line and sustainability: A literature review. *Business and Management Studies*, *1*(2), 6. 10.11114/bms.v1i2.752

Arviana, N., & Wibisono, M. (2023). The effect of fundamental factors on triple bottom line and firm value. *Jurnal Riset Akuntansi Kontemporer*, *15*(1), 12–18. https://doi.org/10.23969/jrak.v15i1.5474

Ateş, S. (2022). The credibility of corporate social responsibility reports: Evidence from the energy sector in emerging markets. *Social Responsibility Journal*, *19*(4), 756–773. https://doi.org/10.1108/srj-04-2021-0149

Barrett, M., Jones, G., Bunds, K., Casper, J., & Edwards, M. (2021). Teamwork makes the net-work: Participant-governed networks and athletics sustainability collaboration. *International Journal of Sustainability in Higher Education*, *23*(5), 1090–1106. https://doi.org/10.1108/ijshe-05-2021-0188

Berning, A., & Venter, C. (2015). Sustainable supply chain engagement in a retail environment. *Sustainability*, *7*(5), 6246–6263. https://doi.org/10.3390/su7056246

Blaha, S., Lambrechts, W., & Mampaey, J. (2021). Walk your talk: On the (non)-formative influence of corporate social responsibility communication on supply chain sustainability measures. *Sustainability*, *13*(17), 9754. https://doi.org/10.3390/su13179754

Boley, B., & Uysal, M. (2013). Competitive synergy through practicing triple bottom line sustainability: Evidence from three hospitality case studies. *Tourism and Hospitality Research*, *13*(4), 226–238. https://doi.org/10.1177/1467358414528528

Business Tech. (2022). Battle of the retail giants: Woolworths vs Shoprite vs Pick n Pay vs Spar vs Massmart. *BusinessTech*.

Business Tech. (2024). The biggest retailers in South Africa: Shoprite vs Woolworths vs Spar vs Pick n Pay. *BusinessTech*.

Choi, M., & Hong, S. (2022). Another form of greenwashing: The effects of chaebol firms' corporate governance performance on the donations. *Sustainability*, *14*(6), 3373. https://doi.org/10.3390/su14063373

Dakora, E., & Slabbert, A. (2014). Strategic options for the expansion of South African retail businesses within Africa: Consequences on operational management. *Journal of Economics and Behavioral Studies*, *6*(3), 218–231. https://doi.org/10.22610/jebs.v6i3.485

Del Río, C., González-Álvarez, K., & López-Arceiz, F.J. (2023). Examining greenwashing and SDG-washing: An analysis of corporate engagement with the SDGs. *Sustainability Accounting, Management and Policy Journal*, *15*(2), 412–456. https://doi.org/10.1108/sampj-02-2023-0080

Engelbrecht, L., & Rensburg, A. (2018). A strategic model of macro-economic and supply chain factors that influence the decision-making process of SA retailers when expanding their footprint in African countries. *The South African Journal of Industrial Engineering, 29*(4). https://doi.org/10.7166/29-4-1956

Eyk, M., Amoah, F., & Yase, T. (2022). Sustaining the township economy: An investigation into the factors influencing the shopping experience of spaza shop customers in South Africa. *Journal of Economics and Behavioral Studies, 14*(3(J)), 20–32. https://doi.org/10.22610/jebs.v14i3(j).3304

Filho, W., Shiel, C., Paço, A., Mifsud, M., Ávila, L.V., Brandli, L.L., Molthan-Hill, P., Pace, P., Azeiteiro, U.M., Vargas, V.R., & Caeiro, S. (2019). Sustainable development goals and sustainability teaching at universities: Falling behind or getting ahead of the pack? *Journal of Cleaner Production, 232*, 285–294. https://doi.org/10.1016/j.jclepro.2019.05.309

Ghisellini, P., Cialani, C., & Ulgiati, S. (2016). A review on circular economy: The expected transition to a balanced interplay of environmental and economic systems. *Journal of Cleaner Production, 114*, 11–32. https://doi.org/10.1016/j.jclepro.2015.09.007

Guerreiro, J., & Pacheco, M. (2021). How green trust, consumer brand engagement and green word-of-mouth mediate purchasing intentions. *Sustainability (Switzerland), 13*(14), 7877. https://doi.org/10.3390/su13147877

Hahn, T., Pinkse, J., Preuss, L., & Figge, F. (2014). Tensions in corporate sustainability: Towards an integrative framework. *Journal of Business Ethics, 127*(2), 297–316. https://doi.org/10.1007/s10551-014-2047-5

Hare, C., & Walwyn, D. (2019). A qualitative study of the attitudes of South African spaza shop owners to coopetitive relationships. *South African Journal of Business Management, 50*(1). https://doi.org/10.4102/sajbm.v50i1.1295

Haywood, L., Lange, W., Musvoto, C., & Wet, B. (2021). What do South African supermarket retailers disclose about the sustainability of suppliers? *Journal of Contemporary Management, 18*(1), 173–193. https://doi.org/10.35683/jcm20087.101

Heidig, W., Dobbelstein, T., Mason, R., & Jooste, W. (2017). First choice or fallback option? The attractiveness of South Africa's retail industry for gen y members. *Problems and Perspectives in Management, 15*(2), 110–123. https://doi.org/10.21511/ppm.15(2).2017.11

Heyes, G., Sharmina, M., Mendoza, J., Gallego-Schmid, A., & Azapagic, A. (2018). Developing and implementing circular economy business models in service-oriented technology companies. *Journal of Cleaner Production, 177*, 621–632. https://doi.org/10.1016/j.jclepro.2017.12.168

Idowu, A., & Ndidiamaka, A. (2018). The influence of supervisor/subordinate social support on employee performance in the South African retail sector. *Journal of Business & Retail Management Research, 12*(3). 10.24052/jbrmr/v12is03/art-01

Ismail, R., Jeffery, R., & Belle, J. (2011). Using ICT as a value adding tool in South African SMEs. *Journal of African Research in Business & Technology, 2011*, 1–12. https://doi.org/10.5171/2011.470652

Jibril, A., Kwarteng, M., Pilík, M., Botha, E., & Osakwe, C. (2020). Towards understanding the initial adoption of online retail stores in a low internet penetration context: An exploratory work in Ghana. *Sustainability*, *12*(3), 854. https://doi.org/10.3390/su12030854

Kaur, A., & Lodhia, S. (2014). The state of disclosures on stakeholder engagement in sustainability reporting in Australian local councils. *Pacific Accounting Review*, *26*(1–2), 54–74. https://doi.org/10.1108/par-07-2013-0064

Kenny, B. (2014). Walmart in South Africa: Precarious labor and retail expansion. *International Labor and Working-Class History*, *86*, 173–177. https://doi.org/10.1017/s0147547914000167

Kurnia, I. (2023). Influence of the triple bottom line concept on sustainability and success of tofu SMEs in Magelang. *Applied Accounting and Management Review (AAMAR)*, *2*(2), 70. https://doi.org/10.32497/aamar.v2i2.5095

Lorgnier, N. (2023). Consumers' perceptions of partnering brands: The roles of perceived sustainable development goals and partnership authenticity in a cross-national context. *Sustainable Development*, *32*(4), 3518–3540. https://doi.org/10.1002/sd.2858

Malgas, M., & Zondi, W. (2019). Differences between migrant owned and locally owned small business retail stores: The case of Delft and Eindhoven. *Journal of Business & Retail Management Research*, *14*(1). 10.24052/jbrmr/v14is01/art-02

Malgas, M., & Zondi, W. (2020). Challenges facing small business retailers in selected South African townships. *The Southern African Journal of Entrepreneurship and Small Business Management*, *12*(1). https://doi.org/10.4102/sajesbm.v12i1.202

Malik, A., Lenzen, M., & Geschke, A. (2015). Triple bottom line study of a lignocellulosic biofuel industry. *GCB Bioenergy*, *8*(1), 96–110. https://doi.org/10.1111/gcbb.12240

Manetti, G. (2011). The quality of stakeholder engagement in sustainability reporting: Empirical evidence and critical points. *Corporate Social Responsibility and Environmental Management*, *18*(2), 110–122. https://doi.org/10.1002/csr.255

Marzouk, J. (2023). The interplay among green absorptive capacity, green entrepreneurial, and learning orientations and their effect on triple bottom line performance. *Business Strategy and the Environment*, *33*(3), 1962–1976. https://doi.org/10.1002/bse.3588

Massmart. (2024a). *Responsible business: Environmental sustainability*. https://www.massmart.co.za/responsible-business/environmental-responsibility

Massmart. (2024b). *Company news and information: Massmart launches sixth solar plant becoming the biggest producer of renewable energy in the South African retail sector*. https://www.massmart.co.za/news/2021/05/massmart-launches-sixth-solar-plant-becoming-the-biggest-produce.

Morseletto, P. (2022). Environmental principles for modern sustainable economic frameworks including the circular economy. *Sustainability Science*, *17*(5), 2165–2171. https://doi.org/10.1007/s11625-022-01208-w

Munjal, P., & Sharma, D. (2021). Determining the managerial perception on triple bottom line performance. *Journal of Financial Reporting and Accounting*, *21*(2), 322–343. https://doi.org/10.1108/jfra-03-2021-0084

Nandkeolyar, O., & Chen, F. (2022). Credibility, transparency, and sustainability in fashion: A game-theoretic perspective. *Agricultural and Resource Economics Review*, *52*(1), 43–70. https://doi.org/10.1017/age.2022.24

Niehaus, G., Feiboth, H., & Goedhals-Gerber, L. (2018). Investigating supply chain sustainability in South African organisations. *Journal of Transport and Supply Chain Management*, *12*. https://doi.org/10.4102/jtscm.v12i0.335

Parviainen, T., Lehikoinen, A., Kuikka, S., & Haapasaari, P. (2017). How can stakeholders promote environmental and social responsibility in the shipping industry? *WMU Journal of Maritime Affairs*, *17*(1), 49–70. https://doi.org/10.1007/s13437-017-0134-z

Perepelkin, J., & Wilson, G. (2018). Incentivizing choice of community pharmacy. *Journal of Pharmaceutical Health Services Research*, *9*(4), 347–359. https://doi.org/10.1111/jphs.12242

Pick n Pay. (2024). *Pick n Pay sustainability report 2024*. pnp-2024-sustainability-report-and-databook-single-pages.pdf.

Pinheiro, A.B., Oliveira, M.C., Freitas, G.A., & d., & Lozano, M. B. (2023). Board attributes and environmental disclosure: What is the nexus in liberal economies? *Revista De Administração De Empresas*, *63*(4). https://doi.org/10.1590/s0034-759020230402

Ridge, M., Johnston, K., & O'Donovan, B. (2015). The use of big data analytics in the retail industries in South Africa. *African Journal of Business Management*, *9*(19), 688–703. https://doi.org/10.5897/ajbm2015.7827

Rubio, L., Valderrama-Hernández, R., Solís-Espallargas, C., & Ruíz-Morales, J. (2022). The implementation of the SDGs in universities: A systematic review. *Environmental Education Research*, *28*(11), 1585–1615. https://doi.org/10.1080/13504622.2022.2063798

Ruiz–Real, J., Uribe-Toril, J., Gázquez–Abad, J., & Valenciano, J. (2018). Sustainability and retail: Analysis of global research. *Sustainability*, *11*(1), 14. https://doi.org/10.3390/su11010014

Sari, N. (2023). The triple bottom line accounting from Scientia sacra perspective. *Riset Akuntansi Dan Keuangan Indonesia*, *8*(1), 23–31. https://doi.org/10.23917/reaksi.v8i1.20893

Seridaran, S., & Noor, M. (2021). The development of a conceptual framework for customer loyalty among users of mobile loyalty programmes. *International Journal of Academic Research in Business and Social Sciences*, *11*(7). https://doi.org/10.6007/ijarbss/v11-i7/10544

Sharma, R., & Gautam, A. (2017). Redefining retail store formats – A South African study. *Applied Finance and Accounting*, *3*(2), 61. https://doi.org/10.11114/afa.v3i2.2518

Shen, M., Chen, H., Lin, C., Hsu, P., & Shen, T. (2019). Sustainable business model to reduce food waste of agricultural products in the retail chain. *International Journal of e-Education e-Business e-Management and e-Learning*, *9*(4), 373–380. https://doi.org/10.17706/ijeeee.2019.9.4.373-380

Shoprite holdings Ltd. (2023). *Shoprite holdings sustainability report 2024 for the year ended 2 July 2023*. Checker sustainability report-2023.pdf.

Shoprite holdings Ltd. (2024). *Shoprite holdings sustainability report 2024 for the year ended 30 June 2024*. Shoprite Holdings – Sustainability Report 2024.pdf.

Stauropoulou, A., Sardianou, E., Malindretos, G., Evangelinos, K., & Nikolaou, I. (2022). The role of customers' awareness towards the sustainable development goals (SDGs) of banks on their behavior. *Environmental Science and Pollution Research*, *30*(5), 13495–13507. https://doi.org/10.1007/s11356-022-23111-8

Strydom, J. (2013). Retail patronage of Sowetan consumers after 1994. *African Journal of Business Management*, *7*(29), 2863–2871. https://doi.org/10.5897/ajbm12.1133

The Spar Group Ltd. 2023. *The Spar Group Ltd environmental, socio-economic and governance report 2023*. https://thespargroup.com/pdf/Environmental_and_Social_Report_2023.pdf.

Tuffour, P. (2023). Toward sustainable development: Developing a decision-making framework for cross-sectoral engagement in green procurement. *Sustainable Development*, *32*(3), 2233–2252. https://doi.org/10.1002/sd.2772

Upadhyay, A., & Akter, S. (2021). An analysis of UK retailers' initiatives towards circular economy transition and policy-driven directions. *Clean Technologies and Environmental Policy*, *24*(4), 1209–1217. https://doi.org/10.1007/s10098-020-02004-9

Urbinati, A., Chiaroni, D., & Chiesa, V. (2017). Towards a new taxonomy of circular economy business models. *Journal of Cleaner Production*, *168*, 487–498. https://doi.org/10.1016/j.jclepro.2017.09.047

Wahid, F. (2012, July 7–8). The green e-procurement infrastructure in the Indonesian public sector: Its antecedents and impacts on the triple bottom line. *2012 International Conference on Green and Ubiquitous Technology*, Bandung, Indonesia. https://doi.org/10.1109/gut.2012.6344190

Wang, C., Li, X., & Mao, Z. (2013). Green retailing practices: An exploratory comparison between Chinese and British retailers. *Journal of General Management*, *39*(2), 35–60. https://doi.org/10.1177/030630701303900203

Wang, X., Hawkins, C., & Berman, E. (2014). Financing sustainability and stakeholder engagement. *Urban Affairs Review*, *50*(6), 806–834. https://doi.org/10.1177/1078087414522388

Wegerif, M. (2020). "Informal" food traders and food security: Experiences from the covid-19 response in South Africa. *Food Security*, *12*(4), 797–800. https://doi.org/10.1007/s12571-020-01078-z

Woolworths holdings limited. (2022). Woolworths holding limited 2022 annual sustainability report: Goodness Business Journey Report. *Woolworths*. 2022-Good-Business-Journey-Report.pdf.

Xu, W., Li, M., & Xu, S. (2023). Unveiling the "veil" of information disclosure: Sustainability reporting "greenwashing" and "shared value". *PLoS One*, *18*(1), e0279904. https://doi.org/10.1371/journal.pone.0279904

Yeow, K.E., & Ng, S. (2021). The impact of green bonds on corporate environmental and financial performance. *Managerial Finance*, *47*(10), 1486–1510. https://doi.org/10.1108/mf-09-2020-0481

Zhang, G., Wang, L., & Meng, H. (2023). Environmental knowledge level and consumer behavior regarding green fashion: A moderated mediation model. *Asia Pacific Journal of Marketing and Logistics*, *36*(4), 977–992. https://doi.org/10.1108/apjml-07-2023-0637

13

Exploring Blockchain Technology Diffusion in Supply Chain Management to Enhance Sustainability Performance: Insights from Egypt an Emerging Economy

Esraa Osama Zayed[1,2] *and Ehab A. Yaseen*[3]

[1] *Faculty of Economics and Business Administration, German International University, Cairo, Egypt*
[2] *College of Business Administration, American University in the Emirates, Dubai, UAE*
[3] *Faculty of Management Technology, German University in Cairo, Cairo, Egypt*

13.1 Introduction

Technologies associated with Industry 4.0 are recently gaining more recognition and forcing multiple changes to the current business model and work systems (Queiroz et al., 2020). BCT in specific is becoming increasingly popular, attracting the interest of scholars, managers, and practitioners (Chen et al., 2020; Wamba & Queiroz, 2020). A significant amount of research focuses on the role of BCT in achieving sustainable development (Mulligan et al., 2024; Park & Li, 2021). BCT is a groundbreaking innovative platform transforming and introducing new ways of thinking in the supply chain field (Liu & Li, 2020).

BCT can be defined as "*a fully distributed system for cryptographically capturing and storing a consistent, immutable, linear event log of transactions between networked actors. Such distributed ledger is consensually kept, updated, and validated by the parties involved in all the transactions within a network*" (Queiroz et al., 2020). Sustainable supply chain management (SSCM) is defined as managing materials, information, and flow of capital to improve organizations' economic, environmental, and social performance, taking into consideration all stakeholders' expectations (Kannan et al., 2014).

Green Innovations in Supply Chain Management: Case Studies and Applications,
First Edition. Edited by Marcia Mkansi, Godfrey Mugurusi, Aaron L. Nsakanda,
Temidayo Akenroye, and Frank Tietze.

Published 2026 by John Wiley & Sons, Inc.

BCT can help achieve various SCM goals including those related to sustainability (Gurtu & Johny, 2019; Vaio & Varriale, 2020). For instance, significant challenges for sustainability integration in SCM remain due to the limitations of current technologies in accurately tracking sustainability performance across the SC (Kshteri, 2021). Sometimes, the opacity of modern SCs allows for unethical practices to go undetected or unpunished (LeBaron, 2020). As BCT introduces multiple opportunities for better traceability and control across the SC (Ozturk & Yildizbasi, 2020; Reddy, 2019), it enables an easy tracking and verification of a product's origin and its entire journey along the SC (Azzi et al., 2019; Casino et al., 2019), hence, the opacity of modern SC can be significantly reduced.

Additionally, unequal power distribution among SC partners as well as low degrees of integration and responsiveness, such as using different types of documentation systems to track products, might lead to information silos, which are commonly pronounced in developing countries. Such information silos hinder the effective monitoring of sustainability implementation across the SC (Kshteri, 2021). Such challenges could be dealt with by implementing BCT. This allows partners to collaborate through a peer-to-peer decentralized network, leading to equality as well as trust between partners. Additionally, participants will jointly access all information through the decentralized feature of BCT (Lohmer et al., 2022; Niu & Li, 2018).

Although adopting BCT into SC will improve the entire process and sustainability performance, it was found that the actual implementation of BCT in SC is still in its infancy, which limits the understanding of the actual value of BCT (Lim et al., 2021). Ghode et al. (2021) stated that most of the current literature focuses on the benefits with less attention on the adoption challenges of BCT in SCM practices. Moreover, the successful adoption of BCT in SCM must begin with a detailed analysis and identification of all challenges/enablers to develop a systematic transition process for such adoption (Ghode et al., 2021; Ozturk & Yildizbasi, 2020).

Accordingly, as mentioned in the abstract, this chapter aims to understand the significant TOE factors that influence BCT diffusion (as an innovation supporting sustainability performance) in SCM. This is achieved through integrating the TOE framework and DoI theory using an exploratory research approach focusing on Egypt as an example of an emerging economy in the MENA region. The research question is "What are the factors influencing the diffusion of blockchain technology in supply chain management in Egypt?"

DoI and TOE integration: The integration of the DoI theory alongside the TOE framework is useful in enhancing the understanding of BCT diffusion in SC in terms of demonstrating how various technological, organizational, and environmental factors influence the different stages of diffusion. TOE is one of the predominant theories for IT adoption at the organizations' level (Chen et al., 2020), as it categorizes factors that affect the ability to implement technological innovations (Clohessy & Acton, 2019; Karamchandani et al., 2021). It can be adopted in broad conditions, and the choice of factors varies according to different innovations, cultures, and contexts (Wong et al., 2020). It is also a suitable tool for identifying factors that influence BCT adoption at the organizational level (Orji et al., 2020; Wong et al., 2020). Additionally, multiple scholars used it to identify challenges for BCT adoption within the SSCM context (Kouhizadeh et al., 2021).

Regarding technological challenges, TOE highlights challenges related to data security, complexity of setup and use, cost of technology adoption, and maturity of BCT—availability and performance (Bouthkhoum et al., 2021; Kouhizadeh et al., 2021; Nayak & Dhaigude, 2019). TOE also highlights challenges related to the top management support, awareness about BCT benefits, organization culture, and people readiness (Bag et al., 2021; Farooque et al., 2020; Lanzini et al., 2021). Finally, environmental challenges are further divided into SC-related challenges such as collaboration, coordination, communication between SC partners, cultural differences among SC partners, and customer influence (Kouhizadeh et al., 2021; Lanzini et al., 2021; Ozturk & Yildizbasi, 2020), and other external challenges related to government policies, support, and competition (Nayak & Dhaigude, 2019; Orji et al., 2020).

On the other hand, DoI is considered useful in analyzing the organization's intention to adopt new technology, focusing on both technology and organizational characteristics (Mathivathanan et al., 2021). It is worth noting that the implementation process of BCT in SCM is considered a complicated process that includes a large number of stakeholders—SC members—who play a significant role in the decision-making process (Chen et al., 2020).

Focusing on emerging economies/Egypt: By reviewing previous literature, Kamble et al. (2020) have stated that country contexts might lead to different significant results regarding BCT implementation in SCM. Although some emerging economies like India and Malaysia are currently implementing BCT. It is noticed that these economies still need to exert more effort to understand various BCT adoption stages (Wamba & Querioz, 2020). Thus, the existence of differences across countries, as well as the complexity of BCT diffusion stages, motivated this study to focus on emerging economies, with Egypt as an example of an emerging economy from the MENA region.

Furthermore, according to the International Trade Administration, the information and communication technology (ICT) sector in Egypt is emerging with a growth rate higher than Egypt's overall GDP growth level, where the Egyptian government is making a series of investments and infrastructure updates to support its ICT 2030 strategy. This is especially after the COVID-19 outbreak, which accelerated the digital transformation plan of Egypt (Egypt— Information and Communications Technology and Digital Economy, 2021).

To achieve this chapter's objective, multiple methodological approaches were taken on three sequential stages. The first stage involved systematically reviewing previous literature about factors influencing BCT diffusion in SCM. The second stage was a survey developed to capture SC experts'—working in organizations operating in Egypt—opinions and views about factors that might hinder or facilitate the successful diffusion of BCT into SCM practices. The last stage was to investigate the interrelationships between significant technological, organizational, and environmental factors influencing BCT diffusion in SC using ISM analysis, followed by partial least squares structural equation modeling (PLS-SEM) analysis for the ISM final model.

This chapter is structured as follows: the first section is a state-of-the-art explaining relevant literature about BCT and SSCM as well as relevant theories used to identify factors influencing such diffusion. The second section, the case section, which presents the research approaches, is conducted to understand these factors, mainly through a deductive exploratory research approach focusing on Egypt as an example for an emerging economy from the MENA region.

The third section, the learning outcomes are mentioned, as based on the data collected, a practical road map for BCT (as an innovation supporting sustainability performance) diffusion in SCM in Egypt. Finally, a plan for further research development in this area is suggested.

13.2 State of the Art

13.2.1 Blockchain Technology and Supply Chain Sustainability

Recently, BCT started to extend outside the realm of cryptocurrency and enter various business practices, such as financial services (e.g., Ripple and Libra), food and agricultural distributions (e.g., IBM Food Trust and Bumble Bee Foods), and healthcare and pharmaceutical supply chains (e.g., ProCredEx and MediLedger) (Hooper & Holtbrugge, 2020; Park & Li, 2021). BCT is an emerging technology, which is predicted to have a 5–10 year maturity cycle (Koh et al., 2020). The key attractive features of BCT include its use as a distributed ledger, decentralized control, transparency, permanency of transactions, robust makeup preventing tampering or manipulation, and openness of accessibility (Bager et al., 2022; Batwa & Norrman, 2020; Upadhyay et al., 2021). BCT is thought to be a new business collaboration tool supporting a shared secure network of data between trusted partners sharing a digital ledger history with no need for trusted intermediaries (Kayikci et al., 2022).

It is evident that every component of the supply chain has an impact on the environment. Environmental sustainability practices within the supply chain are related to energy and water consumption, resource management (consumables), recycling, and waste management (Bhinge et al., 2015). Also, social sustainability within the supply chain examines socioeconomic conditions regarding involved employees in the supply chain, such as safety and human rights. Finally, economic sustainability within the supply chain is often referred to as governance in terms of how a company develops a robust management structure that ensures overall transparency, traceability, and accountability, and ultimately can strengthen relations with external stakeholders and attract potential investors (Park & Li, 2021).

BCT can be a problem solver for many challenges currently faced by supply chains in terms of traceability and security of information (Lim et al., 2021). Sustainability is considered a stimulating issue in the application of BCT in SCM (Park & Li, 2021).

Previous literature suggests that the enhanced visibility, traceability, as well as synchronized transactions offered by BCT can promote environmental sustainability (Saberi et al., 2019). BCT is expected to enhance three indicators in environmental sustainability within SC. The first one is environmental emission abatement by tracking locations and amounts of emissions. While the second indicator is resource management by monitoring the origins of raw materials, avoiding excessive extraction and utilization of natural resources. Finally, the third indicator is waste management, as BCT can allow businesses to track their waste, making recycle and reuse efforts more possible (Kouhizadeh & Sarkis, 2018; Saberi et al., 2019). Accordingly, implementing BCT can make it easier for focal firms to collect information from their supply chain partners lowering economic and reputational harm, thus improving SSCM performance (Saho, et al., 2024). This confirms that applying BCT in SCM can protect the environment in addition to cost reduction benefits (Park & Li, 2021).

As for the social sustainability, which is always the rarely studied pillar in SCM (Seuring & Müller, 2008; Venkatesh et al., 2020), BCT can maintain a secure and immutable information protecting all involved partners—individuals, governments, or organization—from corruption (Park & Li, 2021). For instance, since BCT keeps a transparent record of transactions for the whole processes, the unethical sourcing processes that potentially abuse human rights can easily be detected, promoting healthy and safe business environments in the supply chains (Cartier et al., 2018; Di Vaio & Varriale, 2020; Saberi et al., 2019). Furthermore, organizations can deploy a smart contract system powered by BCT to exchange employees' details—compensations, work conditions, and contract periods—with all members of the upstream and downstream SC (Hasan et al., 2019).

Finally, BCT could improve the economic performance of SC in many aspects. First, as BCT provides a powerful solution to the asymmetry information issue

Table 13.1 Potential benefits of BCT on supply chain sustainability performance (Dutta et al., 2020; Park & Li, 2021; Tsolakis et al., 2021)

Environmental sustainability	• Tracking the environmental performances of suppliers. • Environmental emission abatement. • Monitor environmental impacts of transportation. • Enhance transparency of the operations. • Support tracing the location of the materials and authenticating all actors involved in the recycling process. • Resource management.
Social sustainability	• Enhance social sustainability through a transparent information system, which prevents the corruption of all involved supply chain participants. • Protect human rights and promote safe and healthy business environments.
Economic sustainability	• Help in supplier development. • Remove intermediaries. • Improve supply chain governance performance. • Provides a powerful solution to the asymmetric information issue using the smart contract. • Establish trust and collaboration among stakeholders.

using the smart contract, hence, avoiding possible corruption or errors (Venkatesh et al., 2020). Second, BCT can help track the historical sustainability performance of suppliers, hence minimizing the organizations' effort to audit suppliers and making it easier for the organization to evaluate its current suppliers and to plan suppliers' development programs (Dutta et al., 2020; Van Hoek, 2020). Therefore, BCT is expected to promote supply chain environmental, social, and economic sustainability in various ways as summarized in Table 13.1.

13.2.2 Technology–Organization–Environment (TOE) Framework and Diffusion of Innovation (DoI) Theory

Multiple frameworks and theories are used to understand the determinants for BCT implementation in SCM as revealed by systematically reviewing previous articles. Topics were explored, such as the unified theory of acceptance and use of technology (UTAUT), technology acceptance model (TAM), institutional theory (Francisco & Swanson, 2018; Queiroz & Wamba, 2019), or through categorizing the factors into categories according to the TOE framework (Kouhizadeh et al., 2021; Orji et al., 2020; Wong et al., 2020). Yet, Saurabh and Dey (2021) lately criticized TAM for having a narrow focus on adoption constructs, leading to its inability to fully identify potential drivers/barriers for technology adoption. Also, the sole use of UTAUT in technology adoption studies was criticized, as it is more concerned with users' perceptions and would not help much in actual system designs. Additionally, Wang et al. (2019) criticized the sole use of the institutional theory to explain why organizations used BCT, as doing so will not offer insights about the unique technology aspects.

Caldarelli et al. (2021) confirmed the usefulness of the TOE framework to analyze the BCT adoption factor, stating that TOE is practitioner and academically friendly, readable, and exploitable by multiple audience, and it can easily be applied to organizations of different sizes and sectors as well as at different BCT development stages. Furthermore, the TOE framework can be adopted in broad conditions as the choice of organizational, technological, and environmental factors varies according to different innovations as well as different cultures and contexts (Wong et al., 2020). Additionally, Saberi et al. (2019) had introduced a division for the environmental view into external and SC view, which was followed by other researchers afterward such as Lohmer and Lasch (2020) and Kouhizadeh et al. (2021). Furthermore, the interplay between technological, environmental, and organizational factors is significantly important to be considered for the implementation and diffusion process of a new technology into any organizational function (Wamba & Querioz, 2020). A summary of the most frequently mentioned factors that might impact/hinder BCT adoption/implementation into SCM practices is presented in Table 13.2. Saberi et al. (2019) categorize factors following the traditional TOE framework categories as well as the further modified categories for the environmental factors including the SC view and external view.

Multiple articles recommended the use of DoI theory to study and analyze factors that might influence the diffusion process of BCT in SCM (Kamble et al., 2019; Querioz & Wamba, 2019). DoI theory is a significant theory when considering the diffusion of a new idea, process, or technology and its dissemination through multiple organizations or the society (Wamba & Queiroz, 2020). Simply an innovation is defined as "an idea, practice or object that is perceived as new by an individual or other unit of adoption" (Clohessy & Acton, 2019). DoI theory is used to explain the spread of innovation across the ecosystem, unlike adoption theories that explain how a single party in this ecosystem adopts such innovation (Chen et al., 2020). The concept of diffusion relates to the process by which an innovation is spread across members of a social system over time (Wamba & Queiroz, 2020). According to DoI theory, there is a five-stage process, which the decision unit, either individuals or organizations, go through—presented in Figure 13.1—starting from knowledge about the innovation, then forming an attitude toward this innovation, followed by reaching an adoption or rejection decision, then implementation, and finally confirmation (Rogers, 2003).

Table 13.2 Most significant TOE factors influencing BCT adoption in SCM stated in previous literature

Technological factors	Data security	The ability of BCT to protect users' information and ensure full integrity during transmission as well as keeping data secure across the supply network.	Bouthkhoum et al. (2021), Kouhizadeh et al. (2021), Lanzini et al. (2021); Ar et al. (2020), Kamble et al. (2020), Ozturk and Yildizbasi (2020), Tayal et al. (2020), Yadav and Singh (2020a, 2020b)
	Complexity of setup and use	Degree of the usage difficultness of the innovative technology and is usually negatively related to its rate of adoption.	Bouthkhoum et al. (2021), Farooque et al. (2020), Orji et al. (2020), Ozturk and Yildizbasi (2020)
	Cost technology adoption	Costs of new hardware and software for the organization and partners, as well as costs of implementation phases and trainings.	Kamble et al. (2020), Ozturk and Yildizbasi (2020), Nayak and Dhaigude, 2019
	Maturity of BCT (availability and performance)	Related to BCT actual adoption in business practices, as well as BCT scalability, usability, and interoperability.	Kouhizadeh et al. (2021), Farooque et al. (2020), Shoaib et al. (2020)
Organizational factors	Top management support	Top management team is primary responsible for providing support, directions, and resources for successful BCT implementation.	Bag et al. (2021), Lanzini et al. (2021), Kouhizadeh et al. (2021), Farooque et al. (2020)
	Awareness about BCT benefits	One of the main motivators to adopt a certain innovative technology is its relative advantage over other existing innovations.	Bag et al. (2021), Mathivathanan et al. (2021), Tayal et al. (2020)
	Organization culture, people readiness	Complex set of values and beliefs that outline how an organization operate.	Lanzini et al. (2021), Nayak and Dhaigude (2019)

Environmental factors	SC view	Collaboration, coordination, communication between SC partners	Bringing all SC partners together to implement BCT requires massive coordination and collaboration across the partners to achieve an optimal interoperable and compatible platform.	Bag et al. (2021), Kouhizadeh et al. (2021), Farooque et al. (2020), Ozturk and Yildizbasi (2020)
		Culture difference among SC partners	Supply chains are getting more complex and global oriented, connecting various partner of different culture.	Bag et al. (2021), Kouhizadeh et al. (2021), Nayak and Dhaigude (2019)
		Customer influence	Customers currently demand to know the provenance of the products they buy as well as full information related to it.	Kouhizadeh et al. (2021), Lanzini et al. (2021), Tayal et al. (2020)
Environmental factors	External view	Government policies and support	Governmental support refers to encouragement by government for BCT adoption.	Bouthkhoum et al. (2021), Kouhizadeh et al. (2021), Orji et al. (2020), Tayal et al. (2020), Yadav and Singh (2020b), Yadav et al. (2020)
		Competition	The desire to keep up with other organizations dealing with the same customer segment and produce similar products.	Lanzini et al. (2021), Nayak and Dhaigude (2019)

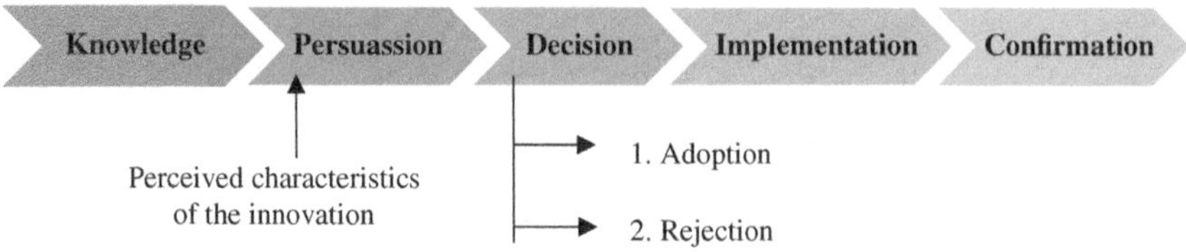

Figure 13.1 Five-stage model of the innovation decision process (*Source:* Adapted from Rogres, 2003).

Accordingly, this chapter analyzed the integration of both TOE framework and DoI theory, as DoI will be useful to understand the organizations' intention to adopt a new technology based on technology characteristics, while TOE will be useful to identify the factors that will influence the decision to leverage the technological innovation such as BCT in SCM. Such integration of DoI theory alongside the TOE framework is useful to enhance the understanding of BCT diffusion in SC in terms of demonstrating how various technological, organizational, and environmental factors influence different stages of diffusion.

13.3 The Case

> "What are the factors influencing blockchain technology diffusion in supply chain management?"

To achieve this chapter's objective, multiple methodological approaches were taken on multiple sequential stages. The first stage was a survey developed to capture SC experts'—working in organizations operating in Egypt—opinions and views about factors that might hinder or facilitate the successful diffusion of BCT into SCM practices. The following stage was to investigate the interrelationships between significant technological, organizational, and environmental factors influencing BCT diffusion in SC using the ISM analysis. Finally, PLS-SEM analysis for the ISM final model was conducted.

Hence, after systematically investigating previous literature, the following research framework—Figure 13.2—was constructed by integrating the most significant factors from previous literature that influence BCT adoption in SCM.

13.3.1 Data Collection

Survey: For this research, a quantitative method is applied, with a survey approach used to collect data. The survey approach is suitable for this research interest, which is analyzing the most significant factors that will influence BCT

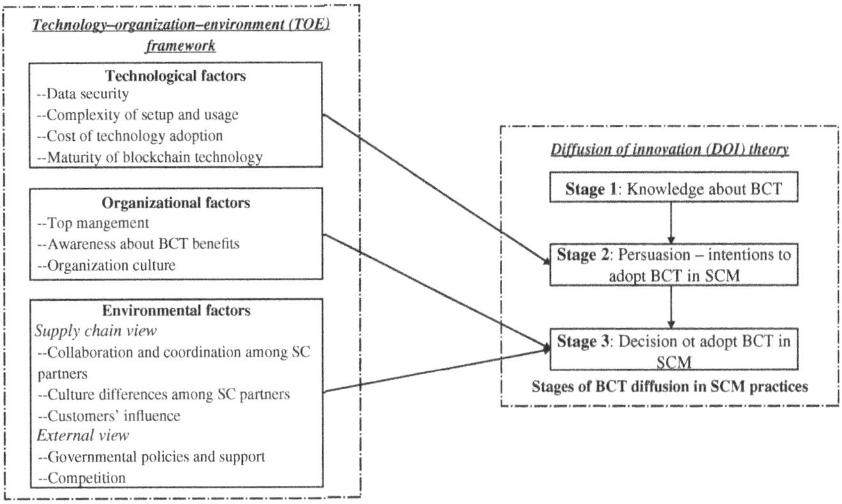

Figure 13.2 Conceptual framework for TOE factors influencing BCT diffusion in SCM.

diffusion in SCM as suggested by previous literature (Jain et al., 2020; Kamble et al., 2019; Philsoophian et al., 2022; Queiroz et al., 2021; Saurabh and Dey, 2021; Wamba et al. 2020). This research is of exploratory purpose, with the intention to investigate a relatively unexplored field among the professionals and industry experts in organizations operating in Egypt rather than yielding generalizable findings.

A survey was developed to capture SC experts'—working in organizations operating in Egypt—opinions and views about factors that might hinder or facilitate the successful diffusion of BCT into SCM practices. This survey was developed in English language. The survey instrument was developed based on validated items collected from the previous literature as well as some other items developed by the researcher for the purpose of this research based on reviewed literature as shown in Appendix 1.

Due to novelty of BCT and scarcity of similar research in Egypt, sample size considerations for this study differ from a regular quantitative study in well-developed topic. Philsoophian et al. (2022) confirmed that BCT is a new innovative technology, which has a small number of experts in the field of SCM who have sufficient knowledge about BCT. In addition, Bouthkhoum et al. (2021) confirmed that large-scale studies might be limited.

The population was composed of SC practitioners and experts with working experience in organizations operating in Egypt from various industries with a minimum of two years of working in the field of SCM. With a mixture of the two non-probability sampling techniques, convenience and snowball sampling were followed in this research. Great efforts were made to ensure that the sample size for this research is acceptable and will lead to a valuable statistical analysis. These efforts started by an initial sampling that was conducted through the professional network platform "LinkedIn," by searching for potential experts with SC experience in their profiles.

13.3.2 Data Analysis

Survey Analysis: The initial sample consisted of ***200 respondents***, and ***152 respondents*** participated in this research study, a response rate of ***76%***. This sample size exceeds the minimum requirement of 10 observations per construct according to the rule of thumb. A power analysis for this sample size was conducted, and results showed a strong estimated power of 1.0000. All participants were either actively involved in BCT projects or indicated that their organizations were currently considering BCT adoption, ensuring that they had direct experience with the technology in SC context.

This study has respondents from varied demographic profiles, with 82% male respondents, while females were 18%. Age ranged from 25 to 50 years old, with various working experiences in the field of SCM with the majority of 7–10 and over 10 years of experience. The respondents had various job titles, such as supply and demand planning manager, logistics manager, material planning manager, operations manager, procurement manager, or SC director.

A nonresponse bias was tested through categorizing responses into two groups—early respondents and late respondents—according to their timestamp, independent sample T-tests (including mean and standard deviations) for each group were performed using IBM-SPSS 19 software. Findings confirmed that there were no statistically significant differences between the scale items for the two groups, indicating the absence of a nonresponse bias. Furthermore, to demonstrate the existence or nonexistence of common method bias and its influence on the research model, Harman's single-factor test was employed. Findings confirmed that the data is not affected by the common method bias.

Pilot testing was conducted after collecting ***30 responses***, exceeding 10% of the targeted sample size as recommended by Kim and Shin (2019). A reliability test using Cronbach's alpha was performed to ensure inter-item consistency for the first ***30 responses*** to ensure that the developed instrument is reliable, hence proceeding with the data collection. Results showed that the instrument used is suitable for the research objective and statistical proportions do exist.

Hence, further investigation for reliability after finalizing the data collection phase was performed using the whole data set of ***152 responses***. Cronbach's alpha values range from 0.7 to 0.8, which confirms the reliability of these constructs as well as the appropriateness of adding these items together to formulate the construct.

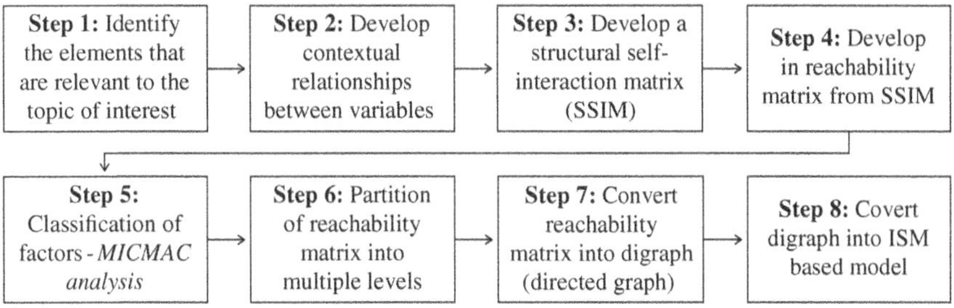

Figure 13.3 Interpretive Structural Modeling procedures (Ghode et al., 2021; Kamble et al. 2020; Zayed & Yaseen, 2021).

To test the significance of relationships between independent and dependent variables, a multiple regression analysis was conducted using the IBM-SPSS 19 Statistical package. Multiple regression is considered a multivariate technique, which is commonly used in business research, providing a method to objectively assess the degree and the character of the relationship between the independent variables and the dependent variable. The regression coefficients indicate the relative importance of each of the independent variables in the prediction of the dependent variable (Sekaran & Bougie, 2013).

Interpretive structural model (ISM): Afterward, to analyze the interrelationships among factors influencing the successful diffusion of BCT in SCM, the interpretive structural modeling (ISM) analysis technique was utilized. ISM was previously used by researchers to identify factors influencing the successful adoption of BCT in SCM (Elhidaoui et al., 2022). For example, Ghode et al. (2021) used ISM to model the challenges of BCT adoption in SC after identifying these factors from the literature and verifying them by experts' opinions. Like any modeling analysis, ISM includes multiple steps to be done as suggested by Warfield in 1974 represented in Figure 13.3.

13.3.3 Data Findings

13.3.3.1 Survey Findings

Statistical findings of the most significant factors influencing Stage 2 "Persuasion/Intentions" and Stage 3 "Decision" of BCT in SCM diffusion stages are presented in Tables 13.3–13.8.

Regression analysis on factors influencing Stage 2 "Persuasion/Intentions" highlights that only two factors—complexity of the BCT setup and knowledge about BCT—significantly influence adoption intentions. These two variables explain about 47.5% of the variation, suggesting that other examined factors are not statistically significant.

Table 13.3 Regression model 1 summary[a]—using a stepwise method

Model	R	R Square	Adjusted R square	Std. error of the estimate	Durbin–Watson
1	0.674[b]	0.454	0.450	0.9766	—
2	0.689[c]	0.475	0.468	0.9610	1.986

[a] Dependent variable: Intention.
[b] Predictors: (Constant), Knowledge.
[c] Predictors: (Constant), Knowledge, Complexity.

Table 13.4 Regression model 1—ANOVA[a]—stepwise regression

Model		Sum of squares	*df*	Mean square	*F*	Sig.
1	Regression	118.925	1	118.925	124.694	0.000[b]
	Residual	143.060	150	0.954	—	—
	Total	261.985	151	—	—	—
2	Regression	124.390	2	62.195	67.350	0.000[c]
	Residual	137.595	149	0.923	—	—
	Total	261.985	151	—	—	—

[a] Dependent variable: Intention.
[b] Predictors: (Constant), Knowledge.
[c] Predictors: (Constant), Knowledge, Complexity.

Table 13.5 Regression model 1—coefficients[a]—stepwise regression

Coefficients								
		Unstandardized coefficients		Standardized coefficients			Collinearity statistics	
Model		*B*	Std. error	Beta	*t*	Sig.	Tolerance	VIF
1	(Constant)	1.898	0.258	—	7.367	0.000	—	—
	Knowledge	0.653	0.058	0.674	11.167	0.000	1.000	1.000
2	(Constant)	2.708	0.418	—	6.471	0.000	—	—
	Knowledge	0.612	0.060	0.631	10.187	0.000	0.919	1.088
	Complexity	−0.161	0.066	−0.151	−2.433	0.016	0.919	1.088

[a] Dependent variable: Intention.

Table 13.6 Regression model 2 summary[a] —stepwise method

Model	R	R square	Adjusted R square	Std. error of the estimate	Durbin–Watson
1	0.676[b]	0.457	0.453	1.055	—
2	0.761[c]	0.579	0.573	0.932	—
3	0.769[d]	0.592	0.583	0.921	2.059

[a] Dependent variable: Decision to adopt BCT in SCM.
[b] Predictors: (Constant), INTI.
[c] Predictors: (Constant), INTI, AWAR.
[d] Predictors: (Constant), INTI, AWAR, COOR.

Table 13.7 Regression model 2—ANOVA[a]—stepwise regression

Model		Sum of squares	*df*	Mean square	*F*	Sig.
1	Regression	140.320	1	140.320	126.124	0.000[b]
	Residual	166.884	150	1.113	—	—
	Total	307.204	151	—	—	—
2	Regression	177.891	2	88.945	102.486	0.000[c]
	Residual	129.313	149	0.868	—	—
	Total	307.204	151	—	—	—
3	Regression	181.753	3	60.584	71.474	0.000[d]
	Residual	125.451	148	0.848	—	—
	Total	307.204	151	—	—	—

[a] Dependent variable: Decision to adopt BCT in SCM.
[b] Predictors: (Constant), INTI.
[c] Predictors: (Constant), INTI, AWAR.
[d] Predictors: (Constant), INTI, AWAR, COOR.

Regression analysis on factors influencing Stage 2 "Persuasion/Intentions" highlights that only two factors—complexity of the BCT setup and knowledge about BCT—significantly influence adoption intentions. These two variables explain about 47.5% of the variance, suggesting that other examined factors are not statistically significant.

Likewise, results showed that only three independent variables are making a statistically significant unique contribution, which are awareness about BCT benefits, collaboration and coordination among SC partners, and intentions to adopt BCT in SCM. Meanwhile, the remaining six independent variables have *p* values greater than 0.05, which indicate that they do not have a statistically significant unique contribution.

Table 13.8 Regression model 2—coefficients[a]—stepwise regression

Model		Unstandardized coefficients		Standardized coefficients			Collinearity statistics	
		B	Std. error	Beta	t	Sig.	Tolerance	VIF
1	(Constant)	2.036	0.314	—	6.484	0.000	—	—
	INTE	0.732	0.065	0.676	11.230	0.000	1.000	1.000
2	(Constant)	0.968	0.321	—	3.014	0.003	—	—
	INTE	0.530	0.065	0.489	8.119	0.000	0.778	1.285
	AWAR	0.390	0.059	0.396	6.580	0.000	0.778	1.285
3	(Constant)	1.997	0.577	—	3.460	0.001	—	—
	INTE	0.495	0.066	0.457	7.453	0.000	0.733	1.365
	AWAR	0.321	0.067	0.327	4.809	0.000	0.598	1.672
	COOR	−0.152	0.071	−0.143	−2.135	0.034	0.612	1.634

[a] Dependent variable: Decision to adopt BCT in SCM.

13.3.4 ISM Findings

To investigate the interrelationships between technological, organizational, and environmental factors influencing BCT diffusion in SC, the ISM analysis was conducted.

Step 1: Identify the elements that are relevant to the topic of interest: For this research, factors influencing BCT diffusion were identified according to the TOE framework by systematically reviewing previous literature and concluding a list of the most significant factors. This was explained previously. Although some factors were found to be slightly insignificant in the survey analysis, they were still included in the ISM analysis to provide a comprehensive understanding of the potential interrelationships between all factors. Their inclusion can reveal underlying patterns and insights that may not be immediately apparent, ultimately enhancing the robustness of the overall analysis and contributing to a more nuanced interpretation of the diffusion process.

Step 2: Develop contextual relationships between variables: Five experts were interviewed to identify the nature of contextual relationships among factors influencing BCT diffusion in supply chain; two experts were from the academia with similar previous research experience and three experts from the industry.

Step 3: Develop a structural self-interaction matrix (SSIM): Contextual relationships between variables mean that either those variables "influence" each other or "lead to" each other; contextual relationships among 10 barriers for sustainable SCM according to experts' opinions. According to the ISM methodology, four symbols are used to identify the relationship between any two barriers, where *V* is used if there is an influence of *i* on *j*, *A* is used if there is an influence of *j* on *i*, *X* is used if both factors are influencing each other, and *O* if barriers are not related.

Step 4: Develop in reachability matrix (RM) from SSIM: Developing RM is the process of transferring SSIM into a binary matrix; such process is composed of two steps: initial and final reachability matrix development. The first step is the creation of initial reachability matrix based on SSIM previously developed by substituting the four symbols (*V*, *A*, *X*, *O*) by 1 or 0 values. Next, the final reachability matrix is developed through updating the initial reachability matrix with transitivity that requires adjusting some of initial reachability matrix cells according to researcher's inference and in accordance with the basic assumption of ISM, final reachability is presented in Table 13.9.

Step 5: Classification of factors—MICMAC analysis: To identify key factors for the topic being studied, factors are clustered according to driving power and dependence power, which are calculated and shown in Table 13.4. Dependence and driving power calculations help in classifying factors into driver factors, dependent factors, linkage factors, and autonomous factors known as the cross-Impact Matrix Multiplication Applied to Classification (MICMAC) analysis. Driver factors are factors with strong driving power, yet weak dependence power. Linkage factors are factors with high driving and dependence power; thus, any action about such factors would significantly affect the overall structure. Yet, dependent factors are factors with low driving power yet high dependence power. Finally, autonomous factors can easily be disconnected from the structure. Factors falling in the category of linkage factors or independent/drive factors are considered among the key factors.

Step 6: Partition of reachability matrix into multiple levels: This is performed by establishing reachability set, antecedent set, and intersection of sets. The reachability set includes the factor itself and other factors it may affect, while the antecedent set includes the factor itself and other factor that may affect it, and the intersection set includes the mutual factors between reachability and antecedent sets.

Step 7: Convert reachability matrix into digraph (directed graph): Afterward, using information from the final reachability matrix, level partitions digraphs and ISM-based model are finally developed. Digraphs are formed by placing high level factors on the highest position, followed by second level factors in second positions and so on.

Step 8: Covert digraph into Interpretive Structural Modeling (ISM) based model: Lastly, the ISM-based model is created by replacing the nodes of digraphs with the statement of the factors included is presented in Figure 13.4.

Table 13.9 Final reachability factors influencing BCT diffusion in SCM

	Barriers	T1	T2	T3	T4	O1	O2	O3	ESC 1	ESC 2	ESC 3	EE1	EE2	Driving power
T1	Data security	1	1	1	0	1	0	0	0	1	1	0	1	7
T2	Complexity of setup and use	1	1	1	0	1	1	1	1	0	0	0	0	7
T3	Cost technology adoption	0	0	1	1	1	0	0	1	0	0	1	0	5
T4	Maturity of BCT	1	1	1	1	1	0	1	1	0	1	0	0	9
O1	Top management	1	0	0	0	1	0	1	0	1	1	1	1	8
O2	Awareness about BCT benefits	0	0	1	0	1	1	0	1	0	0	0	1	5
O3	Organization culture, people readiness	0	1	1	0	0	1	1	1	0	0	0	0	5
ESC 1	Collaboration, coordination, communication between SC partners	1	0	1	0	1	1	1	1	1	1	0	1	9
ESC 2	Culture difference among SC partners	0	1	0	1	1	1	1	1	1	0	1	0	8
ESC 3	Customer influence	0	0	0	0	1	1	1	1	0	1	0	1	6
EE1	Government policies and support	0	1	0	1	1	0	1	1	1	1	1	1	9
EE2	Competition	0	0	0	0	0	1	0	0	0	1	0	1	3
	Dependence power	5	6	7	4	10	9	8	9	5	7	4	7	—

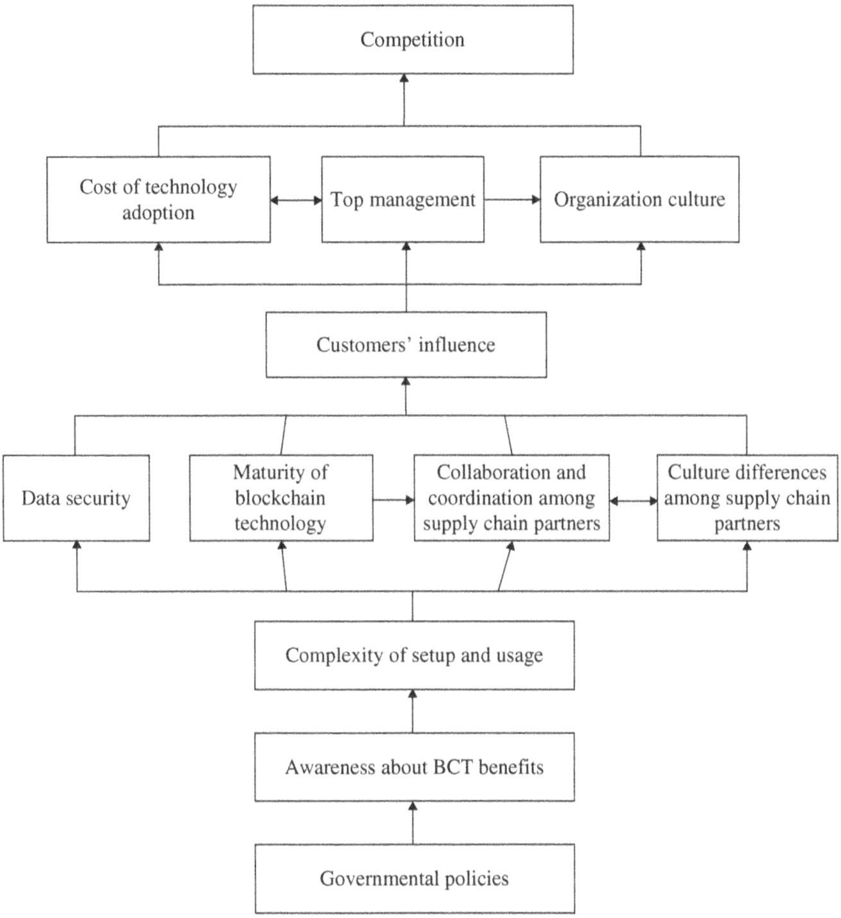

Figure 13.4 ISM model for interrelationships among factors influencing BCT in SCM.

According to the ISM model developed in Figure 13.4, governmental policies, awareness about BCT benefits, and complexity of setup and use are factors at the base of the ISM model. This means that these factors are of high influence on all other factors and are not or are weakly influenced by other factors. Consecutively, intermediate factors are considered, which are data security, maturity of BCT, collaboration and coordination among SC partners, and cultural differences. These factors are highly interconnected and represent the core technical and operational challenges of implementing BCT in SCM. Then, the intermediate factors, customer demand and influence can accelerate the adoption of BCT by pushing companies to improve security, maturity, and collaboration.

Immediately, three factors are considered, which are cost of technology adoption, top management support, and organizational culture. These factors are critical for decision-makers for the adoption of BCT. While high costs can represent a significant barrier, they were found to be less critical as organizations may be more willing to invest when there is a strong perception of customer influence along with other organizational benefits. Furthermore, the culture within the organization can determine the probability of success of BCT implementation but not a challenge for adoption. Finally, at the top of the hierarchy, competition can be considered as the ultimate driver of BCT adoption. Organizations adopting BCT can gain a competitive advantage, pushing others in the industry to follow. It is influenced by all the underlying factors and serves as the final motivator.

13.4 Learning Outcomes

BCT offers significant potential for transforming SCM by enhancing transparency, efficiency, and sustainability performance. Based on the analysis of factors influencing the BCT diffusion in SCM, findings suggest that the adoption of the BCT technology in SCM is a multilayered process, beginning with external factors like governmental policies and moving through awareness, technical challenges, and organizational considerations, all culminating in the competitive pressures that ultimately drive adoption.

Accordingly, this chapter aims to provide a roadmap, which can be a practical and strategic guide for companies, policymakers, and related stakeholders to navigate these factors effectively, facilitating the successful implementation of BCT in SCM. The suggested roadmap—presented in Figure 13.5—outlines 10 interconnected steps that address the various factors influencing BCT adoption. Each step builds on the preceding one, creating a comprehensive approach to successfully implement BCT in SCM.

Following this structured roadmap will allow stakeholders to address the various factors influencing BCT adoption in SCM systematically. This approach not only facilitates effective BCT implementation but also ensures that its adoption advances sustainability goals, providing a significant competitive edge in the marketplace.

The initial step is establishing a supportive regulatory environment primarily targeted at policymakers, who need to establish and enforce clear regulations that support BCT adoption in SCM, focusing on data protection, digital identity, and cross-border transactions. Additionally, policymakers can provide tax incentives, grants, or subsidies for companies that adopt BCT to enhance sustainability, encouraging early adoption and innovation.

The second step is building awareness of BCT benefits. This step is crucial for ensuring that all stakeholders, including SC partners, customers, and internal

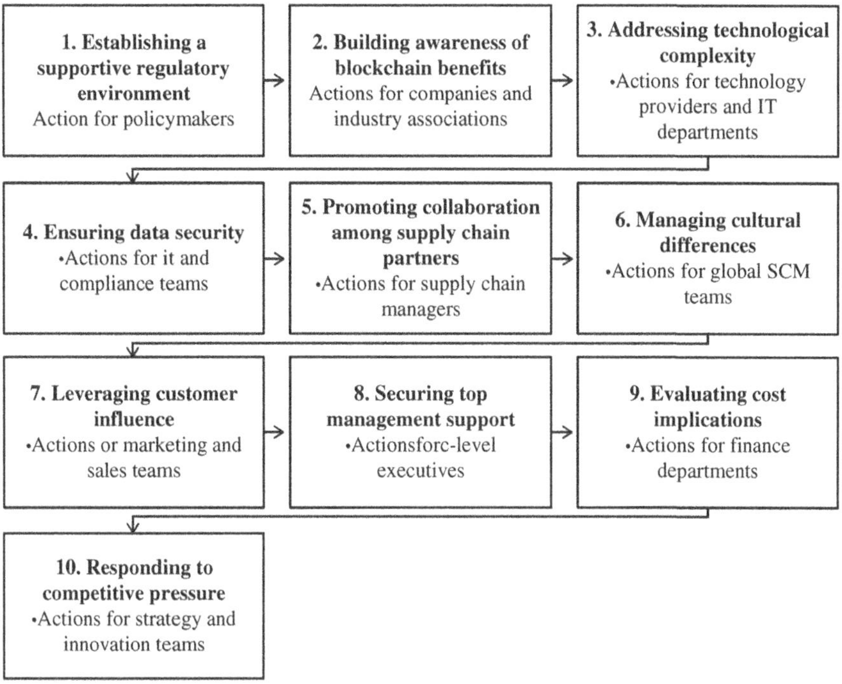

Figure 13.5 Ten steps to BCT in the SCM roadmap.

teams, understand the potential enhancements BCT can bring, particularly in driving sustainability through improved traceability and reduced fraud. Additionally, forming partnerships with educational institutions is vital. These partnerships aim to develop comprehensive training programs that not only highlight the strategic importance of BCT in sustainable SCM practices but also equip participants with the knowledge and skills needed to implement and leverage BCT effectively. The third step involves addressing technological complexity, where technology providers and IT departments play a key role in focusing on developing scalable, user-friendly BCT solutions that simplify the adoption process.

The sixth step is managing cultural differences, especially in globally distributed SCs. Global SCM teams are encouraged to provide cultural sensitivity training and adapt BCT solutions to meet the specific needs of different cultural regions. The seventh step is leveraging customer influence, as marketing and sales teams are encouraged to highlight the sustainability benefits of BCT, using it as a selling point to differentiate products in the market. BCT adoption can be promoted as a means of ensuring ethical and sustainable sourcing practices.

The eighth step is securing top management support that involves aligning BCT adoption with the company's strategic goals. This top-level endorsement not only facilitates the necessary investment in BCT but also fosters a culture of innovation and adaptability within the organization, which is crucial for a long-term success. The ninth step involves evaluating the financial implications of adopting BCT. This financial scrutiny helps in understanding the economic viability of BCT projects and in exploring funding options, such as partnerships, grants, or investments from venture capitalists focused on sustainability, to support the technology adoption.

The last step, responding to competitive pressure, involves using industry dynamics as a catalyst for advancing BCT adoption. Strategy and innovation teams are encouraged to continuously monitor competitor activities and benchmark their own BCT strategies to ensure the organization remains at the forefront of innovation in SCM. Finally, it is imperative to emphasize that the effective adoption of BCT in SCM necessitates the concerted collaboration of diverse stakeholders as explained in each step and presented in Figure 13.5.

13.5 Plans to Further Develop the Initiative

To further develop this initiative, the aim is to understand and enhance BCT diffusion in SCM to improve sustainability practices. It is recommended to *expand the research scope by conducting cross cultural studies* to capture a broader range of insights on BCT diffusion and might reveal unique regional challenges and opportunities. Furthermore, *longitudinal studies* are proposed to observe changes and development over time, providing a more dynamic understanding of how BCT diffusion evolves.

Moreover, *practical application and testing* are highly suggested through pilot programs. This will test the practical application of the structured roadmap and allow for real-time adjustments and improvements, or feedback loops to gather continuous feedback from organizations implementing BCT based on the roadmap, using this data to refine and enhance the model. Finally, *develop and integrate specific sustainability metrics into the BCT diffusion assessment framework* to explicitly measure the environmental and social impact of BCT implementations in SCM.

Another limitation for this research is treating SCM as a broad, holistic concept without differentiating the significance of influencing factors across various sectors or industries. While this approach offers a wide-ranging perspective, it overlooks potential variations in how these factors impact specific industries. Acknowledging this limitation is important, as sector-specific dynamics may introduce unique challenges or opportunities that the generalized approach does not fully address. These recommendations do advance not only the research initiative but also the practical utility and relevance of the findings, ensuring the BCT diffusion in SCM contributes positively and efficiently to sustainability performance within SCs.

A Appendix 1: Survey Items

A.1 Section 1: Technology–Organization–Environment Framework

Data security

- **DSEC1**: Data security will impact the diffusion of BCT in SCM in my company.
- **DSEC2**: BCT has enough safeguards to make me feel comfortable using it (Wamba & Queiroz, 2020).

Complexity of BCT setup and usage

- **CLEX1**: Complexity of BCT setup and usage will impact the diffusion of BCT in SCM in my company.
- **CLEX2**: It will be difficult to understand and apply the concept of blockchain in supply chains (Kamble et al., 2019).

Cost of BCT

- **COST1**: Cost of BCT will impact the diffusion of BCT in SCM in my company.
- **COST2**: Adopting BCT in supply chains will increase hardware and facility cost (Wong et al., 2020).

Maturity of BCT

- **MATU1**: Maturity of BCT will impact the diffusion of BCT in SCM in my company.
- **MATU2**: BCT are not compatible with other systems I use (Queiroz & Wamba, 2019).

Top management

- **TMGT1**: Top management will impact the diffusion of BCT in SCM in my company.
- **TMGT2**: Our top management encourages innovation (Wong et al., 2020).
- **TMG3**: Our top management will inspire employees to apply BCT (Wong et al., 2020).

Awareness about BCT benefits

- **AWAR1**: Awareness about BCT benefits will impact the diffusion of BCT in SCM in my company.
- **AWAR2**: Using BCT will improve our SC performance (Kamble et al., 2019).
  Organizational culture
- **CULT1**: Organizational culture will impact the diffusion of BCT in SCM in my company.

Collaboration and coordination policy SC partners

- **COOR1**: Collaboration and coordination among SC partners will impact the diffusion of BCT in SCM in my company.
- **COOR2**: Our SC partners can be trusted to carry out blockchain transactions faithfully (Queiroz & Wamba, 2019).

Culture differences in supply chain

- **CDIF1**: Culture differences among SC will impact the diffusion of BCT in SCM in my company.
- **CDIF2**: Due to cultural differences, our SC partners might resist BCT. Customers' influence.
- **CUST1**: Customers' influence will impact the diffusion of BCT in SCM in my company.
- **CUST2**: Our main customers have participated or are likely to participate in a BCT system (Hew et al., 2020).

Governmental policies

- **GOVE1**: Governmental policies will impact the diffusion of BCT in SCM in my company.
- **GOVE2**: Relevant policies are introduced by the government to boost BCT use in supply chains (Wong et al., 2020).

Competition

- **COMP1**: Competition will impact the diffusion of BCT in SCM in my company.
- **COMP2**: Competitive pressures force my firm to investigate BCT to gain competitive edge (Wong et al., 2020).

A.2 Section 2: Diffusion of Innovation Theory Stages

Knowledge about BCT

- **KNOW1**: I have the knowledge necessary to use BCT (Queiroz & Wamba, 2019).
- **KNOW2**: My firm has the necessary knowledge to learn and implement blockchain enabled SC applications (Wamba & Queiroz, 2020).
- **KNOW3**: Having necessary knowledge about BCT impacts the intentions to adopt BCT in SCM.

Intention to adopt BCT

- **INTE1**: I think my firm intends to digitally transform operations and SCM through BCT (Wong et al., 2020).
- **INTE2**: My intentions to adopt BCT impact my decision to adopt BCT in SCM.

Decision to adopt BCT

- **DECI1**: Using BCT in SCM is a good idea (Venkatesh et al., 2003).

References

Ar, I.M., Erol, I., Peker, I., Ozdemir, A.I., Medeni, T.D., & Medini, I.T. (2020). Evaluating the feasibility of blockchain in logistics operations: A decision framework. *Expert Systems with Applications, 158*, 113543.

Azzi, R., Chamoun, R.K., & Sokhan, M. (2019). The power of a blockchain-based supply chain. *Computers & Industrial Engineering, 135*, 582–592.

Bag, S., Dmitriev, A.V., Sahu, A.K., & Sahu, A.K. (2021). Barriers to adoption of blockchain technology in green supply chain management. *Journal of Global Operations and Strategic Sourcing, 14*, 104–133.

Bager, S.L., Singh, C., & Persson, M. (2022). Blockchain is not a silver bullet for agro-food supply chain sustainability: Insights from a coffee case study. *Current Research in Environmental Sustainability, 4*, 100163.

Batwa, A., & Norrman, A. (2020). A framework for exploring blockchain technology in supply chain management. *Operations and Supply Chain Management: An International Journal, 13*(3), 294–306.

Bhinge, R., Moser, R., Moser, E., Lanza, G., & Dornfeld, D. (2015). Sustainability optimization for global supply chain decision-making. *Procedia CIRP, 26*, 323–328.

Bouthkhoum, O., Hanine, M., Nabil, M., El-Barakaz, F., Lee, E., Rustam, F., & Ashraf, I. (2021). Analysis and evaluation of barriers influencing blockchain implementation in moroccan sustainable supply chain management: An integrated IFAHP-DEMATEL framework. *Mathematics, 9*.

Caldarelli, G., Zardini, A., & Rossignoli, C. (2021). Blockchain adoption in the fashion sustainable supply chain: Pragmatically addressing barriers. *Journal of Organizational Change Management, 34*(2), 507–524.

Cartier, L.E., Ali, S.H., & Krzemnicki, M.S. (2018). Blockchain chain of custody and trace elements: An overview of tracking and traceability opportunities in the gem industry. *The Journal of Gemology, 36*, 212–227.

Casino, F., Dasaklis, T.K., & Patsakis, C. (2019). A systematic literature review of blockchain-based applications: Current status, classification and open issues. *Telematics and Informatics, 36*, 55–81.

Chen, S., Liu, X., Yan, J., Hu, G., & Shi, Y. (2020). Processes, benefits, & challenges for adoption of blockchain technologies in food supply chains: A thematic analysis. *Information Systems and e-Business Management, 19*, 909–935.

Clohessy, T., & Acton, T. (2019). Investigating the influence of organizational factors on blockchain adoption an innovation theory perspective. *Industrial Management and Data Systems, 119*(7), 1457–1491.

Di Vaio, A., & Varriale, L. (2020). Blockchain technology in supply chain management for sustainable performance: Evidence from the airport industry. *International Journal of Information Management, 52*, 102014.

Dutta, P., Choi, T., Somani, S., & Butala, R. (2020). Blockchain technology in supply chain operations: Applications, challenges & research opportunities. *Transportation Research Part E*, *142*, 1–33.

Egypt – Information and Communications Technology and Digital Economy. (2021). *International trade administration | Trade.Gov*. https://www.trade.gov/country-commercial-guides/egypt-information-and-communications-technology-and-digital-economy

Elhidaoui, S., Benhida, K., El-Fezazi, S., Kota, S., & Lamalem, A. (2022). Critical success factors of blockchain adoption in green supply chain management: Contribution through an interpretive structural model. *Production and Manufacturing Research*, *10*(1), 1–23.

Farooque, M., Jain, V., Zhang, A., & Li, Z. (2020). Fuzzy DEMATEL analysis of barriers to blockchain-based life cycle assessment in China. *Computers and Industrial Engineering*, *147*, 106684.

Francisco, K., & Swanson, D. (2018). The supply chain has no clothes: Technology adoption of blockchain for supply chain transparency. *Logistics*, *2*, 2.

Ghode, D.J., Yadav, V., Jain, R., & Soni, G. (2021). Blockchain adoption in the supply chain: An appraisal on challenges. *Journal of Manufacturing Technology Management*, *32*(1), 42–62.

Gurtu, A., & Johny, J. (2019). Potential of blockchain technology in supply chain management: A literature review. *International Journal of Physical Distribution and Logistics Management*, *49*, 881–900.

Hasan, H., AlHadhrami, E., AlDhaheri, A., Salah, K., & Jayaraman, R. (2019). Smart contract-based approach for efficient shipment management. *Computers & Industrial Engineering*, *136*, 149–159.

Hew, J.-J., Wong, L.-W., Tan, G.W.-H., Ooi, K.-B., & Lin, B. (2020). The blockchain-based Halal traceability systems: A hype or reality? *Supply Chain Management*, *25*(6), 863–879.

Hooper, A., & Holtbrugge, D. (2020). Blockchain technology in international business: Changing the agenda for global governance. *Review of International Business and Strategy*, *30*, 183–200.

Jain, G., Singh, H., Chaturvedi, K.R., & Rakesh, S. (2020). Blockchain in logistics industry: In fizz customer trust or not. *Journal of Enterprise Information Management*, *33*(3), 541–558.

Kamble, S., Gunasekaran, A., & Arha, H. (2019). Understanding the blockchain technology adoption in supply chains-Indian context. *International Journal of Production Research*, *57*(7), 2009–2033.

Kamble, S.S., Gunasekaran, A., & Sharma, R. (2020). Modelling the blockchain enabled traceability in agriculture supply chain. *International Journal of Information Management*, *52*, 101967.

Kannan, D., Jabbour, A.B.L.S., & Jabbour, C.J.C. (2014). Selecting green suppliers based on GSCM practices: Using fuzzy TOPSIS applied to a Brazilian electronics company. *European Journal of Operational Research*, *233*(2), 432–447.

Karamchandani, A., Srivastava, S.K., Kumar, S., & Srivastava, A. (2021). Analyzing perceived role of blockchain technology in SCM context for the manufacturing industry. *International Journal of Production Research*, *59*, 3398–3429.

Kayikci, Y., Gozacan-Chase, N., Rejeb, A., & Mathiyazhagan, A. (2022). Critical success factors for implementing blockchain-based circular supply chain. *Business Strategy and the Environment*, *31*, 3595–3615.

Kim, J.-S., & Shin, N. (2019). The impact of blockchain technology application on supply chain partnership and performance. *Sustainability*, *11*(21), 6181.

Koh, L., Dolgui, A., & Sarkis, J. (2020). Blockchain in transport and logistics – Paradigms and transitions. *International Journal of Production Research*, *58*, 2054–2062.

Kouhizadeh, M., & Sarkis, J. (2018). Blockchain practices, potentials, and perspectives in greening supply chains. *Sustainability*, *10*, 3652.

Kouhizadeh, M., Saberi, S., & Sarkis, J. (2021). Blockchain technology & the sustainable supply chain: Theoretically exploring adoption barriers. *International Journal of Production Economics*, *231*, 1–21.

Kshetri, N. (2021). Blockchain & sustainable supply chain management in developing countries. *International Journal of Information Management*, *60*, 102376.

Lanzini, F., Ubacht, J., & De Greeff, J. (2021). Blockchain adoption factors for SMEs in supply chain management. *Journal of Supply Chain Management Science*, *2*(1–2), 47–68.

LeBaron, G. (2020). *Combatting modern slavery: Why labour governance is failing and what we can do about it*. Cambridge, UK: Polity.

Lim, M.K., Li, Y., Wang, C., & Tseng, M.L. (2021). A literature review of blockchain technology applications in supply chains: A comprehensive analysis of themes, methodologies and industries. *Computers and Industrial Engineering*, *154*.

Liu, Z., & Li, Z. (2020). A blockchain-based framework of cross-border e-commerce supply chain. *International Journal of Information Management*, *52*, 102059.

Lohmer, J., & Lasch, R. (2020). Blockchain in operations management and manufacturing: Potential and barriers. *Computers and Industrial Engineering*, *149*, 106789.

Lohmer, J., Silvia, E.R., & Lasch, R. (2022). Blockchain technology in operations & supply chain management: A content analysis. *Sustainability*, *14*, 6192.

Mathivathanan, D., Mathiyazhagan, K., Rana, N.P., Khorana, S., & Dwivedi, Y.K. (2021). Barriers to the adoption of blockchain technology in business supply chains: A total interpretive structural modelling (TISM) approach. *International Journal of Production Research*, *59*(11), 3338–3359.

Mulligan, C., Morsfield, S., & Cheikosman, E. (2024). Blockchain for sustainability: A systematic literature review for policy impact. *Telecommunications Policy*, *48*, 102676.

Nayak, G., & Dhaigude, A.S. (2019). A conceptual model of sustainable supply chain management in small and medium enterprises using blockchain technology. *Cogent Economics and Finance*, *7*, 1667184.

Niu, X., & Li, Z. (2018). Research on supply chain management based on blockchain technology. *Journal of Physics: Conference Series*, *1176*. https://doi.org/10.1088/1742-6596/1176/4/042039

Orji, I.J., Kusi-Sarpong, S., Huang, S., & Vazquez-Brust, D. (2020). Evaluating the factors that influence blockchain adoption in the freight logistics industry. *Transportation Research Part E*, *141*.

Ozturk, C., & Yildizbasi, A. (2020). Barriers to implementation of blockchain into supply chain management using an integrated multi-criteria decision-making method: A numerical example. *Soft Computing*, *24*, 14771–14789.

Park, A., & Li, H. (2021). The effect of blockchain technology on supply chain sustainability performances. *Sustainability*, *13*, 1726.

Philsoophian, M., Akhavan, P., & Namvar, M. (2022). The mediating role of blockchain technology in improvement of knowledge sharing for supply chain management. *Management Decision*, *60*(3), 784–805.

Queiroz, M.M., & Wamba, S.F. (2019). Blockchain adoption challenges in supply chain: An empirical investigation of the main drivers in India and the USA. *International Journal of Information Management*, *46*, 70–82.

Queiroz, M.M., Telles, R., & Bonilla, S.H. (2020). Blockchain and supply chain management integration: A systematic review of the literature. *Supply Chain Management: An International Journal*, *25*, 241–254.

Queiroz, M.M., Wamba, S.F., Bourmont, M.D., & Telles, R. (2021). Blockchain adoption in operations and supply chain management: Empirical evidence from an emerging economy. *International Journal of Production Research*, *59*(20), 6087–6103.

Reddy, V.P.V. (2019). Enhancing supply chain management using blockchain technology. *International Journal of Engineering and Advanced Technology*, *8*(6), 4657–4661.

Rogers, E.M. (2003). *Diffusion of innovations*. New York: Free Press.

Saberi, S., Kouhizadeh, M., Sarkis, J., & Shen, L. (2019). Blockchain technology and its relationships to sustainable supply chain management. *International Journal of Production Research*, *57*, 2117–2135.

Saho, S., Kumar, S., Sivarajah, U., Lim, W.M., Westland, C., & Kumar, A. (2024). Blockchain for sustainable supply chain management: Trends and ways forward. *Electronic Commerce Research*, *24*, 1563–1618.

Saurabh, S., & Dey, K. (2021). Blockchain technology adoption, architecture, and sustainable agri-food supply chains. *Journal of Cleaner Production*, *284*, 124731.

Sekaran, U., & Bourgie, R. (2013). *Research methods for business: A skill-building approach*. Chichester, West Sussex: Wiley.

Seuring, S., & Müller, M. (2008). From a literature review to a conceptual framework for sustainable supply chain management. *Journal of Cleaner Production, 16*(15), 1699–1710.

Shoaib, M., Lim, M.K., & Wang, C. (2020). An integrated framework to prioritize blockchain-based supply chain success factors. *Industrial Management & Data Systems, 120*(11), 2103–2131.

Tayal, A., Solanki, A., Kondal, R., Nayyar, A., Tanwar, S., & Kumar, N. (2020). Blockchain-based efficient communication for food supply chain industry: Transparency and traceability analysis for sustainable business. *International Journal of Communication Systems, 34*(4), e4696.

Tsolakis, N., Niedenzu, D., Simonetto, M., Dora, M., & Kumar, M. (2021). Supply network design to address United Nations sustainable development goals: A case study of blockchain implementation in Thai fish industry. *Journal of Business Research, 131*, 495–519.

Upadhyay, A., Mukhuty, S., Kumar, V., & Kazancoglu, Y. (2021). Blockchain technology and the circular economy: Implications for sustainability and social responsibility. *Journal of Cleaner Production, 293*, 126130.

Van Hoek, R. (2020). Developing a framework for considering blockchain pilots in the supply chain – Lessons from early industry adopters. *Supply Chain Management, 25*(1), 115–121.

Venkatesh, V., Kang, K., Wang, B., Zhong, R.Y., & Zhang, A. (2020). System architecture for blockchain based transparency of supply chain social sustainability. *Robotics and Computer-Integrated Manufacturing, 63*, 101896.

Venkatesh, V., Morris, M.G., Davis, G.B., & Davis, F.D. (2003). User acceptance of information technology: Toward a unified view. *MIS Quarterly, 27*(3), 425–478.

Wamba, S.F., & Queiroz, M.M. (2020). Blockchain in the operations and supply chain management: Benefits, challenges and future research opportunities. *International Journal of Information Management, 52*, 102064.

Wamba, S.F., Queiroz, M.M., & Trinchera, L. (2020). Dynamics between blockchain adoption determinants and supply chain performance: An empirical investigation. *International Journal of Production Economics, 229*, 107791.

Wang, Y., Han, J.H., & Beynon-Davies, P. (2019). Understanding blockchain technology for future supply chains: A systematic literature review and research agenda. *Supply Chain Management: An International Journal, 24*, 62–84.

Wong, L., Leong, L., Hew, J., Tan, G.W., & Ooi, K. (2020). Time to seize the digital eruption: Adoption of blockchain in operations and supply chain management among Malaysian SMEs. *International Journal of Information Management, 52*, 101997.

Yadav, S., & Singh, S.P. (2020a). An integrated fuzzy-ANP and fuzzy-ISM approach using blockchain for sustainable supply chain. *Journal of Enterprise Information Management, 34*, 54–57.

Yadav, S., & Singh, S.P. (2020b). *Blockchain critical success factors for sustainable supply chain* (p. 152). Resources, Conservation & Recycling.

Yadav, V.S., Singh, A.R., Raut, R.D., & Govindarajan, U.H. (2020). Blockchain technology adoption barriers in the Indian agricultural supply chain: An integrated approach. *Resources, Conservation and Recycling*, 161, 104877.

Zayed, E.O., & Yaseen, E.A. (2021). Barriers to sustainable supply chain management implementation in Egyptian industries: An interpretive structural modeling (ISM) approach. *Management of Environmental Quality*, *32*(6), 1192–1209.

14

A Guide to Securing Funding for Scaling Green Innovations in Africa

Aaron L. Nsakanda[1], Marcia Mkansi[2], Ernest Marfo Asiedu[3], and Frank Ranganai Matenda[3]

[1] *Sprott School of Business, Carleton University, Ottawa, Ontario, Canada*
[2] *Office of Graduate Studies and Research, University of South Africa, Pretoria, South Africa*
[3] *Department of Operations Management, University of South Africa, Pretoria, South Africa*

14.1 Introduction

Several authors have indicated that there is a dire need to decrease greenhouse gas (GHG) emissions and decarbonize to limit the negative impacts of global warming. Among other issues, global warming has resulted in increased global temperatures, rising sea levels, and more severe weather events. The Paris Agreement is a legally binding international treaty on climate change accepted by 196 Parties at the United Nations Climate Change Conference (COP21) on 12 December 2015 in Paris, France. The main aim of this Agreement is to ensure that the rise in the average global temperature is maintained at below 2 °C above preindustrial levels and follows attempts to restrict the temperature rise to 1.5 °C above pre-industrial levels. Several economies made attempts to hasten the development of green industries and GI to achieve these targets (Yang et al., 2020).

In today's world, the significance of GI cannot be exaggerated. GI are innovations that alleviate the adverse influence on the environment as a result of operations, implementing enhanced technologies, management practices, and systems (Roper & Tapinos, 2016; Singh et al., 2020). Abbas and Khan (2023) articulated that "GI refers to the introduction of significantly improved or new organizational

Green Innovations in Supply Chain Management: Case Studies and Applications, First Edition. Edited by Marcia Mkansi, Godfrey Mugurusi, Aaron L. Nsakanda, Temidayo Akenroye, and Frank Tietze.

Published 2026 by John Wiley & Sons, Inc.

products, processes, technologies or management systems that significantly decrease natural resource consumption, emission of toxic substances and the negative impact on the natural environment." GIs mitigate environmental damage, encourage sustainable development, and promote the attainment of economic benefits through advancing cleaner production, encouraging resource efficiency, and plummeting pollution. Ikram and Sadki (2024) propounded that green technologies are vital in attaining the United Nations sustainable goals. Several stakeholders have recognized the significance of promoting GIs since climate change and environmental sustainability continue to gain momentum. This has resulted in the emergence of green funding, which is essential in financing eco-friendly initiatives and projects.

In practice, there is a general consensus that financial sectors play a crucial role in the green technology innovation deployment. In support of this, Hossain et al. (2024) postulated that ecological restoration is a main agenda among stakeholders like corporate and financial sectors. Wang and Zhang (2022) proffered that the green finance (GF) reform has a positive impact on corporate GI. Thus, the availability and presence of GF can be an enticement for corporates to invest in and further promote GI initiatives (Hossain et al., 2024). Huang et al. (2022) indicated that GF has significant positive direct effects and indirect influence on GI, showing that GF augments GI.

Investments in green technology are a high priority way to environmental degradation mitigation and sustainable development. Green funding for eco-friendly initiatives and projects can come in the form of grants, loans, crowdfunding, and impact investing. Eco-friendly initiatives and projects emphasize environmental sustainability and strive to diminish their influence on the environment by reducing carbon emissions, protecting biodiversity, promoting renewable energy, and conserving natural resources. Therefore, green funding promotes a greener and more resilient environment. Green funding allows eco-friendly projects and ventures to address their financial barriers and chase their objectives. They can also design and adopt innovative technologies, enhance their operations, and perform research and development. Further, green funding comes with mentorship, technical expertise, and networking opportunities, thereby improving the success rates of these eco-friendly projects and initiatives.

Even though GF is complementary to the GI process for environmental protection, GI industries are limited by financial limitations (Agrawal et al., 2024). Agrawal et al. (2024) postulated that the capacity of industries in GI is constrained by financial restrictions. These challenges in getting financial resources can interrupt investments in eco-friendly technologies (Andersen, 2017). Agrawal et al. (2024) further opined that the GI of industries is exposed to financial restrictions because it is associated with a high degree of uncertainty and less returns. These financial restrictions need to be resolved in order to develop a favorable financial atmosphere for green growth.

In developing countries, studies on GF are limited, with Africa not being an exception (Ozili, 2022). The majority of studies on GF are developed to developed countries. Given that developing countries have their own unique political, economic, and social features, which are different from those of developed countries, it is challenging to generalize findings from developing countries to developed countries. For instance, GF markets in developing countries like African countries are not well developed as compared to those in developed markets. Also, GI depends on successfully applying green growth models in diverse economies (Hossain et al., 2024). Destek et al. (2024) opined that developed countries can implement green growth models since they are associated with great shock absorption abilities and have zero premature deindustrialization risk. Nevertheless, implementing green growth models in developing economies can be calamitous since they need to substantially modify a number of things like investment options (Destek et al., 2024). Thus, an in-depth study that proposes a guide to securing funding for scaling GIs in Africa is required.

Motivated by the increasing prominence of GIs, the significance of green funding in promoting eco-friendly projects and initiatives and the associated financial restrictions, this study aims to propose a guide to securing funding for scaling GIs in Africa. This guide may substantially enhance the chances of securing funding by African corporations, which find it challenging to raise green funding. Designing this framework for African corporations is important since African countries face a myriad of financial barriers, lag behind the rest of the world when it comes to green technology adoption, and lack awareness about GF. Several African green ventures usually are not able to scale. Nonetheless, these ventures need to grow swiftly to promote economic development on the continent in a sustainable manner. Further, various African funding streams attempt to mobilize financial sectors to fund green investments. However, a possibly disturbing facet is that most African companies are not able to secure funding to scale GIs.

The rest of the chapter is structured as follows. Section 14.2 outlines the review of literature and Section 14.3 presents the methodology. Findings are presented in Section 14.4, and Section 14.5 discusses the findings. Section 14.6 provides the conclusions of the study.

14.2 Literature Review

GF and GI are becoming increasingly important in reality as the world is attempting to deal with environmental dilapidation and climate change (Vaka et al., 2020). These two aspects, GF and GI, are vital in attaining a sustainable future (Agrawal et al., 2024). GF is "financing of investments that provide environmental benefits" (International Finance Corporation, 2017). Liu et al. (2022) and Soundarrajan and

Vivek (2016) indicated that GF is the offering of finance to back environmentally sustainable initiatives and projects. This indicates that GF is a financing tool to finance initiatives and projects that encourage environmental sustainability.

GF is critical for many reasons. Massive investments are needed for a shift to a low-carbon economy (Agrawal et al., 2024). Nonetheless, traditional sources of finance may not be adequate enough to finance initiatives and projects that encourage environmental sustainability and cater to a relatively low-carbon economy (Campiglio, 2016). GF can assist to mobilize extra capital needed for a shift to this low-carbon economy (Dorfleitner & Braun, 2019). Several environmentally sustainable initiatives and projects are associated with significant risks as a result of their comparatively untried nature and absence of a trail record (Agrawal et al., 2024). GF may assist in alleviating these significant risks by offering patient and long-term capital (Bhatnagar et al., 2022). Wang and Zhi (2016) indicated that GF describes the financial regulations in investment and loans to back environmentally sustainable development goals. Basically, Agrawal et al. (2024) postulated that GF generates prospects for diverse institutions to pursue climatic and environmental activities.

Further, GF has a positive influence on GI (Wang et al., 2022). GF promotes the designing and implementation of green technologies, which encourage environmental sustainability. Green technologies can reduce GHG emissions and reduce the influence of human actions on the environment. GI is of paramount significance since conventional technologies, processes, and products may not be adequate to deal with the climate change challenges and attain environmental sustainability (Rennings, 2000). Further, existing literature has indicated that GI can create new business opportunities, stimulate economic growth, and create new jobs and new markets (Agrawal et al., 2024; Hsu et al., 2016). This discussion indicates that GF has vital policy implications (Zhang et al., 2019).

Even though GF and GI are critical in attaining environmental sustainability, the capacity of corporates in GI is limited by financial challenges due to their uncertainty and less returns. Andersen (2017) opined that the problems in getting funding disturb investments in eco-friendly technologies. The financing restrictions hamper the effective application and scaling of green initiatives and projects, which are usually long-term, sophisticated, high-risk, and capital-intensive in nature. Several corporations, especially financially strapped corporates, find it difficult to raise enough financial resources to fund GIs (Dong et al., 2024; Yu et al., 2021). Interestingly, there is a financing gap for GIs globally. In developing countries, this gap continues to broaden. Ozili (2022), Gilchrist et al. (2021), Falcone and Sica (2019), and Berensmann and Lindenberg (2016) articulated that GF is associated with several challenges like absence of consciousness concerning GF, nonexistence of policy coordination for green financing, varying definitions of GF, unreliable policies, and deficiency of profitable inducements to investors who want to invest in climate change alleviation. This puts numerous companies at a dilemma when they want to scale up their sustainable initiatives and projects.

Most GF studies are devoted to developed countries (Breitenfellner et al., 2020; Falcone & Sica, 2019). In developing economies, particularly African economies, research on GF is limited (Akomea-Frimpong et al., 2022; Jawadi et al., 2025; Taghizadeh-Hesary et al., 2022). The characteristics of developed countries and developing countries are not similar (Jawadi et al., 2025). Otek Ntsama et al. (2021) postulated that financial markets for low- and middle-income countries are not similar to those in developed economies. Further, the authors (Otek Ntsama et al., 2021) opined that, divergent to developed economies, the financial, institutional, and political blockades are keeping the green market in low- and middle-income economies in a developing state. Wang et al. (2022) discovered that, in the emerging economies and economies with lower GF levels, GF has a positive influence on GI, whereas, in economies associated with better GI or environmental performance, GF has an adverse influence on GI. Therefore, we cannot generalize findings generated in developed countries to developing countries. Also, this indicates that there is no harmony in the existing literature when it comes to the influence of GF on GI.

In Africa, several sectors lag behind in complying with environmental regulations (Jawadi et al., 2025). An extant literature also indicated that, in Africa, there is the absence of an essential infrastructure that is required for shifting to sustainable practices. Without complying with the existing environmental regulations and in the absence of adequate infrastructure, it is challenging to secure green funding. Jawadi et al. (2025) indicated that, in spite of the availability of green funds, several economic players in Africa are associated with restricted access to funding because of the absence of collateral, bureaucracy, and supposedly high risk of investing in Africa.

14.3 Methodology

14.3.1 Research Design and Approach

In this study, a qualitative case study research design was adopted. Case study research is a usually qualitative approach that offers an in-depth comprehension of a particular case or phenomenon within its context implementing several techniques (Alam, 2021; Yin, 2013). Simons (2009) articulated "case study is an in-depth exploration from multiple perspectives of the complexity and uniqueness of a particular project, policy, institution, program, or system in a real-life." Yin (2013) postulated that a case study approach is one of the best approaches for displaying evidence in a linear manner. Further, Alam (2021) indicated that in a case study, researchers are openly implicated in data gathering, therefore making the research of interest ethically justifiable. A case study is good for describing, comparing, evaluating, and understanding diverse facets of a research problem. Researchers implementing a case study design have flexibility in choosing multiple or single cases, and they may define cases premised on geographical location or more abstract criteria.

In this study, face-to-face interviews were conducted. Face-to-face interviews improve rapport between interviewees and interviewers when it comes to interviewee recruitment and motivating them to partake in research (Horsfall et al., 2021). Interviews are cheap and fast to execute and are detail oriented (Jain, 2021). When conducting face-to-face interviews, researchers can interpret nonverbal cues in a better way and gather in-depth data, resulting in more precise and inclusive collection of information.

Purposive sampling was used when selecting cases (i.e., in this case companies of interest) and the respective participants to interview. A purposive sample refers to the one whose features are defined for an aim that is pertinent to the study (Andrade, 2021). Purposive sampling allowed researchers to focus on relevant cases and respondents and to execute an in-depth analysis. Andrade (2021) indicated that purposive samples permit researchers to only study the population of specific interest, make the sample homogeneous (resulting in improved statistical significance), and eliminate participants who are in danger of grave adverse events.

Thematic analysis was used to analyze the gathered qualitative data. The thematic analysis was used to detect repeated themes or patterns. These themes or patterns are then explained and organized in detail to comprehend the data. Thematic analysis is flexible and simple to execute (Braun & Clarke, 2006; Javadi & Zarea, 2016) and has the capacity to reveal clear, rich perspectives from data. Therefore, thematic analysis is a useful technique for examining sophisticated disciplines and comprehending varied perceptions

14.3.2 Case Descriptions

In this study, six participants were selected from five cases, i.e., institutions (Table 14.1). We believe that this research sample is representative of the study population. Therefore, the study findings can be generalized to the whole population.

Table 14.1 Demographic profile

Case	Nature of the case	Number of participants	Participant's role	About the participant
A	Case 1 is a private equity manager based in Johannesburg, South Africa, with offices in Nairobi and Mauritius. It was founded in the early 2000s. Originally, it managed a single large private equity fund investing in mid-cap companies across all industries in South Africa (Capital Growth Fund). In 2012, it launched the Metier Sustainable Capital Fund (later Loreco Metier Sustainable Capital), focused on renewable energy, energy efficiency, and resource-efficient infrastructure (including waste and water management). Case 1 is now fundraising for its second follow-on fund, which has a similar mandate but with a strategic focus on distributed and embedded generation.	1	Private Equity Investor	Principal at the Sustainable Capital Fund at Metier. He has a background in environmental science and environmental economics. He has been working at Metier for approximately 12 years in the field of green infrastructure and low-carbon investment. He was a principal on the first Metier Sustainable Capital Fund and is now a principal on the second fund.
B	Case 2 aims to limit the global rise in temperature. It was established in 2010, became operational in 2013, and is still ongoing. The GCF undergoes a fundraising process every four years from donor countries.	1	GCF Reviewer	The interviewee was a reviewer for the Green Climate Fund (GCF). Her area of specialization is water. She is now teaching on climate finance.
C	Case 3 revolves around the broader landscape of sustainable finance and green innovation, with a focus on the financial sector's role.	1	Sustainability Banking Expert	With 25 years of experience in banking and financial services, primarily at Investec, she recently joined the Gordon Institute of Business Science (GIBS) to lead their climate leadership and sustainability program. Her role focuses on consolidating sustainability initiatives, implementing policies, and exploring business opportunities to enhance sustainability through financing.

(Continued)

Table 14.1 (Continued)

Case	Nature of the case	Number of participants	Participant's role	About the participant
D	The case primarily focuses on financing movable assets. They engage with corporates purchasing items like airplanes, ships, trains, or large fleets of vehicles, providing financing or loans based on affordability.	1	World Bank Investment Banker	She previously worked at Absa, financing corporate purchases of movable assets. Now at IFC, she specializes in investment banking for infrastructure projects, evaluating them for affordability, financial viability, and climate alignment.
E	Case focuses on the Development Bank of Southern Africa (DBSA) and its relationship with the Green Climate Fund (GCF) and the Global Environment Facility (GEF). DBSA is an accredited entity with the GCF and GEF, allowing it to access funding for climate-related projects. DBSA has a mandate for infrastructure development, encompassing energy, transport, water, sanitation, ICT, and social sectors.	2	DBSA Officials	He is a senior figure in DBSA's engagement with the GCF. She facilitates access to GCF funding, aligning it with DBSA's infrastructure development mandate.

14.4 Findings

This section presents the findings of the study. These findings are related to the funding systems in scaling GIs in Africa.

14.4.1 Case Analysis

Here, we present the findings from each case. The themes that emerged from the dataset's analysis are categorized into six aggregated themes: (i) funding approach/methodology, (ii) nature of funding business, (iii) funding scheme, (iv) funding access eligibility criteria, (v) impacts of businesses accessing the fund, and (vi) the investment amount involved.

14.4.1.1 Case 1

Table 14.2 presents the main themes that emerged in relation to the aforementioned six (6) aggregated themes, after categorizing the codes created from the participants' narratives in Case 1. These themes are explained in the following.

The Funding Approach/Methodology The first theme focuses on how the private equity fund is structured and how investment authority is managed (Table 14.2). This underlines the commitments made by investors, the concept of a "blind pool," and the process of drawing money from investors when investments are made, as highlighted by the participant:

> *"Our fund is capitalised by commitments from our limited partners, which means they have given us a mandate in what's called a blind pool."*
>
> *"This means we have sole investment authority and can decide what to invest in once they've made that commitment, as long as it's within a certain set of investment guidelines."*

The second theme centers on the investment strategy and objectives of the fund (Table 14.2). It emphasizes the search for projects that align with the fund's mandate, the focus on equity investments without using debt, and the goal of growing companies and exiting the investments within a specified timeframe. The participant's exact words are in the following:

> *"When we make an investment, we call that money from investors, and they are obliged to give it to us for investment. So, we don't do debt; we have to have an equity stake in those companies."*
>
> *"Our investment thesis is then, or criterion really, if we invest equity, can we help grow that company and sell that equity position within the lifespan of the fund, which is ten years, and make a return that meets our investors' hurdle return."*

Table 14.2 Thematic analysis on the funding systems in scaling green innovations from Case 1

Aggregated themes	Themes	Codes
Funding approach/ methodology	Fund structure and investment authority	*Private equity fund capitalized by commitments from limited partners.* *Investors provide commitments to a "blind pool," granting Metier sole investment authority within agreed-upon investment guidelines (mandate).* *Money is drawn from investors when an investment is made.*
	Investment strategy and objectives	*Metier seeks projects within its mandate that offer potential investment returns within the fund's timeframe.* *Equity investments are made (no debt), aiming to grow the company and sell the equity position within ten years.*
Nature of funding business	Renewable energy and efficiency investment	*Sustainable capital fund investing in renewable energy, energy efficiency, and resource-efficient infrastructure in South and sub-Saharan Africa.* *Investments span solar PV and battery storage, energy supply for telecommunications towers, rooftop PV and energy storage for commercial and residential.* *Focus on distributed/embedded generation beyond large-scale utility projects.*
	Resource-efficient infrastructure projects	*Investments span wastewater treatment, and recycling (e-waste, undersea cables).* *A biogas project refurbishing company*
Funding scheme	Utility-scale renewable energy investments	*Invested in large utility-scale renewable energy projects in South Africa fully invested* *Small hydropower project in Uganda fully invested.*
	Pan-African diversified energy and infrastructure investments	*A larger fund of roughly $150 million with a Pan-African mandate fully invested* *Focus of 50% outside DAC 4 countries (middle and high-income countries in Africa) and 50% in other countries across Africa.*

Table 14.2 (Continued)

Aggregated themes	Themes	Codes
Funding access eligibility criteria	Investment risk assessment criteria	*Mandate alignment* *Commercial viability* *Investment returns* *Negative screening*
	Investment strategy alignment criteria	*Equity stake* *Business size* *Business model innovation* *Project stage*
Impacts of businesses accessing the fund	Sustainable and ethical business practices	*Greenhouse gas emission reductions* *Promoting sustainable practices and circular economy principles* *Adoption of supply chain due diligence regarding forced labor and ethical sourcing*
	Economic and operational performance	*Employment creation* *Financial performance* *Access to expertise and growth support* *Improve data gathering to retain customers and incentivize sales in areas where people don't have normal credit scores that you would have in more developed countries* *Innovations in business models to better manage credit scores for seasonal workers*
Typical investment amount	Investment in medium to large enterprises	*Investment in medium to large category companies.* *Investment range $150 million*

The Nature of Funding Business Case 1 ensures sustainable capital fund investment in renewable energy, energy efficiency, and resource-efficient infrastructure in South and sub-Saharan Africa (Table 14.2). Its investments span solar photovoltaic (PV) and battery storage, energy supply for telecommunications towers, and rooftop PV and energy storage for commercial and residential use. Additionally, it focuses on distributed/embedded generation beyond large-scale utility projects. As outlined by the participant:

> *"We launched a fund called the Metier Sustainable Capital Fund. That fund had a very specific mandate to invest in renewable energy, energy efficiency, and generally resource-efficient infrastructure."*

> *" ... solar PV and battery storage. One entity that does energy supply for telecommunications towers and moves them from diesel to hybrid solar diesel and generators in West Africa."*
>
> *"And inside South Africa, we are invested in rooftop PV and energy storage in the commercial industrial space in a platform that provides those services to customers in a company that does similar work for the retail market."*

Lastly, Case 1 invests in resource-efficient infrastructure projects (Table 14.2). These project investments span wastewater treatment, recycling (electronic and electrical waste, undersea cables), and a biogas project refurbishing company. The participant asserted that:

> *" ... which included waste management explicitly and water management in South and sub-Saharan Africa ... "*

The Funding Scheme As shown in Table 14.2, the first theme highlights the funding scheme's focus on large-scale renewable energy projects, including significant investments in South Africa and a hydropower project in Uganda. It emphasizes the scale and impact of the investments. The participant indicated that:

> *"That fund invested mainly in large utility-scale renewable energy projects in South Africa ... We also did our first project outside South Africa, which was a small hydropower project in Uganda ... "*

The second theme focuses on a more diversified investment strategy across the African continent. It highlights the fund's significant size, its Pan-African mandate, and the balanced focus on both middle and high-income countries and other countries across Africa. The participant's exact words are in the following:

> *"That had a Pan-African mandate with the focus of 50% outside South Africa, well outside what's called the DAC (Development Assistance Committee) four countries. So, the middle and high-income countries in Africa and 50% in other countries across Africa and we are now fully invested."*

The Funding Access Eligibility Criteria The first theme is investment risk assessment criteria. Case 1 ensures that projects fit within the renewable energy, energy efficiency, or resource-efficient infrastructure mandate, reducing the risk of investing in areas outside the fund's expertise and strategic focus (Table 14.2). The participant highlighted that:

> *"We launched a fund called the Metier Sustainable Capital Fund. That fund had a very specific mandate to invest in renewable energy, energy efficiency, and generally resource-efficient infrastructure."*

Case 1 requires that the technology is commercially proven to eliminate the risk associated with untested or experimental technologies, thereby avoiding commercial risk and research and development investment uncertainties, as stated in the participant's narrative:

> *"So, if it was using some new technology that was not commercially proven, we would very quickly say, sorry, we probably can't advance because you need to demonstrate to us that this is commercially proven. We don't take commercial risk. That's important to note."*

Case 1 guarantees that projects must offer returns that meet investors' hurdle rates within the fund's timeframe. The participant asserts that:

> *"We select ones that look like they meet our mandate and could give us the returns within the timeframes of the fund."*

Case 1 ensures negative screening by excluding investments in arms trade, coal, pornography, gambling, etc. This helps them avoid reputational and ethical risks, ensuring that the fund's investments align with socially responsible and sustainable practices. The participant's exact words are in the following:

> *"First of all, our first screen is a negative screen, which is probably publicly available because it's the same kind of negative screening that the IFC (International Finance Corporation) and other DFIs (Development Finance Institutions) use. So, for example, we won't do anything related to the arms trade, coal, pornography, gambling, etc."*

The final theme is investment strategy alignment criteria (Table 14.2). Case 1 ensures an equity stake in the company, whereby they have a significant ownership interest and influence over the company's strategic decisions. It looks for companies in the medium to large categories. This focus on the company size ensures that investments are made in businesses with sufficient scale and resources to achieve growth and profitability. The participant outlines that:

> *"Mostly, I would say in the medium category, let's say medium to large categories."*
>
> *"There are lots of criteria we look at: are they the right size, is the timing right, what are the risks, country risk, etc. We only provide equity. So, we don't do debt; we have to have an equity stake in those companies."*

Similarly, it prefers investing in early-stage companies with innovative business models rather than new technologies. This approach targets companies with unique and scalable business strategies that can disrupt markets and drive growth. The participant asserts this in this narrative:

> *"I think we do actually invest in quite early-stage companies compared to, let's say, conventional private equity in that some of our business models, like the biogas one I mentioned, have really not been done before in South Africa."*

Lastly, Case 1 likewise prefers companies that have matured past the very early stages (venture capital stage). This preference reduces the risk associated with very early-stage ventures and focuses on businesses that have already demonstrated some level of market validation and operational stability.

The Impacts by Businesses Accessing the Fund The first theme focuses on the fund's commitment to sustainability and ethical business practices (Table 14.2). It highlights the importance of reducing GHG emissions, promoting sustainable and circular economy principles, and ensuring ethical sourcing and labor practices in the supply chain. The participant cites that:

> *"For example, we report for every project on employment, GHG emission reductions, and obviously financial performance."*
>
> *"... asking our portfolio companies to get confirmation from their suppliers that they don't have any slave labour, child labour, or forced labour in their supply chains at the minimum."*

The second theme centers on the economic and operational aspects of the funding business (Table 14.2). It includes employment creation, financial performance, access to expertise and growth support, and innovative approaches to managing credit scores and customer retention in challenging markets. The participant queries:

> *"How do you use data gathering to retain customers and incentivise sales in areas where people don't have normal credit scores that you would have in more developed countries?"*
>
> *"How do you deal with people's payment schemes where people have seasonal payments, where they have money at certain times of the year when they've been able to sell crops but at other times may not have money?"*

The Typical Investment Amount The typical investment amount highlights the fund's focus on investing in medium to large category companies, as well as the substantial size of US$150 million. It underlines the scale and scope of the fund's investment strategy, targeting substantial and impactful business ventures, as cited by the participant:

> *"It was roughly $150 million."*

14.4.1.2 Case 2

Table 14.3 illustrates the main themes that emerged from the six aggregated themes, subsequent to classifying the codes created from the participants' narratives in Case 2. These themes are presented in Table 14.3.

Table 14.3 Thematic analysis on the funding systems in scaling green innovations from Case 2

Aggregated themes	Themes	Codes
Funding approach/ methodology	Funding commitment and needs assessment	*Every four years, the GCF solicits funding commitments from developed countries.*
		Presents assessment of developing countries' climate mitigation/ adaptation needs.
	Proposal submission and accountability	*Developing countries submit proposals after GCF secures commitments.*
		Proposals require partnership between a government department and an accredited organization.
		Accreditation ensures accountability and responsible fund management.
Nature of funding business	Climate change mitigation and adaptation	*Reduce greenhouse gas emissions & improve climate change adaptation.*
		Mitigation of climate change.
		Adaptation to climate change impacts
	Resource and technology transfer for sustainable development	*Transfer resources/technology from developed to developing countries.*
		Sectors include water, sanitation, healthcare, energy, transportation, and others.
Funding scheme	Climate and environmental funding mechanisms	*Adaptation Fund*
		Global Environment Facility
		Climate Investment Funds (World Bank)
	Direct access and sector-specific support	*Direct access entities (larger NGOs).*
		Funds are provided to support green transportation programs and water treatment initiatives.

(Continued)

Table 14.3 (Continued)

Aggregated themes	Themes	Codes
Funding access eligibility criteria	Impact and development potential	*Impact potential*
		Paradigm shift potential
		Sustainable development potential
	Recipient needs and project viability	*Needs of the recipient*
		Country ownership
		Efficiency & effectiveness
Impacts of businesses accessing the fund	Resource provision and climate adaptation	*Direct resource provision (e.g., water systems)*
		Climate change adaptation training programs
	Innovation and financial efficiency	*Innovative business models*
		Co-financing & cost-effective methods
Typical investment amount	Project size (GCF) and adaptation funds	*Ranges from $10 million to $250 million*
		The adaptation funds has smaller packages for up to $5 million

The Funding Approach/Methodology The first theme focuses on the process of securing funding commitments from developed countries and assessing the climate mitigation and adaptation needs of developing countries (Table 14.3). This suggests the cyclical nature of funding solicitation and the importance of understanding the specific needs of developing nations. The participant highlighted:

> *"The way that it raises money is that every four years it goes through a fund-raising process, or what they call a replenishment process, in which they approach the donor countries with their assessment of the needs of the developing countries for funding for climate mitigation and adaptation and make a pitch for how the fund can make a difference towards both adaptation and mitigation."*

The second theme centers on the submission and evaluation of proposals, emphasizing the requirement for partnerships between government departments and accredited organizations (Table 14.3). This also highlights the role of accreditation in ensuring accountability and responsible management of funds. The participant's exact words are in the following:

> *"Once they've attracted these commitments from the donor countries, they then invite developing countries to submit proposals for financing."*
>
> *"Now, these proposals tend to be submitted by a government department in collaboration with an organization that is accredited to the fund."*

The Nature of Funding Business The first theme focuses on the core objectives of the business, which are to mitigate climate change by reducing GHG emissions and to adapt to the impacts of climate change (Table 14.3). It highlights the dual approach of addressing both the causes and effects of climate change. The participant highlighted that:

> *"The purpose of this fund is both to mitigate climate change by funding projects that reduce emissions and to support adaptation to climate change, which is to improve people's capacity to deal with the impacts of the extremes that we are experiencing."*

The second theme emphasizes the transfer of resources and technology from developed to developing countries to support sustainable development (Table 14.3). It covers various sectors such as water, sanitation, healthcare, energy, and transportation. This showcases a comprehensive approach to improving the infrastructure and quality of life in developing regions, as alluded to by the participant:

> *"The Green Climate Fund and similar organizations have been established to ensure the transfer of finances and technology from developed to developing countries to help reduce GHG emissions and improve the ability to adapt to climate change."*

The Funding Scheme The first theme focuses on the various funding mechanisms dedicated to climate and environmental projects (Table 14.3). It highlights the Adaptation Fund, Global Environment Facility, and Climate Investment Funds, which share similar investment criteria aimed at supporting sustainable development and climate resilience. This is highlighted by the participant:

> *"There is also the Adaptation Fund, the Global Environment Facility, and the Climate Investment Funds under the World Bank, and all of these have quite similar investment criteria."*

The second theme emphasizes the role of direct access entities, such as larger non-governmental organizations (NGOs), in managing and utilizing funds

(Table 14.3). It also highlights the specific sectors that receive support, including green transportation programs and water treatment initiatives, showcasing the targeted impact of the funding. The participant's observation is in the following:

> *"Both of these funds have what they call direct access entities. These, in some cases, are NGOs, but they tend to be larger NGOs ... In Pakistan, the GCF (Green Climate Fund) funded a green bus rapid transport system. There is also a project around e-mobility in Costa Rica, but this is more of a light rail transit system for the capital city."*

The Funding Access Eligibility Criteria The first theme focuses on the potential impact and broader development benefits of the projects. It includes criteria related to the number of beneficiaries, GHG emission reductions, the ability to catalyze further impact and innovation, and the social, environmental, and economic co-benefits of the projects. The second theme centers on the specific needs of the recipient countries and the viability of the projects. It includes criteria related to the development status and vulnerability of the target population, government commitment and alignment with national policies, and the economic and financial soundness, cost-effectiveness, and co-financing aspects of the projects. The specific criteria outlined by the participant are in the following:

> *"So, this is really about what impact this project is going to make in terms of primarily the number of people that the project is going to benefit or the amount by which GHG emissions are going to be reduced by the project, or both."*
>
> *"The second criterion is called paradigm shift potential, and this is about the extent to which the proposed project will catalyze impact beyond just the target beneficiaries or the target area."*
>
> *"The third criterion is called sustainable development potential. So, it looks at the dimensions of sustainability—the social, environmental, and economic dimensions."*
>
> *"The fourth criterion is called needs of the recipient, and this is both in terms of the recipient country as well as the actual people who are going to benefit from the project."*
>
> *"The fifth criterion is called country ownership, and this is about the extent to which the project proponents can demonstrate that the government of the beneficiary country owns the project and will sustain the interventions beyond the lifespan of the project."*
>
> *"The last criterion is about efficiency and effectiveness, and this one is focused on the economic or financial soundness of the project."*

The Impacts by Businesses Accessing the Fund The first theme focuses on the direct provision of essential resources and the implementation of training programs to help communities adapt to climate change. It suggests the tangible impacts on resource availability and the enhancement of local capacities to cope with climate-related challenges. The participant indicated:

> *"This becomes a bit of a grey area because, you know, somebody who receives training on adapting to climate change is counted as a beneficiary in the same way as somebody who's receiving a whole new water supply in India—a rainwater harvesting unit on his house. You're counting these two people as beneficiaries."*
>
> *" ... and of course it has to either reduce carbon emissions or it has to support climate adaptation, or both, yeah."*

The second theme emphasizes the role of innovation in business models and the importance of financial efficiency. It includes the development of new and creative business approaches, as well as the use of co-financing and cost-effective methods to maximize the impact of the investments. The participant asserted:

> *" ... it just needs to be innovative for the context in which it's going to be applied. And we also have to show that everything that we're proposing is cost-effective compared to other possible options that would have been applied."*

The Typical Investment Amount This highlights the fund's focus on investing in project sizes (GCF) and adaptation funds. The typical project size investment amount ranges from US$10 million to US$250 million. The adaptation funds have smaller packages for up to US$5 million. The participant's observations are in the following:

> *" ... these projects tend to range between $10 million to $250 million per project, some are even bigger than that."*
>
> *" ... even though smaller packages tend to be around, you know, $4 million, so it's not really something that an individual or a small NGO would be able to manage."*
>
> *" ... where project proponents can access perhaps up to $5 million to fund innovations that support adaptation to climate change ... "*

14.4.1.3 Case 3

Table 14.4 delineates the themes that emerged during the categorization of codes derived from participant's narratives in Case 3. These themes are centered on the six aggregated themes, as elaborated thereof.

Table 14.4 Thematic analysis on the funding systems in scaling green innovations from Case 3

Aggregated themes	Themes	Codes
Funding approach/ methodology	Investment and return strategies	*Take equity stakes in companies.* *Seek shorter-term returns (5–10 years).* *Use ESG screening, focus on lending, not investing.*
	International and institutional funding mechanisms	*International Funding (e.g., Green Climate Fund); funds move through agencies to governments/entities for allocation.* *MDBs & DFIs de-risk projects.* *Attracts commercial banks.*
Nature of funding business	Sustainable finance and innovation	*Sustainable finance* *Green innovation* *Concerns of prioritizing climate- and environment-related innovation*
	Addressing sustainable development goals	*Challenges of allocating capital to address Sustainable Development Goals.* *Renewable energy, green building, decarbonizing initiatives.*
Funding scheme	Retail loans with ESG incentives	*Retail loans* *Attach ESG criteria to incentivize sustainability*
	Green climate funds and debt swaps	*Green climate funds* *Varying standards of sustainability with debt swaps*
Funding access eligibility criteria	Investor risk appetite	*Investor risk appetite* *A deterrent to international funds*
	Government regulations and policies	*Government regulations & policies* *Enables an efficient system*
Impacts of businesses accessing the fund	Structured and compliant financing	*Tightly structured to meet objectives* *Noncompliance can result in losing premium benefits*
	Integration of sustainability	*Need for integration* *Sustainability needs to be integrated into all financing*
Typical investment amount	N/A	*No specific amounts*

The Funding Approach/Methodology The first theme focuses on strategies related to investment and returns (Table 14.4). It includes taking equity stakes in companies, seeking shorter-term returns, and using ESG (Environmental, Social, and Governance) screening with a focus on lending rather than investing.

These strategies highlight the fund's approach to managing investments and achieving financial returns within a specified timeframe. The exact words of the participant are in the following:

> *"What I do understand is that the local banks are doing this, but not all of them, only a few. I know that Standard Bank, which has quite a broad African footprint, will do ESG screening. Access Bank, which is the largest Nigerian bank, will also do ESG screening, but these banks are predominantly lending, not investing. What you need here is more equity investment."*
>
> *" ... private equity is inherently taking an equity stake ... usually within 5 to 10 years ... "*

The second theme centers on the mechanisms of international and institutional funding (Table 14.4). It includes the flow of funds from international sources like the GCF through agencies to governments and entities, the role of Multilateral Development Banks (MDBs) and DFIs in de-risking projects, and the attraction of commercial banks, as highlighted by the participant:

> *"The GCF ... would accumulate the finances and then be responsible for allocating them accordingly."*
>
> *"There's a whole range of risks that first need to be addressed, hence the whole model around blended finance where you need to get the MDBs and the DFIs involved to de-risk, and then it helps open it up for the other banks, you know, for the retail banks to get involved. So, my next thought was, well, why is that not happening first of all, and it is happening. So, there is a lot of work that's being done by the MDBs and the DFIs in terms of development and sustainable finance across all emerging markets, but specifically in Africa, they are doing great work."*

The Nature of Funding Business The first theme focuses on the financial and innovative aspects of the business (Table 14.4). It highlights the commitment to sustainable finance and GI, as well as the concerns and considerations involved in prioritizing climate and environmental innovations. The participant indicated that:

> *"Understanding sustainable finance is extremely broad and bringing it down to GI takes you down a completely different alleyway."*

The second theme centers on the efforts to address Sustainable Development Goals (SDGs) through capital allocation (Table 14.4). It includes the challenges faced in this process and specific initiatives such as renewable energy, green building and decarbonizing efforts that contribute to achieving these goals. The participant alluded that:

> *"... helping them meet their sustainability ambitions, and there are loads of examples in that space. Renewable energy obviously has been critical for Africa, and that's just taken off."*

The Funding Scheme The first theme focuses on the use of retail loans that incorporate ESG criteria to promote and incentivize sustainable practices among borrowers (Table 14.4). It highlights the integration of sustainability considerations into retail lending. The participant asserted that:

> *"Loans that are attached to meeting certain ESG criteria can have a positive impact, or can reduce a negative impact."*

The second theme centers on the role of GCFs and the use of debt swaps with varying standards of sustainability (Table 14.4). This underlines the financial mechanisms and standards employed by GCFs to support sustainability initiatives. The exact words of the participant are in the following:

> *"Banks will give a premium, a premium, they'll give you a lower interest rate if you meet certain criteria."*
>
> *"... governments are aware of this, and they're all trying to negotiate their own path, you know, doing climate debt swaps, for example."*

The Funding Access Eligibility Criteria The first theme focuses on the risk appetite of investors and how it can act as a deterrent to attracting international funds (Table 14.4). This stresses the challenges and considerations related to investor risk tolerance in the context of funding schemes. The participant's exact words are in the following:

> *"The country risk is one of the biggest deterrents of international funds coming into these types of projects."*
>
> *"So, the country risk appetite, the currency, the political appetite—there were just so many risk factors that when they went back to their companies to say, well, can we, almost every single one came back saying no, we wouldn't actually be able to allocate any capital towards this initiative."*

The second theme centers on the role of government regulations and policies in creating an efficient system for funding (Table 14.4). This accentuates how supportive and well-structured regulations can facilitate the smooth operation and success of funding schemes. As alluded to by the participant:

> *"I think for me, a large element of this is around the government and creating the right regulatory environment and policy environment is a massive part."*

The Impacts by Businesses Accessing the Fund The first theme focuses on the importance of structured and compliant financing (Table 14.4). This highlights the need for financing to be tightly structured to meet specific objectives and the consequences of non-compliance, such as losing premium benefits, as affirmed by the participant:

> *"The financing is structured so strictly that they have to meet those objectives ... it just means you don't get the premium, you don't get the benefits of the improved financing or the better financing."*

This theme emphasizes the necessity of integrating sustainability into all financing activities (see Table 14.4). It underlines the importance of ensuring that sustainability principles are embedded in the financial practices of businesses accessing the fund. As highlighted by the participant:

> *"... it almost needs to become part of the norm where a sustainability element is incorporated into all types of financing."*

The Typical Investment Amount In Case 3, there is no specific investment amount, as it varies based on business models.

14.4.1.4 Case 4

Table 14.5 outlines the themes that emerged during the categorization of codes derived from participant's narratives in Case 4 under the six aggregated themes. These themes are elucidated as follows.

The Funding Approach/Methodology The first theme focuses on the financing of movable assets and providing loans to corporate entities (Table 14.5). This includes climate financing for asset purchases that contribute to reducing GHG emissions, such as electric vehicles (EVs). The participant's words are in the following:

> *"We would engage with corporates that were, you know, purchasing movable assets such as airplanes, ships, trains, or large fleets of vehicles, and we would give them financing or a loan to purchase those movable assets based on affordability, etc."*
>
> *"We will essentially look at the project, see if it meets our climate criteria ... we tend to look at things that are green or focused on renewable energy."*

Table 14.5 Thematic analysis on the funding systems in scaling green innovations from Case 4

Aggregated themes	Themes	Codes
Funding approach/ methodology	Asset and Corporate Financing	*Financing movable assets (planes, ships, vehicles).* *Loans to corporates.* *Climate financing if asset purchase reduced greenhouse gas emissions (e.g., EVs).*
	Infrastructure and climate-focused projects	*Focus on infrastructure projects (renewable energy).* *Projects meet financial thresholds, then climate assessment.* *Assesses emissions reduction & alignment with Paris Agreement.* *Preferential interest rates for climate-friendly projects.*
Nature of funding business	Sustainable investment banking	*Investment banking for sustainable/ climate-friendly projects.* *Impact on reducing greenhouse gas emissions.* *Alignment with climate goals.*
Funding scheme	Climate-focused loan programs	*Loans with preferential rates for projects meeting climate criteria* *World Bank Donor Funding with no-interest loans to qualifying companies*
	Sector-specific access to funding	*Primarily for companies in the climate sectors.* *Companies in the renewable energy sectors.*
Funding access eligibility criteria	Financial viability	*Project must be financially viable first.*
	Climate Impact and alignment criteria	*Level of greenhouse gas emissions reduction.* *Percentage of transaction contributing to climate focus.* *Alignment with the Paris Agreement*
Impacts of businesses accessing the fund	Incentives for climate-friendly	*Access to more competitive financing* *Potentially lower interest rates*
	Climate performance tracking	*Ensures climate targets are met* *Failure to meet targets can result in loss of preferential rates*
Typical investment amount	N/A	*Focus on large corporates, not startups.*

The second theme centers on the funding of infrastructure projects with a focus on renewable energy and climate impact (Table 14.5). It includes the assessment of projects based on financial thresholds and climate criteria, the role of the climate team in evaluating emissions reduction and alignment with the Paris Agreement, and the provision of preferential interest rates for climate-friendly projects. As indicated by the participant:

> *"So, every time I look at a project, let's say for instance I get a project to invest in a renewable energy plant somewhere in, I don't know, Vereeniging or something, I will take a look at the project, assess it for affordability in terms of how they will be able to refinance the loan, etc. If they meet the minimum financial thresholds, then we will look at it from a climate angle, which will involve the climate teams. Essentially, what the climate team will do is conduct their own separate assessment to determine the level of reduction in GHG emissions, what percentage of the transaction contributes to climate focus, or what percentage of it is aligned with the Paris Agreement."*

The Nature of Funding Business The first theme focuses on the core business activity of providing investment banking services specifically for sustainable and climate-friendly projects (Table 14.5). This highlights the commitment to financing initiatives that promote environmental sustainability. The participant asserted:

> *"... primarily focused on infrastructure. So, the way that the company is subdivided, let's say I have an infrastructure project to invest in, like a renewable energy plant, be it hydro, wind, or solar."*

The second theme centers on the criteria used to assess projects for investment (Table 14.5). This includes evaluating the financial viability of projects, their impact on reducing GHG emissions, and their alignment with broader climate goals. The participant highlighted:

> *"In the event that the project or the corporate's purchase was aligned with the reduction of GHG emissions, for instance, if it's a company that normally buys buses and now they are converting their buses to electric vehicles, that would reduce GHG emissions, and it would fall under our climate or sustainable finance department."*

The Funding Scheme The first theme focuses on the loan programs designed to support climate-friendly projects (Table 14.5). It includes loans with preferential rates for projects that meet specific climate criteria and no-interest loans provided through World Bank donor funding for qualifying companies. The exact words of the participant are in the following:

> *" ... we can give them specific preferential interest rates only for the portion that is related to renewable energy ... "*
>
> *" ... it comes from the donor funding, it doesn't come from the World Bank pool of funds specifically, so that's when I interact with donor funding ... "*

The second theme highlights the targeted access to funding for companies operating in the climate and renewable energy sectors (Table 14.5). It stresses the focus on supporting businesses that contribute to environmental sustainability and renewable energy initiatives. As indicated by the participant:

> *"Then based on that, you can either get an approval saying that because there's an element of renewable energy, we can give them specific preferential interest rates only for the portion that is related to renewable energy, but for the coal-fired power plants, we can't give them anything."*

The Funding Access Eligibility Criteria The first theme focuses on the financial aspect of the eligibility criteria (Table 14.5). It emphasizes the importance of ensuring that a project is financially viable before considering other factors. This ensures that the project is economically sound and capable of delivering the expected financial returns. As highlighted in the participant's narrative:

> *" ... first and foremost, the project needs to be financially friendly, so it needs to be financeable before we look at climate. So, is it a project that is financially viable?"*

The second theme centers on the climate-related criteria for funding access (Table 14.5). It includes the level of GHG emissions reduction and the extent to which the transaction contributes to climate-focused goals. This ensures that the projects not only meet financial standards but also contribute positively to climate change mitigation and adaptation. The participant asserted:

> *" ... then we will look at it from a climate angle, which will involve the climate teams. Essentially, what the climate team will do is conduct their own separate assessment to determine the level of reduction in GHG emissions ... "*
>
> *"So, if they see that the strategy is not fully aligned with moving towards renewable energy, but is just primarily focused on having something in the mix, then the project can still be rejected entirely."*

The Impacts by Businesses Accessing the Funding The first theme focuses on the financial incentives provided to businesses that undertake climate-friendly initiatives (Table 14.5). It highlights access to more competitive financing options and the potential for lower interest rates as rewards for meeting climate-related criteria. The participant highlighted:

> *"The reason our corporates or the companies were eager to get climate financing or sustainable finance is because it essentially gave them an opportunity to receive a reduced interest rate ... "*

The second theme centers on the continuous monitoring and tracking of climate performance. It emphasizes the importance of meeting climate targets and the consequences of failing to do so, such as losing preferential financing rates, as alluded to by the participant:

> *"So, on an annual basis, you'll probably, as an investment banking team, have to submit reports to people like the climate team on those targets that the company would have given us. In the event that people are deviating or are no longer meeting the criteria or have now moved back from renewable to just coal or something, they will lose their preferential rate because, as you know, a loan is paid, I guess it depends, monthly, quarterly, or twice a year."*

The Typical Investment Amount In Case 4, there is no specific investment amount because it focuses on large corporates, not startups.

14.4.1.5 Case 5

Table 14.6 presents the themes that emerged from the six aggregated themes, following the classification of the codes created from the participants' narratives in Case 5. These themes are elaborated later.

The Funding Approach/Methodology The first theme focuses on the role of the DBSA as an accredited entity for the GCF and its program design approach (Table 14.6). This suggests DBSA's efforts to align funding with its infrastructure mandate, attract private sector participation, provide concessional funding and grants and de-risk projects to make them bankable. This was asserted by Participant 1:

> *"The DBSA program will be the first private sector climate finance facility in Africa using a pioneering green bank model. It will de-risk projects to make them bankable."*

Table 14.6 Thematic analysis on the funding systems in scaling green innovations from Case 5

Aggregated themes	Themes	Codes
Funding approach/ methodology	DBSA's role and strategies as a green climate fund accredited entity	*DBSA is an accredited entity for the Green Climate Fund (GCF)* *Accesses funding aligned with DBSA's infrastructure mandate* *DBSA utilizes a program unit approach* *Designs programs to crowd in the private sector* *Offers concessional funding and grant support* *Helps de-risk projects & prepare them for bankability*
	Managing foreign currency risk and leveraging private sector financing	*GCF does not take on foreign currency risk, an agreement had to be reached to accommodate this.* *The facilities had to be restructured so the projects didn't take on the forex risk.* *GCF looks for opportunities to leverage and bring in private sector financing.*
Nature of business	Sustainable infrastructure development	*Infrastructure development (energy, transport, water, ICT)* *Aims to create infrastructure to assist climate change* *Seeks to develop the development impact of projects*
	Climate change mitigation and business adaptation	*Projects reducing emissions and adapting to climate change* *Large corporations & SMEs*
Funding scheme	Leveraging sovereign guarantees for green transportation	*Climate Finance Facility (CFF) and Embedded Generation Investment Program (EGIP) (GCF approved).* *Uses funding to make use of sovereign guarantees in green transportation programs.*
	Financing and grant support for investor connectivity	*Programs can offer both debt financing* *Grants to connect investors*

Table 14.6 (Continued)

Aggregated themes	Themes	Codes
Funding access eligibility criteria	Climate impact and mitigation	*Demonstrating strong climate impacts is crucial* *GHG emissions reductions (mitigation projects)*
	Assessment and collaborative financing models	*DBSA assesses GCF's six criteria* *Private developers who have private sector offers can work with DBSA through blended finance models to achieve sustainable projects*
Impacts of businesses accessing the fund	Transformative policy influence and private sector engagement	*Fund is intended to be transformative and influence even policy within a sector* *Aims to bring in private sector to be invested in businesses* *To be more involved as members of Community Trusts*
	Enhancing infrastructure and sustainable private sector operations	*Increasing access to water and supporting infrastructure* *To meet certain financial goals* *Support the private sector so they have sustainable operations*
Typical investment amount	Strategic project financing and public fund mobilization	*DBSA has a minimum project size of 150 million Rand.* *mobilizes 4% in public funds for implementation*

The second theme centers on the management of foreign currency risk and the efforts to leverage private sector financing (Table 14.6). It highlights the need to restructure facilities to avoid foreign currency risk and the GCF's focus on finding opportunities to attract private sector investment. As stated by Participant 2:

> *"We had a challenge with the Green Climate Fund that the Green Climate Fund doesn't take any foreign currency risk."*

The Nature of Funding Business The first theme focuses on sustainable infrastructure development. DBSA's focus is on infrastructure development across sectors like energy, transport, water, sanitation, and information and communications technology (ICT) but can be flexible to large corporations and small-to-medium enterprises (SMEs) (see Table 14.6). The words of Participant 2 are in the following:

> *"We access GCF funding because it aligns with DBSA's infrastructure development mandate. As we know, we're an infrastructure DFI, we finance projects in energy, transport, water, sanitation, ICT, and also social sectors."*

The second theme highlights climate change mitigation and business adaptation. It provides finance to projects that aim to reduce emissions as well as support climate change adaptation, as asserted by Case 5 Official 1:

> *"Finance facility, which is more cross-cutting amongst mitigation and climate change adaptation."*

The Funding Scheme The first theme focuses on leveraging sovereign guarantees for green transportation (see Table 14.6). DBSA utilizes funding mechanisms such as the Climate Finance Facility (CFF) and the Embedded Generation Investment Program (EGIP), both approved by the GCF. DBSA often works to help the municipal infrastructure through a program that assists in water reuse. The exact words from Participant 1 are in the following:

> *"... And the EGIP, which is a renewable energy program, submitted a climate adaptation-focused initiative called the Water Reuse Program, which has been approved."*
>
> *"But people know what the DBSA has been doing. A lot of renewable energy projects. So, the team that is currently implementing the EGIP, which is supported by GCF."*

The second theme highlights financing and grant support for investor connectivity (see Table 14.6). DBSA makes use of sovereign guarantees to finance independent power producers and green transportation efforts, as indicated by Participant 2:

> *"But that program is dependent on sovereign guarantees."*

The Funding Access Eligibility Criteria The first theme highlights climate impact and mitigation. The primary hurdle is the climate rationale, demonstrating that strong climate impacts are crucial. For mitigation projects, calculating GHG emissions reductions is fundamental, as indicated by Participant 2:

> *"Calculating the GHG emissions reductions would be, you know, fundamental. ... So, if a project, even though it may have, you know, very good climate impacts."*

The second theme highlights assessment and collaborative financing models. DBSA assesses GCF's six criteria. From a private sector perspective, private developers who have a private sector offers can work with DBSA through blended finance models to achieve sustainable projects. The exact words from Participant 1 are in the following:

> *" ... Or seven criteria, but six criteria. But I think out of those six criteria, you know, the key ones, of course, being a climate fund, you know, the climate rationale is often the one that ... "*

The Impacts by Businesses Accessing the Funding The first theme explains that the GCF requires its funds to be transformative and influence policy within a sector. Success involves attracting private sector investment in businesses. Additionally, GCF financing can help cover the costs of new bus systems and allow small businesses to become involved as members of Community Trusts. As indicated by Participant 1:

> *"Project under the Green Fund at the time that we feel could be scaled up with GCF support. So, at the time we submitted a concept note based on the work we did with the GCF and our concept note was shortlisted and we were requested to then submit."*

The second theme focuses on enhancing infrastructure and sustainable private sector operations. This funding increases access to infrastructure in parts of South Africa to address potential issues like water scarcity. As highlighted by Participant 1:

> *"Where you set up a dedicated office to manage, you know, the program. So, the impacts are generally looked at in terms of assessing, addressing issues around water scarcity in South Africa."*

The Typical Investment Amount This theme captures the essence of DBSA's criteria for project size and its approach to leveraging public funds for implementation. On the global front, DBSA has a minimum project size of 150 million Rand for the projects that it can support. To further encourage this impact in larger programs, 4% is generally used to mobilize this funding from public funds to be spent through development projects. This is indicated by Participant 1:

> *"We can, as DBSA, set a minimum project size that we can potentially support over 150 million Rand. ... the 4% million dollars, two and a half million dollars, would be looking at financing."*

14.4.2 Cross-Case Analysis

This section presents the findings from the dataset across the cases. Table 14.7 illustrates the differences and similarities of the funding concepts (themes) across all the cases.

The funding methodologies of various entities show both similarities and differences. All entities rely on external funding sources, such as donor countries, limited partners/investors, or countries set frameworks. Additionally, all entities perform due diligence and assessment of projects to ensure their viability and alignment with objectives (see Table 14.7).

However, there are notable differences. Case 1 operates as a private equity fund, while others involve public funds or blended finance approaches. Case 1 primarily takes equity stakes, whereas others use loans (often with preferential rates) and grants. Case 2 GCF solicits proposals from developing countries, while others are more proactive in seeking investments. Case 1 also operates within a blind pool, providing sole investment authority (see Table 14.7).

In terms of the nature of business, all entities focus on sustainable development, climate change mitigation, and environmental protection. They specialize in sectors such as renewable energy, energy efficiency, water management, waste management, and sustainable infrastructure. Case 2 GCF has the broadest mandate, covering multiple sectors, while others are more targeted. Case 1 focuses on South and sub-Saharan Africa, whereas others operate globally with a focus on developing countries. Case 1 operates as a private equity fund, while the others operate as public programs (see Table 14.7).

Regarding funding schemes, specific programs or facilities are dedicated to climate finance. Some entities emphasize the availability of program documents and funding proposals, while others are less explicit. Geographic scope also varies, with some entities focusing on specific regions and others functioning more internationally (see Table 14.7).

When it comes to funding access eligibility criteria, all entities prioritize financially viable projects that align with broader sustainability or climate objectives and evaluate the potential for GHG emission reductions. However, some prioritize commercially proven technologies and market-based returns, while others consider broader development impacts. Preferences for project size and stage also vary, with some focusing on larger projects and others on startups (Table 14.6).

The impacts of businesses accessing the fund show similarities in aiming for environmental benefits like GHG emission reductions and promoting renewable energy. Many also consider social and economic development impacts, including job creation and community benefits. However, methods for measuring impacts vary, with some facing challenges in long-term measurement and others using specific criteria (Table 14.7).

Table 14.7 Cross-case analysis on the funding systems in scaling green innovations

Funding aggregated themes	Funding concepts	Case 1	Case 2	Case 3	Case 4	Case 5
Funding methodology	Reliance on external funding/partnerships	x	x	x	x	x
	Project-based or program-based approach	x	x	x	x	x
	Due diligence/viability assessment	x	x	x	x	x
	Private focus	x	—	—	—	—
	Public/blended	—	x	x	x	x
	Equity	x	—	—	—	—
	Debt/grants	—	x	x	x	x
Nature of funding business	Focus on sustainability/ green initiatives	x	x	x	x	x
	Targeting specific sectors	x	x	x	x	x
	Broadest mandate	x	—	—	—	—
	More targeted	—	x	x	x	x
	Focus on South and sub-Saharan Africa	x	—	—	—	—
	Globally with a focus on developing countries	—	x	x	x	x
	Private focus	x	—	—	—	—
	Public focus	—	x	x	x	x
Funding scheme	Program-based or facility-based	x	x	x	x	x
	Publicly available program proposals	—	x	x	x	x
	Regional geographic scope	x	—	—	—	—
	Internationally geographic scope	—	x	x	x	x
Funding access eligibility criteria	Financial viability	x	x	x	x	x
	Alignment with sustainability goals	x	x	x	x	x
	Climate-specific criteria	x	x	x	x	x

(*Continued*)

Table 14.7 (Continued)

Funding aggregated themes	Funding concepts	Case 1	Case 2	Case 3	Case 4	Case 5
Impacts of businesses accessing the fund	Greenhouse gas emission reductions	x	x	x	x	x
	Promoting renewable energy	x	x	x	x	x
	Social and economic impacts	x	x	x	x	x
Typical investment/ funding amounts	Investment range variation	x	x	x	x	x
	Program level variation	x	x	x	x	x

Finally, typical funding amounts differ widely, ranging from US$10 million to US$250 million or more. Some entities specify minimum amounts, while others refer to total program amounts. Case 1 focuses on typical project investments, while others refer to total program funding (see Table 14.7).

14.5 Discussion

The study reveals that, in the African landscape, there are a variety of funding options available, including private equity, blended finance, grant-based funding, and debt financing. Nevertheless, accessing these different funding streams often requires navigating complex eligibility criteria and understanding the varying risk tolerance of different funders.

The study findings also revealed that the geographic focus and specific mandates of funding organizations significantly influence the types of projects they support. Some organizations focus specifically on Africa, while others have a global mandate. Furthermore, some funders prioritize commercial viability, while others focus on development impacts.

Further, the study revealed that due diligence is paramount, but impact measurement is challenging. All funding entities emphasize the importance of due diligence to ensure the viability of projects. However, measuring the long-term impacts of GIs, particularly social and environmental impacts, remains a significant challenge.

Lastly, the right funding mix is critical. Combining different sources of funding to create a blend of capital can be highly effective. This approach can help de-risk investments while ensuring they operate in a climate-positive manner.

Considering the aforementioned six aggregated themes, we propose a framework dubbed "A Guide to Securing Funding for Scaling Green Innovation in Africa" as shown in Figure 14.1. This framework aims to provide a practical advice for entrepreneurs, project developers, and fund managers seeking funding to scale GIs in Africa.

This funding guide has six stages as discussed in the following:

Stage 1: Understanding funding methodologies

To successfully navigate the diverse funding landscape, it is essential to familiarize yourself with the various funding methodologies available. Private equity involves direct investment in your company in exchange for equity, suitable for established companies with high growth potential. Debt financing or loans can be obtained from commercial banks, DFIs, or specialized green lenders. Blended finance combines concessional (below-market) financing from DFIs or philanthropic sources with commercial capital, reducing risk for private investors. Grants are non-repayable funds from governments, international organizations or philanthropic foundations, typically requiring alignment with specific development goals.

Assessing the needs stage is crucial. For early-stage startups, seek seed funding or venture capital that provides early-stage funding and support. For growth-stage companies, it is vital to target private equity, venture capital or debt financing to scale operations. Mature-stage companies should explore debt financing for larger projects. Building partnerships with experienced partners who have a track record of securing funding in your sector is also vital.

Stage 2: Defining sustainable nature of business

Clearly defining and quantifying the environmental and social benefits of the GI is essential. This includes GHG emission reductions, resource efficiency (water, energy, materials), biodiversity conservation, job creation and economic development, and community benefits (health, education, access to clean energy/water). Aligning the project with the SDGs is increasingly important for attracting impact investors and grant funding. Focus on sectors with high potential for GI and strong demand in Africa, such as renewable energy (solar, wind, hydro), energy efficiency (green buildings, efficient appliances), water management and sanitation, and waste management and recycling.

Stage 3: Decoding funding schemes

Explore the available programs by identifying specific funding programs or facilities offered by DFIs, international organizations, and private investors. Carefully review the eligibility criteria for each funding scheme, including project size and stage, geographic focus, sector priorities, financial requirements, and environmental and social safeguards. Leverage publicly available program

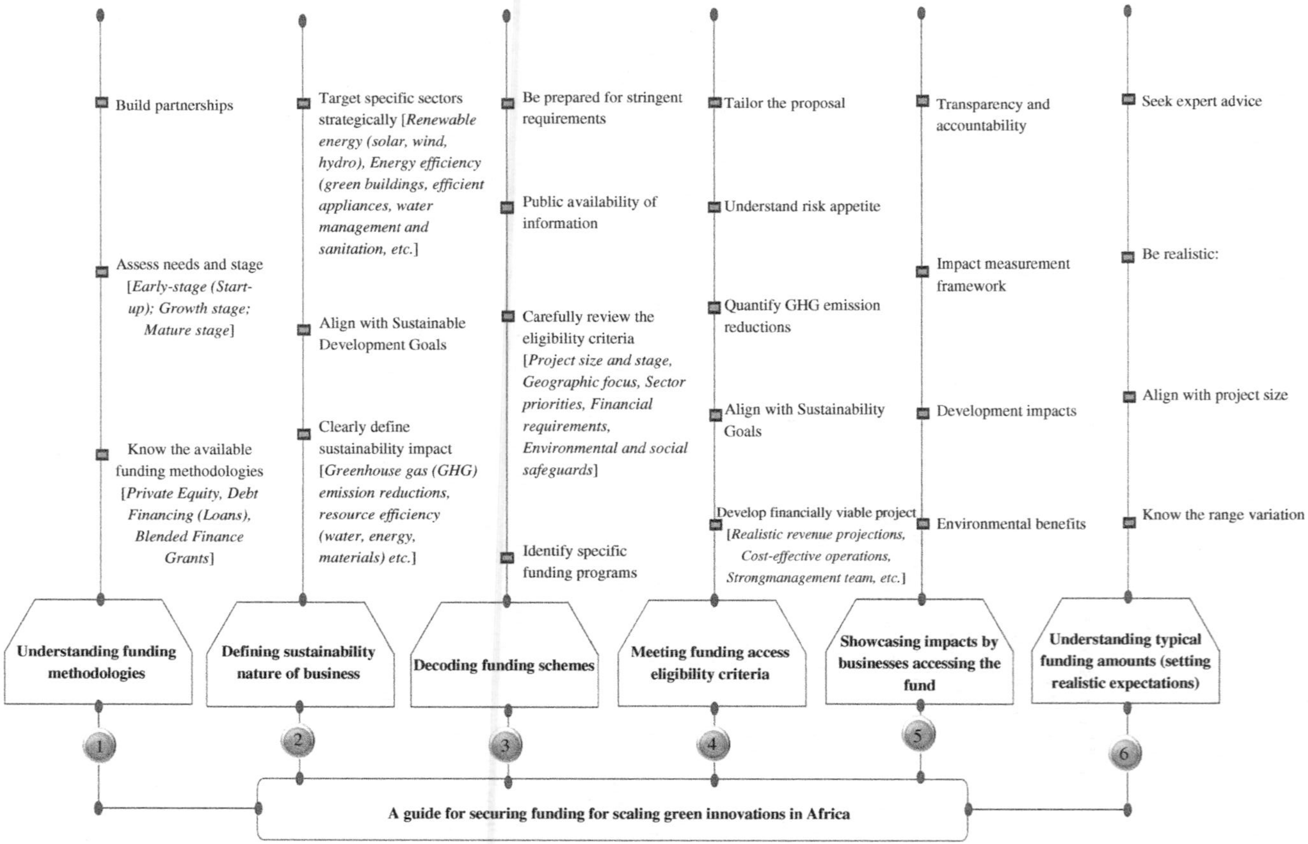

Figure 14.1 A guide to securing funding for scaling GI in Africa. *Source:* Authors' compilation (2025).

documents and funding proposals to understand the application process and evaluation criteria. Be prepared for stringent requirements, as funding schemes often have strict due diligence, impact measurement, and reporting requirements.

Stage 4: Meeting funding access eligibility criteria
Develop a robust business plan that demonstrates the financial viability of the project, including realistic revenue projections, cost-effective operations, a strong management team, and a clear exit strategy. Demonstrate a clear alignment with the fund's broader sustainability or climate objectives and provide a credible estimate of the potential for GHG emission reductions. Understand the risk appetite of different funding sources; private equity investors typically prefer commercially proven technologies and market-based returns, while DFIs may be more willing to accept higher risks for projects with strong development impacts. Tailor the proposal to the specific requirements and priorities of each funding source.

Stage 5: Showcasing impacts of businesses accessing the fund
Clearly outline the environmental benefits of the project, such as GHG emission reductions, improved resource efficiency and promoting renewable energy. Highlight the social and economic development impacts, including job creation, community benefits, and poverty reduction. Develop a robust impact measurement framework that includes clear indicators to measure the environmental, social, and economic impacts of your project, baseline data to track progress over time, a rigorous methodology for collecting and analyzing data, and a plan for reporting your results to investors and stakeholders. Be transparent about your impact measurement methodology and results, and acknowledge the challenges of measuring long-term impacts, developing strategies to overcome these challenges.

Stage 6: Understanding typical funding amounts to set realistic expectations
Be aware of the typical funding amounts available from different sources and target funding sources that are appropriate for the size and stage of your project. Set realistic expectations for the amount of funding to secure and seek expert advice from financial advisors who specialize in GF to develop a funding strategy that aligns with your project's needs and the available funding opportunities.

14.6 Conclusions

This study adopted a qualitative multiple case study approach to suggest a guide to securing funding for scaling GIs in Africa. Face-to-face interviews and thematic analysis were also adopted to attain the study aim. A six-stage guide to securing funding for scaling GIs in Africa with the following stages was designed: (i) understanding funding methodologies, (ii) defining sustainable nature of business, (iii) decoding funding schemes, (iv) meeting funding access eligibility criteria, (v) showcasing impacts of businesses accessing the fund, and (vi) understanding typical funding amounts to set realistic expectations. Since the African funding landscape is diverse and fragmented, this guide developed may enhance the chances of securing funding by African corporations, which find it challenging to raise green funding and raise their awareness about GF. As an avenue for future research, this study can be expanded by conducting a comparative study between developing and developed countries to get new insights.

References

Abbas, J., & Khan, S.M. (2023). Green knowledge management and organizational green culture: An interaction for organizational green innovation and green performance. *Journal of Knowledge Management, 27*(7), 1852–1870.

Agrawal, R., Agrawal, S., Samadhiya, A., Kumar, A., Luthra, S., & Jain, V. (2024). Adoption of green finance and green innovation for achieving circularity: An exploratory review and future directions. *Geoscience Frontiers, 15*(4), 101669.

Akomea-Frimpong, I., Adeabah, D., Ofosu, D., & Tenakwah, E.J. (2022). A review of studies on green finance of banks, research gaps and future directions. *Journal of Sustainable Finance & Investment, 12*(4), 1241–1264.

Alam, M.K. (2021). A systematic qualitative case study: Questions, data collection, NVivo analysis and saturation. *Qualitative Research in Organizations and Management: An International Journal, 16*(1), 1–31.

Andersen, D.C. (2017). Do credit constraints favor dirty production? Theory and plant-level evidence. *Journal of Environmental Economics and Management, 84*, 189–208.

Andrade, C. (2021). The inconvenient truth about convenience and purposive samples. *Indian Journal of Psychological Medicine, 43*(1), 86–88.

Berensmann, K., & Lindenberg, N. (2016). *Green finance: Actors, challenges and policy recommendations.*

Bhatnagar, M., Taneja, S., & Özen, E. (2022). A wave of green start-ups in India—The study of green finance as a support system for sustainable entrepreneurship. *Green Finance, 4*(2), 253.

Braun, V., & Clarke, V. (2006). Using thematic analysis in psychology. *Qualitative Research in Psychology, 3*(2), 77–101.

Breitenfellner, A., Hasenhüttl, S., Lehmann, G., & Tschulik, A. (2020). Green finance – Opportunities for the Austrian financial sector. *Financial Stability Report*, *40*, 45–61.

Campiglio, E. (2016). Beyond carbon pricing: The role of banking and monetary policy in financing the transition to a low-carbon economy. *Ecological Economics*, *121*, 220–230.

Destek, M.A., Hossain, M.R., & Khan, Z. (2024). Premature deindustrialization and environmental degradation. *Gondwana Research*, *127*, 199–210.

Dong, L., Zhang, X., & Chen, J. (2024). Does investor sentiment drive corporate green innovation: Evidence from China. *Sustainability*, *16*(8), 3220.

Dorfleitner, G., & Braun, D. (2019). Fintech, digitalization and blockchain: Possible applications for green finance. In *The rise of green finance in Europe: Opportunities and challenges for issuers, investors and marketplaces* (pp. 207–237). Cham: Springer International Publishing.

Falcone, P.M., & Sica, E. (2019). Assessing the opportunities and challenges of green finance in Italy: An analysis of the biomass production sector. *Sustainability*, *11*(2), 517.

Gilchrist, D., Yu, J., & Zhong, R. (2021). The limits of green finance: A survey of literature in the context of green bonds and green loans. *Sustainability*, *13*(2), 478.

Horsfall, M., Eikelenboom, M., Draisma, S., & Smit, J.H. (2021). The effect of rapport on data quality in face-to-face interviews: Beneficial or detrimental? *International Journal of Environmental Research and Public Health*, *18*(20), 10858.

Hossain, M.R., Rao, A., Sharma, G.D., Dev, D., & Kharbanda, A. (2024). Empowering energy transition: Green innovation, digital finance, and the path to sustainable prosperity through green finance initiatives. *Energy Economics*, *136*, 107736.

Hsu, C.-C., Tan, K.-C., & Mohamad Zailani, S.H. (2016). Strategic orientations, sustainable supply chain initiatives, and reverse logistics: Empirical evidence from an emerging market. *International Journal of Operations & Production Management*, *36*(1), 86–110.

Huang, Y., Chen, C., Lei, L., & Zhang, Y. (2022). Impacts of green finance on green innovation: A spatial and nonlinear perspective. *Journal of Cleaner Production*, *365*, 132548.

Ikram, M., & Sadki, J.E. (2024). Resilient and sustainable green technology strategies: A study of Morocco's path toward sustainable development. *Sustainable Futures*, *8*, 100327.

International Finance Corporation (2017). *Green finance: A bottom-up approach to track existing flows*. Washington, DC: International Finance Corporation.

Jain, N. (2021). Survey versus interviews: Comparing data collection tools for exploratory research. *The Qualitative Report*, *26*(2), 541–554.

Javadi, M., & Zarea, K. (2016). Understanding thematic analysis and its pitfall. *Journal of Client Care, 1*(1), 33–39.

Jawadi, F., Pondie, T.M., & Cheffou, A.I. (2025). New challenges for green finance and sustainable industrialization in developing countries: A panel data analysis. *Energy Economics, 142*, 108120.

Liu, H., Yao, P., Latif, S., Aslam, S., & Iqbal, N. (2022). Impact of green financing, FinTech, and financial inclusion on energy efficiency. *Environmental Science and Pollution Research, 29*(13), 18955–18966.

Otek Ntsama, U.Y., Yan, C., Nasiri, A., & Mbouombouo Mboungam, A.H. (2021). Green bonds issuance: Insights in low- and middle-income countries. *International Journal of Corporate Social Responsibility, 6*, 1–9.

Ozili, P.K. (2022). Green finance research around the world: A review of literature. *International Journal of Green Economics, 16*(1), 56–75.

Rennings, K. (2000). Redefining innovation—Eco-innovation research and the contribution from ecological economics. *Ecological Economics, 32*(2), 319–332.

Roper, S., & Tapinos, E. (2016). Taking risks in the face of uncertainty: An exploratory analysis of green innovation. *Technological Forecasting and Social Change, 112*, 357–363.

Simons, H. (2009). *Case study research in practice*. Sage Publications.

Singh, S.K., Del Giudice, M., Chierici, R., & Graziano, D. (2020). Green innovation and environmental performance: The role of green transformational leadership and green human resource management. *Technological Forecasting and Social Change, 150*, 119762.

Soundarrajan, P., & Vivek, N. (2016). Green finance for sustainable green economic growth in India. *Agricultural Economics/Zemědělská Ekonomika, 62*(1), 35–44.

Taghizadeh-Hesary, F., Zakari, A., Alvarado, R., & Tawiah, V. (2022). The green bond market and its use for energy efficiency finance in Africa. *China Finance Review International, 12*(2), 241–260.

Vaka, M., Walvekar, R., Rasheed, A.K., & Khalid, M. (2020). A review on Malaysia's solar energy pathway towards carbon-neutral Malaysia beyond Covid'19 pandemic. *Journal of Cleaner Production, 273*, 122834.

Wang, W., & Zhang, Q. (2022). Does green finance reform promote corporate green innovation? Evidence from a quasi-natural experiment. *Mathematical Problems in Engineering, 2022*(1), 7503917.

Wang, Y., & Zhi, Q. (2016). The role of green finance in environmental protection: Two aspects of market mechanism and policies. *Energy Procedia, 104*, 311–316.

Wang, Q.-J., Wang, H.-J., & Chang, C.-P. (2022). Environmental performance, green finance and green innovation: What's the long-run relationships among variables? *Energy Economics, 110*, 106004.

Yang, Y.C., Nie, P.Y., & Huang, J.B. (2020). The optimal strategies for clean technology to advance green transition. *Science of the Total Environment, 716*, 134439.

Yin, R.K. (2013). *Case study research: Design and methods*. Sage Publications.

Yu, C.-H., Wu, X., Zhang, D., Chen, S., & Zhao, J. (2021). Demand for green finance: Resolving financing constraints on green innovation in China. *Energy Policy, 153*, 112255.

Zhang, D., Zhang, Z., & Managi, S. (2019). A bibliometric analysis on green finance: Current status, development, and future directions. *Finance Research Letters, 29*, 425–430.

15

Green Innovations in African Health Supply Chains: Case Studies and Policy Implications

Julius Mugwagwa[1] *and Cynthia Siona Ndeh*[2]

[1] *Department of Science, Technology, Engineering and Public Policy (STEaPP), University College London, London, UK*
[2] *London Graduate School, London, UK*

15.1 Introduction

The healthcare systems in Africa are confronted with a multitude of challenges that hinder their effectiveness, including logistical inefficiencies, poor infrastructure, limited resources, and a rapidly growing population. These challenges are compounded by environmental concerns as the health sector consumes vast amounts of energy, produces significant waste, and contributes to carbon emissions through transportation and the disposal of medical equipment. Addressing these environmental impacts through **green innovations**—defined as practices, technologies, and solutions that reduce environmental harm while improving efficiency—has become essential to both enhance healthcare delivery and align with global sustainability goals.

Health supply chains—which include the procurement, storage, transportation, and distribution of medical goods, equipment, and supplies—are often a significant source of environmental degradation. From the energy-intensive cold storage of vaccines to the transportation of medical goods using fossil-fuel-powered vehicles, the environmental footprint of these processes is considerable. However, there is a growing movement toward integrating green innovations into these supply chains not only to address environmental concerns but also to improve the efficiency and resilience of healthcare systems.

Green Innovations in Supply Chain Management: Case Studies and Applications,
First Edition. Edited by Marcia Mkansi, Godfrey Mugurusi, Aaron L. Nsakanda,
Temidayo Akenroye, and Frank Tietze.

Published 2026 by John Wiley & Sons, Inc.

One of the most pressing environmental issues related to health supply chains in Africa is the management of e-waste. E-waste, or electronic waste, includes discarded electronic devices such as computers, medical equipment, mobile phones, and other healthcare technologies that have reached the end of their life cycle. As Africa's health systems increasingly rely on digital tools for managing patient data, telemedicine, and diagnostic technologies, the disposal of outdated or nonfunctional electronic devices has become a significant issue. Improper e-waste management, such as the dumping or incineration of these devices, can lead to the release of hazardous chemicals, which can severely impact both human health and the environment.

One notable initiative addressing e-waste in Africa is eWaste Africa, a leader in the responsible management and recycling of electronic waste across the continent. Based in Ghana, eWaste Africa has developed a model for collecting, refurbishing, and recycling electronic waste from healthcare facilities and beyond. Their initiative not only reduces the harmful environmental impact of e-waste but also promotes the reuse of valuable materials, creating economic opportunities and fostering a circular economy. This initiative has had a direct impact on improving the sustainability of healthcare operations, as obsolete medical equipment, such as old diagnostic machines and hospital computers, are recycled or refurbished for reuse. According to the United Nations Environment Programme (UNEP), e-waste recycling programs like eWaste Africa play a critical role in reducing environmental pollution and conserving resources (UNEP, 2020).

In addition to eWaste Africa, several South African case studies highlight the growing integration of green innovations in health supply chains. For example, South Africa has introduced solar-powered vaccine refrigerators in rural areas, where unreliable electricity grids make it challenging to maintain the cold chain for vaccines. These solar-powered refrigerators provide a sustainable solution to ensure that vaccines remain at the correct temperature, even in off-grid locations. By using renewable energy, these refrigerators reduce dependency on fossil fuels and contribute to decreasing greenhouse gas emissions. According to the South African Department of Health, the introduction of solar-powered refrigeration has greatly improved vaccine storage in remote areas, ensuring better public health outcomes and reducing waste caused by vaccine spoilage (SA Department of Health, 2021).

Another example is the adoption of electric and hybrid vehicles for pharmaceutical distribution. South African pharmaceutical companies have begun transitioning to hybrid and electric vehicles to transport medical goods across the country. This move is a part of broader efforts to reduce carbon emissions, decrease fuel dependency, and minimize the environmental impact of logistic operations. In cities such as Cape Town, hybrid vehicles are already being used in the transportation of critical medical supplies, helping distributors achieve their sustainability goals while also reducing transportation costs over time. According to a 2022 report by GreenCape, an organization focused on supporting sustainable business practices in South Africa, the use of electric vehicles in the medical supply chain could reduce operational costs and emissions by up to 30% (GreenCape, 2022a).

In addition to transportation, waste management is another key area where South African healthcare institutions have made strides. Hospitals and clinics across the country are increasingly adopting sustainable waste management systems that emphasize the reuse and recycling of medical supplies and packaging. For instance, biodegradable packaging is now being used for pharmaceutical products, and some hospitals have introduced initiatives to sterilize and reuse medical instruments rather than discarding them after a single use. These initiatives not only reduce the volume of medical waste but also contribute to the development of a circular economy within the healthcare sector.

Despite these promising initiatives, the widespread adoption of green innovations in African health supply chains remains limited due to several barriers. Financial constraints, inadequate infrastructure, lack of technical expertise, and regulatory gaps continue to hinder the adoption of green technologies. The initial capital required to implement solar-powered systems or purchase electric vehicles is often prohibitive for healthcare facilities in low-resource settings. Furthermore, many African countries still lack the regulatory frameworks necessary to support sustainable practices in the health sector, making it difficult for organizations to invest in long-term solutions.

This research aims to explore the role of green innovations in improving the efficiency, sustainability, and resilience of African health supply chains. By examining key case studies from across the continent, including the e-waste management initiatives of eWaste Africa and South Africa's use of solar-powered refrigeration and green logistics, this paper will provide insight into the practical applications of green technologies in healthcare. The research will also address the policy challenges and propose recommendations for scaling up green innovations in African health systems, with the goal of fostering a more sustainable and resilient healthcare environment for the future.

15.2 Green Innovations in Health Supply Chains: A Global Overview

Green innovations in health supply chains are increasingly viewed as essential to improving both the sustainability and efficiency of healthcare systems worldwide. As healthcare demand continues to rise globally, so does the strain on healthcare systems to balance operational effectiveness with environmental responsibility. The healthcare sector, particularly in low-resource settings like Africa, faces significant challenges such as waste management, inefficient transportation, and energy consumption. Therefore, integrating green innovations—technologies and practices designed to minimize environmental impacts while improving operational outcomes—becomes critical in transforming healthcare supply chains into more sustainable and resilient systems.

15.2.1 Defining Green Innovations in Healthcare

Green innovations in healthcare supply chains cover a wide array of technologies, practices, and strategies aimed at reducing environmental impacts, improving efficiency, and promoting sustainability. The primary components include:

1) **Energy efficiency in healthcare operations**: This includes the adoption of energy-efficient technologies such as light-emitting diode (LED) lighting, energy-efficient air conditioning and heating systems, and automated energy management systems in healthcare facilities. Moreover, transitioning to renewable energy sources, such as solar or wind power, is a growing trend in both medical facilities and supply chains.
   - **Example**: The Mara Group in Rwanda has implemented a solar-powered health center in rural areas where traditional energy infrastructures are unreliable (Mara Group, 2021). Such initiatives ensure sustainable healthcare delivery by reducing reliance on fossil fuels.
2) **Sustainable healthcare logistics**: The movement of medical supplies often relies on vehicles powered by fossil fuels, contributing to carbon emissions and fuel consumption. A shift toward electric and hybrid vehicles in healthcare logistics reduces the carbon footprint associated with the transportation of essential medicines, vaccines, and equipment.
   - **Example**: In South Africa, the pharmaceutical company Medlog has piloted the use of electric vehicles for medical supply distribution in urban centers, thereby reducing emissions and operational costs (GreenCape, 2022b).
3) **Waste management and e-waste recycling**: Healthcare systems generate substantial waste, including paper, plastics, and, notably, electronic waste (e-waste) from outdated medical devices, diagnostic equipment, and IT infrastructure. Proper waste disposal and recycling of such materials are essential to reduce their environmental impact.
   - **Example**: **eWaste Africa** has emerged as a leading example of e-waste recycling, collecting old medical devices from healthcare providers in Ghana, refurbishing them, and safely recycling hazardous components. The initiative not only reduces harmful environmental impacts but also provides valuable materials for new products (eWaste Africa, 2020).
4) **Sustainable procurement**: This involves the sourcing of medical products and equipment that are produced sustainably, with attention to minimizing

environmental harm. Examples include sourcing products with minimal packaging, using biodegradable or recyclable materials, and choosing products with long lifespans.

- **Example**: The World Health Organization (WHO) and other international organizations have developed sustainable procurement guidelines for health products to support countries in making environmentally friendly purchasing decisions (WHO, 2021).

15.2.2 Theoretical Frameworks for Green Innovation Adoption

Adopting green innovations in health supply chains involves overcoming a series of challenges, including financial constraints, infrastructure limitations, and regulatory issues. Several theoretical frameworks help explain why and how green innovations are adopted and integrated into healthcare systems, particularly in resource-constrained environments.

1) **Diffusion of innovations theory** (Rogers, 2003)
   - **Rogers' diffusion of innovations** theory provides a framework for understanding how innovations spread through populations over time. Innovations pass through five stages: knowledge, persuasion, decision, implementation, and confirmation. In the healthcare context, this theory helps explain how green innovations, such as solar-powered medical devices or green logistics solutions, first undergo trials in pilot programs, and then gradually spread as their effectiveness and feasibility are demonstrated. The process is influenced by factors such as perceived relative advantage, compatibility with existing practices, and the complexity of the innovation.
   - **Example**: The introduction of solar-powered refrigeration for vaccines in African rural areas is a case where early adopters (e.g., healthcare NGOs) tested these technologies before widespread implementation across the continent, supported by evidence of their effectiveness in preserving vaccines without electricity (UNICEF, 2020a).
2) **Environmental innovation theory**
   - Environmental innovation theory explores how organizations develop innovations that help reduce environmental impact while increasing profitability or operational efficiency. In the context of healthcare, green innovations can often lead to win-win scenarios, where both environmental sustainability and operational efficiency are achieved. For instance, transitioning to energy-efficient medical equipment or using biodegradable packaging can help healthcare providers reduce operational costs while contributing to environmental conservation.

- **Example**: The Cleveland Clinic in the United States adopted an environmental innovation strategy that reduced waste by over 50% and saved millions of dollars annually. This green strategy involved the replacement of traditional packaging with reusable and recyclable materials, significantly cutting down on disposable items in healthcare settings (Green Healthcare Partnership, 2019).

3) **Resource-based view (RBV)**
   - The RBV suggests that an organization's ability to achieve a competitive advantage is based on its access to and management of valuable, rare, inimitable, and non-substitutable resources. For healthcare systems, green innovations such as renewable energy systems, sustainable waste management technologies, and electric vehicle fleets can be viewed as valuable resources that improve both sustainability and operational performance. The RBV highlights how these innovations can help healthcare organizations differentiate themselves and offer better services.
   - **Example**: By investing in solar-powered cold storage units for vaccines, healthcare systems gain a competitive edge in regions with unreliable electricity, enhancing the system's resilience and ensuring the continued delivery of healthcare services even during power outages.
4) **Stakeholder theory** (Freeman, 1984)
   - Stakeholder theory argues that businesses (and organizations) should consider the interests and welfare of all stakeholders in decision-making processes. In healthcare, this includes government agencies, healthcare providers, patients, suppliers, and environmental organizations. The adoption of green innovations requires collaboration across various stakeholders to address shared challenges, such as e-waste management or the carbon footprint of logistics.
   - **Example**: The eWaste Africa initiative demonstrates how healthcare providers, waste management organizations, and environmental groups can collaborate to create sustainable e-waste management practices. This collaboration ensures that outdated medical equipment is safely disposed of or refurbished, benefiting the environment and society while also creating new economic opportunities.

15.2.3 Global Best Practices in Green Innovations for Healthcare Supply Chains

Across the globe, numerous healthcare systems and organizations are implementing green innovations to reduce their environmental footprint and improve efficiency. These examples offer valuable insights into how healthcare systems can balance sustainability with the practical demands of service delivery.

- **The United States**: The Cleveland Clinic has made significant strides in energy conservation, waste reduction, and sustainable sourcing. The facility has reduced its carbon footprint by investing in green building technologies and waste recycling programs. It has also implemented a comprehensive energy efficiency program that includes the use of solar panels and energy-efficient HVAC (heating, ventilation, and air conditioning) systems to power its healthcare facilities (Green Healthcare Partnership, 2019).

- **The United Kingdom**: The National Health Service (NHS) has set ambitious targets for achieving net-zero carbon emissions by 2045. This includes retrofitting buildings for energy efficiency, using sustainable procurement practices for medical supplies, and reducing waste. NHS hospitals have implemented waste segregation systems to recycle medical packaging and disposable medical instruments (NHS Sustainability, 2021).
- **India**: In India, the Health Ministry is advancing solar-powered cold chains to address the challenges of vaccine storage in remote areas with unreliable power grids. This initiative ensures that vaccines are stored under optimal conditions and reduces dependence on electricity from conventional power sources (UNICEF, 2020b).

15.2.4 Challenges in Implementing Green Innovations in African Health Supply Chains

While green innovations hold tremendous potential for transforming healthcare systems in Africa, several barriers hinder their widespread adoption:

- **Financial barriers**: The initial investment required for green technologies, such as solar energy systems, electric vehicles, and waste management infrastructure, is often high, making it difficult for resource-constrained healthcare systems to implement these solutions without external funding or partnerships.
- **Infrastructure constraints**: In many African countries, the infrastructure required to support green innovations (such as electricity grids for solar energy or reliable roads for electric vehicle transportation) is lacking, making it harder to scale up these technologies.
- **Regulatory challenges**: Weak regulatory frameworks and a lack of incentives for environmental sustainability in healthcare mean that green innovations are often under-supported by governments, hindering their growth and scalability.

Nevertheless, emerging public–private partnerships, international funding for sustainable healthcare practices, and the growth of local green technologies present opportunities for overcoming these barriers.

15.3 Case Studies of Green Innovations in African Health Supply Chains

This section delves into case studies that demonstrate the adoption of green innovations in African health supply chains. These innovations are not only addressing environmental sustainability but also enhancing healthcare delivery, especially in resource-constrained settings. The examples discussed highlight the ways in which various stakeholders, including governments, international organizations, private companies, and local innovators, are transforming health systems with a focus on reducing waste, improving energy efficiency, and utilizing renewable resources.

15.3.1 E-Waste Management in Healthcare: The eWaste Africa Initiative

The rapid technological advancement in medical devices and information technology (IT) has resulted in an increase in e-waste generation across healthcare facilities in Africa. This poses serious environmental and health risks due to the hazardous materials contained in e-waste, such as mercury, lead, and cadmium, which can leach into the environment if not properly managed.

eWaste Africa offers a key example of how healthcare e-waste can be managed sustainably. Headquartered in Ghana, eWaste Africa collects obsolete and broken-down medical equipment from healthcare facilities across West Africa, ensuring that they are refurbished, resold, or recycled in an environmentally responsible way. The organization's e-waste recycling process helps mitigate the harmful impact of discarded electronic waste while providing economic benefits through the resale of refurbished devices.

- **Example**: Ghana Health Service collaborated with eWaste Africa to pilot an e-waste recycling program that targets old medical devices such as diagnostic machines, blood pressure monitors, and medical computers. By refurbishing and recycling parts, this initiative prevents toxic chemicals from leaking into the soil and water, while creating job opportunities in the local economy (eWaste Africa, 2020).
- **Example**: In Nigeria, Health-E-Waste Nigeria, a non-governmental organization, works with hospitals and health centers to collect and safely dispose of e-waste. Their efforts are focused on educating healthcare providers about the dangers of improper disposal and providing safe, sustainable disposal solutions, thus reducing environmental pollution and fostering responsible recycling practices (Health-E-Waste Nigeria, 2021).

15.3.2 Solar-Powered Health Clinics in Rural Areas

In many African countries, access to reliable electricity remains a challenge, particularly in remote rural areas. This lack of power affects the ability of healthcare providers to offer consistent services, including the refrigeration of vaccines, operation of medical equipment, and general facility lighting.

The integration of solar power into health clinics has proven to be a sustainable solution that directly addresses these challenges.

The Mara Group's solar-powered health clinics in Rwanda provide a model for how solar energy can support healthcare delivery in underserved areas. By using solar energy to power medical equipment, lighting, and refrigeration, these clinics ensure that essential healthcare services continue uninterrupted, even in the absence of a stable grid supply.

- **Example**: In Rwanda, the Mara Group has partnered with local communities and international organizations to establish solar-powered health centers. These clinics are equipped with solar-powered cold storage units for vaccines and medical supplies. The renewable energy solution has been instrumental in preventing vaccine spoilage in areas where the electricity grid is unreliable, ensuring that immunization campaigns are not hindered by power outages (Mara Group, 2021).

In Kenya, the Kenya Red Cross has also implemented solar-powered healthcare solutions, particularly for mobile clinics that serve rural and nomadic communities. These solar units allow healthcare workers to provide care in areas that previously lacked electricity, powering equipment such as diagnostic tools and mobile health technology (Kenya Red Cross, 2020).

15.3.3 Sustainable Logistics and Transportation for Health Supplies

The transportation of medical supplies, particularly in rural and remote regions, often requires fuel-intensive vehicles that contribute to carbon emissions and environmental degradation. Sustainable logistics practices are being adopted in various African countries to address these challenges, including the use of electric vehicles (EVs), hybrid vehicles, and more fuel-efficient transport options for the distribution of medical supplies.

Medlog, a logistics company in South Africa, has integrated EVs into its fleet for transporting medical goods within urban centers. This initiative not only reduces greenhouse gas emissions but also cuts fuel costs, contributing to a more sustainable healthcare logistics network.

- **Example**: Medlog's EV fleet has been instrumental in distributing vaccines, medicines, and other critical medical supplies in Johannesburg and Pretoria. The move toward EVs helps reduce the environmental footprint of healthcare logistics, aligning with South Africa's broader sustainability goals. Medlog is also exploring the use of electric motorcycles and drones for last-mile deliveries to further reduce emissions and improve efficiency (GreenCape, 2022b).

Additionally, UPS Healthcare, in partnership with local logistics providers, has launched green delivery solutions across several African countries. Using hybrid and EVs, UPS has optimized supply routes for medical deliveries, reducing both transportation costs and carbon emissions in countries like Kenya and Nigeria (UPS Healthcare, 2021).

15.3.4 Eco-friendly Packaging for Medical Supplies

The packaging of medical supplies, including medicines and medical devices, has a significant environmental impact. Traditional packaging materials, such as plastic and Styrofoam, are not only non-biodegradable but also contribute to excessive waste. In response to this, a growing number of healthcare organizations in Africa are exploring eco-friendly packaging alternatives.

In Kenya, the Kenya Medical Supplies Authority (KEMSA) has spearheaded the transition to biodegradable and recyclable packaging for medical products. KEMSA's efforts have focused on replacing single-use plastics with sustainable alternatives made from plant-based materials that are both eco-friendly and functional for storing medical goods.

- **Example**: KEMSA has worked with local manufacturers to produce eco-friendly packaging that reduces the environmental footprint of healthcare logistics. The use of biodegradable and recyclable materials for drug packaging not only reduces waste but also aligns with national and international sustainability efforts. This shift is also supported by Kenya's ongoing efforts to ban single-use plastics (KEMSA, 2021).

In South Africa, the Green Packaging Initiative is another project that aims to minimize waste generated by medical packaging. The initiative works with local healthcare facilities and suppliers to ensure that medical products are packaged in reusable and recyclable containers. The initiative has contributed to reducing plastic waste in the healthcare sector, especially in urban hospitals.

15.3.5 Renewable Energy for Vaccine Refrigeration

Proper vaccine storage is a critical challenge in many African countries, particularly in remote regions where power outages are frequent and access to reliable electricity is limited. Solar-powered refrigeration units have emerged as an effective solution for ensuring that vaccines are stored at the proper temperature, reducing waste and ensuring the safety and efficacy of vaccines.

UNICEF has been at the forefront of deploying solar-powered refrigerators to African countries, ensuring that vaccines are stored correctly in places with unreliable electricity. This initiative has improved vaccination coverage, particularly in rural and hard-to-reach areas, and has reduced the need for traditional, energy-hungry refrigeration methods.

- **Example**: UNICEF's solar-powered refrigerators in Ethiopia and Uganda have been crucial in expanding immunization efforts. These solar units ensure that vaccines remain viable even in areas that have no access to the national power grid. The use of solar energy for vaccine storage also contributes to sustainability by reducing the carbon footprint associated with conventional refrigeration systems (UNICEF, 2020b).

In Nigeria, the Bill & Melinda Gates Foundation is working with local governments to install solar-powered cold chain systems for vaccines and medical supplies in remote regions. This partnership has expanded vaccination programs, especially in the northern regions of the country, where power shortages are common (Bill & Melinda Gates Foundation, 2021).

15.4 Policy Implications for Green Innovations in African Health Supply Chains

The successful implementation and scaling of green innovations in African health supply chains require supportive policy frameworks that address both environmental sustainability and healthcare system strengthening. Governments, international organizations, and healthcare stakeholders must work together to create an enabling environment that promotes the adoption of green technologies. In the context of Africa, where healthcare systems often face significant challenges, green innovations are not only an environmental priority but also a means of improving operational efficiency, cost-effectiveness, and service delivery. This section outlines the policy implications for fostering these green innovations, focusing on e-waste management, renewable energy integration, green logistics, and sustainable healthcare supply chains.

15.4.1 E-Waste Management Policies

As discussed in earlier sections, e-waste management is a critical issue in African healthcare systems, particularly with the rapid increase in the use of electronic devices in health facilities. The lack of formal e-waste recycling infrastructure and policies poses significant environmental and health risks, as improper disposal leads to hazardous chemicals leaching into the environment. To address this, governments and policymakers must put in place comprehensive e-waste management policies that ensure proper disposal, recycling, and reuse of obsolete medical equipment and electronics.

For instance, Ghana, which is home to eWaste Africa, has developed e-waste management policies that mandate the safe recycling and disposal of e-waste. However, despite these policies, enforcement remains a challenge, and more work is needed to ensure that healthcare facilities across the country comply with these regulations. A key policy recommendation is the integration of extended producer responsibility (EPR) frameworks, where manufacturers of medical devices and electronics are held accountable for the lifecycle management of their products, including end-of-life disposal and recycling.

- **Policy example**: In Ghana, the Environmental Protection Agency (EPA) enforces regulations that require healthcare facilities to dispose of e-waste responsibly. However, challenges such as a lack of collection infrastructure and awareness remain. Expanding partnerships with private e-waste recyclers and establishing a national e-waste fund could incentivize better management practices. Additionally, policies that encourage the refurbishment and reuse of medical devices could further minimize e-waste generation.
- **Policy recommendation**: Governments should incentivize the creation of e-waste recycling infrastructures within health facilities and collaborate with private recyclers like eWaste Africa to establish collection centers and environmentally safe disposal routes. Public–private partnerships can help create a sustainable framework for managing e-waste in Africa's healthcare sector.

15.4.2 Renewable Energy Integration Policies

The integration of renewable energy, particularly solar power, into healthcare facilities is one of the most effective ways to address energy access and sustainability in the healthcare sector. In regions with unreliable or nonexistent power grids, solar energy can provide a reliable source of electricity for essential services such as vaccine refrigeration, diagnostic equipment, and basic healthcare needs.

Countries like Kenya, Uganda, and Rwanda have already taken significant steps in integrating solar power into their health systems. However, the scaling of these efforts requires clear policies that encourage the adoption of renewable energy in healthcare infrastructure. Policies should include financial incentives for solar installations, tax breaks for renewable energy companies, and subsidies for solar-powered medical equipment.

- **Policy example**: The Kenya Solar Energy Association (KSEA) has been instrumental in promoting solar energy adoption within healthcare. The Kenyan government's Energy Act of 2019 supports renewable energy projects by offering tax incentives and subsidies. Solar-powered vaccine refrigerators and solar clinics have been deployed to remote areas, providing healthcare where grid power is unavailable (KSEA, 2020).

- **Policy recommendation**: African governments should implement feed-in tariffs and other financial mechanisms to support private sector investments in renewable energy for healthcare. Additionally, policies should be expanded to include training for healthcare workers on the maintenance of solar-powered equipment, ensuring the sustainability of these green innovations.

15.4.3 Sustainable Procurement and Green Logistics Policies

Efficient and sustainable logistics are critical to ensuring that medical supplies are delivered in an environmentally friendly manner, minimizing carbon emissions and reducing the sector's overall environmental footprint. Policymakers must adopt sustainable procurement policies that prioritize environmentally friendly logistics solutions and encourage the adoption of green transportation methods.

For instance, in South Africa, the government has started integrating green logistics into public procurement by including environmental criteria in the procurement process for medical supplies and transport services. These green logistics initiatives include the use of EVs for delivery, fuel-efficient trucks, and carbon offset programs. South Africa's efforts to reduce emissions from logistics could serve as a model for other African nations.

- **Policy example**: The South African Department of Environmental Affairs (DEA) has introduced the Green Transport Strategy, which includes incentives for logistics companies that adopt sustainable practices. As part of the strategy, Medlog, a logistics company in South Africa, has integrated electric vehicles into its fleet for transporting medical supplies (GreenCape, 2022a).
- **Policy Recommendation**: Policymakers should encourage the use of electric vehicles and other low-carbon transportation options in healthcare logistics. Public–private partnerships can help subsidize the initial costs of adopting electric vehicles, ensuring that these innovations are accessible to healthcare providers across the continent. Additionally, the introduction of carbon pricing and tax incentives for green transport technologies would further encourage healthcare providers to prioritize sustainability.

15.4.4 E-Waste Education and Capacity Building

Effective e-waste management also requires significant capacity building at the national and local levels. Healthcare workers, IT staff, and waste management personnel need to be trained on the risks of improper e-waste disposal and the importance of safe recycling practices. This education can be provided through targeted training programs, public awareness campaigns, and partnerships with local universities.

- **Policy example**: The WHO and UNEP have collaborated with African governments to offer training workshops for healthcare workers on e-waste management. These initiatives are aimed at improving the knowledge and skills of healthcare staff on how to handle and dispose of e-waste responsibly (WHO & UNEP, 2020).
- **Policy recommendation**: Governments should invest in national awareness campaigns and training programs to increase the understanding of e-waste management among healthcare staff. Furthermore, policymakers should incorporate e-waste management into healthcare curricula in universities and vocational institutions, ensuring that future healthcare professionals are equipped to manage e-waste effectively.

15.4.5 Financing Green Innovations

One of the critical barriers to the adoption of green innovations in African health supply chains is the high upfront cost of technologies like solar-powered clinics, electric vehicles, and green logistics systems. To overcome this, governments and development partners must develop financing mechanisms that support the upfront investment in green healthcare technologies.

- **Policy example**: In Kenya, the government has worked with the World Bank to launch green bonds that finance sustainable projects, including those in the health sector. These bonds allow healthcare providers to access capital for purchasing solar-powered equipment and electric vehicles for logistics (World Bank, 2020).
- **Policy recommendation**: African governments should create green finance schemes that provide low-interest loans, subsidies, and tax relief to healthcare providers that adopt environmentally friendly technologies. Furthermore, climate adaptation funds could be allocated to healthcare projects that address both climate change and public health goals.

Green innovations are thus increasingly becoming a cornerstone of sustainable health supply chains, particularly in addressing environmental, economic, and health challenges. Table 15.1 summarizes key policy areas critical to fostering green practices in healthcare supply chains, highlighting their impact and offering actionable recommendations for implementation.

Table 15.1 emphasizes the need for effective e-waste management, integration of renewable energy, sustainable procurement, education on e-waste, and financing green innovations to create environmentally friendly health supply chains.

Table 15.1 Summary table of key policy areas impacting green health supply chains

Policy area	Impact on green health supply chains	Key recommendations
E-Waste management	Effective e-waste management reduces environmental and health risks from hazardous materials in medical devices.	Implement EPR frameworks. Encourage partnerships with e-waste recyclers.
Renewable energy integration	Solar energy ensures reliable power for health facilities in off-grid areas, supporting operations like vaccine storage and medical equipment use.	Provide tax incentives and subsidies for solar adoption. Establish training for solar-powered system maintenance.
Sustainable procurement and logistics	Green logistics and sustainable procurement reduce carbon emissions and transportation costs in healthcare supply chains.	Encourage adoption of electric vehicles and green transport solutions. Introduce carbon pricing and tax incentives.
E-Waste education and capacity building	Capacity building enhances e-waste management in healthcare, increasing awareness of risks and proper disposal techniques.	- Invest in training programs for healthcare workers. - Integrate e-waste management into healthcare curricula.
Financing green innovations	Financial barriers hinder the widespread adoption of green technologies in healthcare. Financing mechanisms can help overcome this.	- Develop green finance schemes, including green bonds and low-interest loans. - Allocate climate adaptation funds for healthcare projects.

Source: Table developed by authors drawing from, among others eWaste Africa (2022), NHS Sustainability (2021), and UNICEF (2020b).

Key recommendations include adopting EPR frameworks, incentivizing renewable energy use, promoting electric vehicle adoption, and investing in education and green financing mechanisms. Together, these strategies aim to reduce environmental risks, enhance operational efficiency, and ensure long-term sustainability in healthcare systems.

15.5 Challenges and Barriers to Implementing Green Innovations in African Health Supply Chains

The integration of green innovations in African health supply chains is not only an environmental necessity but also a socio-economic opportunity. However, as identified through various theoretical lenses, the adoption of these innovations faces multiple barriers. Drawing from Institutional Theory, Diffusion of Innovations Theory, and Systems Theory, we can better understand how these barriers manifest and how they interact to impede the widespread uptake of green technologies in the health sector. Below, we further explore these challenges, referencing the theoretical underpinnings and previous findings, and discuss practical solutions for overcoming these obstacles.

15.5.1 Financial Barriers

The financial barriers discussed earlier can be analyzed through the lens of Institutional Theory, which emphasizes the role of both formal and informal institutions in shaping behaviors and decisions. In the context of green innovations, the institutional framework for healthcare financing in Africa remains underdeveloped, and financial structures do not yet prioritize long-term sustainability over short-term expenditures.

- **Link to theory**: According to the Institutional Theory, the financial challenges in African healthcare systems are rooted in the path dependency of existing institutions, which are often focused on immediate needs (such as urgent healthcare delivery) rather than long-term sustainability goals like environmental resilience (North, 1990). This institutional inertia makes it difficult for healthcare providers to allocate funding to green technologies despite their potential long-term cost savings.
- **Example**: The case of Nigeria illustrates this path dependency. Despite the availability of solar power as a long-term solution, the immediate financial needs of health facilities, driven by existing institutional priorities, often sideline the adoption of green innovations, particularly when these require large initial investments.

Policy Recommendation: To address these financial barriers, African governments should encourage financial reforms within the health sector by aligning healthcare financing mechanisms with sustainable development goals. This could include offering green bonds for healthcare infrastructure or subsidized loans that cover the upfront costs of renewable energy technologies, thus creating a more supportive institutional environment for green investments.

15.5.2 Technical Barriers

The technical barriers to green innovations in African health supply chains can be understood through the Diffusion of Innovations Theory, which emphasizes how new technologies spread within a population and what factors influence their adoption. According to Rogers (2003), innovations that require significant changes in existing practices, such as renewable energy systems or advanced e-waste recycling technologies, face higher barriers to adoption due to their complexity and the uncertainty around their effectiveness.

- **Link to theory**: The complexity of green technologies, a central factor identified in the Diffusion of Innovations Theory, presents a major challenge for health supply chains in Africa. Healthcare workers may lack confidence in using or maintaining renewable energy systems or e-waste management equipment due to limited exposure to these innovations. This can lead to resistance to change, which slows down the diffusion process of green technologies across the healthcare sector.
- **Example**: As noted in Ghana with the eWaste Africa initiative, while significant strides have been made in recycling electronic waste, the need for highly specialized skills to handle medical e-waste means that only a limited number of technicians can maintain such systems. This issue complicates the diffusion of e-waste management innovations, particularly in rural areas (eWaste Africa, 2022).

Policy recommendation: The adoption of green technologies in healthcare requires not only the introduction of new technologies but also the development of comprehensive technical training programs. These programs should be designed to build local expertise and support healthcare workers in the operation, maintenance, and troubleshooting of green technologies. By addressing the perceived complexity of innovations, the rate of adoption can be accelerated.

15.5.3 Institutional and Policy Challenges

Institutional and policy challenges remain a critical barrier to the implementation of green innovations, and these challenges are strongly informed by the findings of the Systems Theory, which emphasizes the interconnectedness and interdependence of various subsystems within a larger system. Within the healthcare system, the fragmented approach to policy implementation often results in ineffective support for green innovations. The lack of integrated policy frameworks that address both environmental sustainability and healthcare delivery can create systemic inefficiencies.

- **Link to theory**: According to the Systems Theory, the healthcare sector cannot achieve a meaningful progress in integrating green innovations unless policies are aligned across multiple subsystems—healthcare delivery, environmental protection, and sustainable development. The failure to recognize the interdependencies between these subsystems often leads to fragmented efforts, where the objectives of one subsystem (e.g., health service delivery) are prioritized over another (e.g., environmental sustainability).

- **Example**: In South Africa, although the government has passed policies promoting green energy, there is no unified strategy that integrates healthcare, energy policy, and sustainability. As a result, hospitals may invest in solar panels or renewable energy, but without adequate regulation or incentives, these efforts are isolated and do not reach their full potential (GreenCape, 2022a).

Policy recommendation: To overcome institutional fragmentation, cross-sectoral policy frameworks should be developed that encourage the integration of green innovations into all facets of healthcare delivery. For example, policies could mandate that health facilities adhere to environmental sustainability standards as a part of their operational requirements. Moreover, a coordinating body should be established to oversee the implementation of these policies, ensuring that both healthcare and environmental goals are met simultaneously.

15.5.4 Social and Cultural Barriers

Social and cultural barriers, including resistance to change and a lack of awareness, can also be examined using the Diffusion of Innovations Theory, which highlights the importance of social networks and the role of early adopters in promoting innovation. Social networks within healthcare facilities, as well as community engagement, can either facilitate or hinder the adoption of green innovations.

- **Link to theory**: The social system in which an innovation is introduced plays a critical role in its adoption. Opinion leaders within healthcare settings—such as senior medical staff or administrative leaders—can either encourage or obstruct the introduction of new technologies. Furthermore, peer influence within the healthcare community significantly impacts adoption rates (Rogers, 2003). Without the active involvement of respected individuals within the healthcare sector, the diffusion of green technologies is often slow.
- **Example**: In Kenya, the reluctance of healthcare workers to adopt solar-powered medical equipment stems from the lack of trust in the new technology and the absence of peer support. This is particularly evident in regions where medical professionals are used to conventional methods and are skeptical of the effectiveness of renewable energy solutions (Kenya Solar Energy Association, 2020).

Policy recommendation: Governments and healthcare organizations should invest in community-level awareness campaigns and create peer support networks to encourage the adoption of green innovations. Identifying and empowering early adopters in healthcare settings who can champion the benefits of sustainable technologies could also facilitate broader acceptance across the system.

15.5.5 Infrastructure Limitations

Infrastructure limitations can be framed using the concept of path dependence from Institutional Theory and Systems Theory, which argues that the existing infrastructure in African healthcare systems was designed without considering future sustainability. The lack of reliable electricity grids, poor transportation networks, and outdated waste management systems are all legacies of the past that constrain the ability of healthcare systems to adopt green technologies.

- **Link to theory**: The path dependency of the healthcare infrastructure results from historical decisions and investments that were not made with long-term sustainability in mind. Existing infrastructure is often inadequate to support the operational requirements of green technologies like solar power or electric vehicles, which create a barrier to their adoption (North, 1990). Additionally, infrastructure deficiencies are often compounded by feedback loops where an inadequate infrastructure leads to further deterioration of the system, making it harder to integrate green innovations.
- **Example**: In Mozambique, the lack of reliable electricity in rural health facilities means that the use of solar energy is not always feasible. Health facilities often operate on unreliable grids or generators, which significantly limit the potential for integrating renewable energy (World Bank, 2020).

Policy recommendation: A holistic approach is needed to address infrastructure challenges. Governments must prioritize infrastructure investment in rural and underserved areas, particularly focusing on electricity grids, roads, and waste management systems that can support the adoption of green technologies. Public–private partnerships can be instrumental in addressing these gaps, especially in sectors like energy and transportation.

The barriers to implementing green innovations in African health supply chains are deeply rooted in both institutional structures and systemic limitations. By drawing upon theoretical frameworks like Institutional Theory, Diffusion of Innovations Theory, and Systems Theory, we gain valuable insights into the challenges faced by the healthcare sector. Financial constraints, technical gaps, fragmented policy environments, cultural resistance, and infrastructure deficiencies all contribute to the slow adoption of green innovations. However, by aligning policies, investing in capacity building, and improving infrastructure, African countries can overcome these challenges and accelerate the transition to more sustainable health supply chains.

15.5.6 Successful Green Innovations in Health Supply Chains: Global Examples and Key Policy Lessons for Africa

Examining global examples of green innovations in health supply chains is crucial for identifying proven strategies that can address environmental and health challenges. By learning from the successes of other countries, African nations can adopt tailored solutions to enhance sustainability, improve healthcare delivery, and tackle pressing issues such as energy access, e-waste management, and carbon emissions. Table 15.2 provides insights from six countries—Germany, the United States, India, Brazil, Japan, and Denmark—highlighting their initiatives and key lessons African nations can adopt to advance sustainable healthcare systems.

The table showcases various strategies, including Germany's e-waste recycling systems, the United States' telemedicine infrastructure, India's solar-powered medical devices, and Brazil's green logistics. Additionally, Japan's regulatory frameworks for green supply chains and Denmark's energy-efficient hospital designs offer models for African countries to emulate. Key policy lessons for Africa include incentivizing renewable energy, developing national e-waste recycling laws, integrating sustainability into healthcare design, and adopting green procurement and logistics practices. These approaches can support Africa's transition to environmentally conscious healthcare systems. By examining the successful cases of green innovations in healthcare supply chains from countries around the world, African nations can draw valuable policy lessons. The key to success lies in creating coherent policies, incentivizing green technologies, and investing in sustainable infrastructure. Adopting these best practices will not only improve environmental sustainability but also strengthen the resilience and efficiency of healthcare systems across Africa, ensuring better health outcomes for future generations.

15.6 Conclusion and Recommendations

The implementation of green innovations in African health supply chains offers substantial opportunities for reducing environmental impacts, improving healthcare efficiency, and enhancing resilience. However, as this paper has shown, the adoption of these innovations is impeded by several barriers, including financial constraints, technical limitations, institutional fragmentation, social resistance, and infrastructure deficiencies. Through theoretical frameworks such as Institutional Theory, Diffusion of Innovations Theory, and Systems Theory, we have gained a deeper understanding of these challenges and how they intersect to hinder the adoption of green technologies in the healthcare sector.

This conclusion synthesizes the key findings of the paper and offers actionable recommendations for overcoming these barriers to create sustainable, green health supply chains in Africa. Additionally, we reflect on the key lessons from the COVID-19 pandemic that are relevant to the development of more resilient, green healthcare systems in Africa.

Table 15.2 Successful green innovations in health supply chains: global examples and key policy lessons for Africa

Country	Green innovation in health supply chains	Key policy lessons for Africa
Germany Hempel and Henze (2020).	Implementation of solar-powered health clinics in remote regions.	Strong policy support for renewable energy: Africa can adopt Germany's approach by incentivizing renewable energy solutions for healthcare facilities in rural areas.
Zeman (2021).	Advanced e-waste recycling systems for medical devices.	Integrated e-waste policies: African countries can develop regulations for proper disposal and recycling of healthcare electronics, modelled after Germany's circular economy initiatives.
	Use of sustainable procurement policies to reduce environmental impact of medical supplies.	Sustainable procurement frameworks: Africa can implement policies that prioritize green procurement in the healthcare sector, reducing waste and encouraging sustainability.
United States Gibson and Black (2020).	Widespread adoption of telemedicine, reducing the need for physical visits and cutting transportation emissions.	Telemedicine infrastructure: African countries can create supportive policies to develop telemedicine services, reducing emissions and increasing healthcare access.
	Green hospitals designed to meet LEED (Leadership in Energy and Environmental Design) standards.	Healthcare infrastructure design standards: African nations can adopt green building codes for hospitals and clinics, ensuring energy efficiency and sustainability.
India Gupta and Sharma (2019).	Implementation of solar-powered medical devices in rural healthcare centers.	Local solar energy solutions: African nations can prioritize solar-powered devices and infrastructure to overcome energy supply challenges, particularly in off-grid areas.
	Low-cost, eco-friendly medical waste management solutions.	Affordable waste management technologies: African countries can develop cost-effective waste management policies, especially for medical waste disposal and recycling.

(*Continued*)

Table 15.2 (Continued)

Country	Green innovation in health supply chains	Key policy lessons for Africa
Brazil Andrade and Silva (2020).	Development of green logistics in the health supply chain, reducing transportation emissions.	Green logistics policies: Africa can benefit from policies promoting green transportation, such as electric vehicles and sustainable supply chain management for medical goods.
	Focus on sustainable public health programs that incorporate environmental considerations in health planning.	Integration of environmental health policies: Africa can integrate sustainability into public health programs, addressing both health outcomes and environmental impacts.
Japan Takahashi and Yamaguchi (2021).	Comprehensive e-waste recycling systems for healthcare devices, reducing environmental harm.	National e-waste recycling laws: African nations can adopt national policies that require the collection, recycling, and disposal of healthcare-related e-waste in an environmentally safe manner.
	Integration of environmental sustainability into medical supply chain management practices.	Regulatory frameworks for green supply chains: Africa can develop laws and incentives for healthcare providers to implement green practices throughout the supply chain.
Denmark Miller and Thompson (2021).	Extensive use of green hospitals that minimize energy consumption through renewable sources and smart technologies.	Green hospital initiatives: African countries can implement energy-saving technologies in hospitals, reducing long-term costs and improving healthcare quality through sustainable design.

Source: Table developed by authors drawing from cited sources.

15.6.1 Key Findings

1) **Financial barriers**: High initial investment costs for green technologies, such as renewable energy systems and waste management solutions, remain a critical challenge. The Institutional Theory framework highlights that the financial structures in place are often focused on short-term needs and lack mechanisms to support long-term sustainability goals.
2) **Technical barriers**: The complexity of green innovations, such as solar energy systems and e-waste recycling, hampers their diffusion. The Diffusion of Innovations Theory underscores that perceived complexity and lack of local

technical expertise are key barriers. Insufficient training and expertise in maintaining these technologies slow the adoption of sustainable solutions.

3) **Institutional and policy challenges**: Many African countries lack integrated policies that link healthcare with environmental sustainability. As the Systems Theory suggests, policy fragmentation hinders the effective implementation of green innovations. Coordination between health, environmental, and energy sectors is crucial for overcoming this challenge.
4) **Social and cultural barriers**: Resistance to change, combined with limited awareness of the benefits of green innovations, creates barriers at the societal level. Diffusion of Innovations Theory indicates that early adopters and opinion leaders in healthcare settings can drive the broader adoption of innovations by reducing resistance and fostering peer support.
5) **Infrastructure limitations**: The lack of an adequate infrastructure—such as reliable electricity grids and waste management systems—prevents the integration of green technologies into existing healthcare systems. The concept of path dependency in Institutional Theory shows that existing infrastructures often lack the flexibility needed to accommodate newer, more sustainable systems.

15.6.2 Recommendations for Overcoming Barriers

Given the key findings, the following recommendations aim to address the challenges to green innovation in African health supply chains:

1) **Develop green finance mechanisms**: Financial mechanisms such as green bonds, subsidized loans, and public–private partnerships should be introduced to provide healthcare facilities with the necessary capital for investing in green technologies. Institutional Theory suggests that these mechanisms can help shift the focus of financial structures to long-term sustainability goals.
2) **Enhance technical training and capacity building**: To overcome technical barriers, healthcare workers, technicians, and engineers must be trained in the operation, maintenance, and repair of green technologies. Diffusion of Innovations Theory emphasizes that increasing local expertise reduces the perceived complexity of these technologies, facilitating their adoption.
3) **Integrate environmental sustainability into healthcare policies**: Governments should adopt cross-sectoral policies that promote environmental sustainability within the healthcare system. The Systems Theory emphasizes the importance of policy integration across subsystems—health, energy, and environment—to create a holistic framework that supports green innovation.
4) **Promote social awareness and leadership**: Efforts to increase awareness about the benefits of green technologies through community engagement and peer influence are essential for overcoming social resistance. The

Diffusion of Innovations Theory underscores the role of early adopters and opinion leaders in facilitating the broader adoption of green innovations within healthcare.

5) **Invest in healthcare infrastructure**: Governments should prioritize infrastructure investments in rural and underserved areas, focusing on electricity grids, roads, and waste management systems that support the integration of green technologies. This is crucial for addressing path dependency and enabling healthcare systems to support sustainable solutions.
6) **Establish regulatory and incentive mechanisms**: Regulatory frameworks should be implemented to enforce environmental standards in healthcare operations. Governments should also provide incentives such as tax breaks and subsidies for facilities that integrate sustainable practices into their operations.

15.6.3 Key Lessons from the COVID-19 Pandemic for Green Health Supply Chains

The COVID-19 pandemic exposed the vulnerabilities and limitations of existing healthcare systems globally, including in Africa. However, it also provided valuable lessons for building more resilient and sustainable health supply chains that can accommodate green innovations. Reflecting on these lessons can guide future efforts to integrate sustainability into African healthcare systems.

1) **Resilience and flexibility in crisis management**: The pandemic revealed the importance of healthcare systems being flexible and adaptable to external shocks. Health systems with a sustainable infrastructure, such as solar-powered facilities, proved more resilient in the face of energy shortages and supply chain disruptions. The adoption of green technologies can contribute to this resilience by reducing dependency on unreliable or expensive energy sources and improving the sustainability of supply chains, as seen in regions where solar-powered health clinics became lifelines during the pandemic (World Health Organization, 2020).
   - **Link to theory**: The application of the Systems Theory highlights the need for healthcare systems to become more interconnected and capable of adapting to disruptions. Green technologies, such as renewable energy systems and eco-friendly supply chains, can help create more flexible, resilient healthcare networks that are less susceptible to external shocks.
2) **E-Waste management and the growth of telemedicine**: The COVID-19 pandemic accelerated the use of digital health solutions, such as telemedicine, to maintain the continuity of care during lockdowns. This led to a significant increase in the generation of e-waste, particularly from medical devices and

personal protective equipment (PPE) used in healthcare settings. The eWaste Africa initiative and other recycling programs highlight the importance of developing robust e-waste management systems to address the environmental impacts of digital health technologies.

- **Link to theory**: Diffusion of Innovations Theory suggests that the adoption of digital technologies, including telemedicine and electronic health records, was accelerated by necessity during the pandemic. However, the resulting surge in e-waste necessitates the development of infrastructures for sustainable disposal and recycling to mitigate the environmental harm caused by improper waste management.

3) **Supply chain vulnerabilities and sustainability**: The pandemic underscored the fragility of traditional supply chains, especially those reliant on global trade for essential medical supplies. Green health supply chains, which integrate local, sustainable solutions such as local production of medical supplies and solar-powered healthcare facilities, can reduce dependency on global supply chains, which are vulnerable to disruption.
   - **Link to theory**: **Institutional Theory** suggests that traditional healthcare institutions must evolve to prioritize local, sustainable practices to mitigate the risks exposed during the pandemic. By shifting focus to green supply chains, healthcare systems can reduce their vulnerability to supply chain disruptions caused by external shocks.

15.6.4 Long-Term Vision for Green Health Supply Chains

The long-term vision for African health supply chains envisions a future where sustainability and resilience are deeply ingrained in the core operations of healthcare systems. This transformative approach requires the widespread adoption of green technologies across multiple sectors. Renewable energy sources, such as solar power, should become the backbone of energy supply in health facilities, ensuring consistent and reliable power for essential services like vaccine storage and life-saving medical equipment. Similarly, sustainable medical waste management must be prioritized to minimize environmental and health risks, while eco-friendly transportation options, such as electric vehicles, can revolutionize the delivery of medical supplies and services.

The COVID-19 pandemic underscored the vulnerabilities in existing healthcare systems and highlighted the urgent need for robust infrastructures capable of withstanding future crises. By integrating green innovations, African healthcare systems can build resilience, improve patient care, and significantly reduce their environmental footprint. Lessons from global leaders in green health supply chains further demonstrate the feasibility of aligning sustainability with operational efficiency.

Incorporating green supply chains into African healthcare systems extends beyond environmental benefits. It creates opportunities for economic growth through job creation in renewable energy, waste management, and sustainable logistics sectors. Moreover, it strengthens the overall resilience of healthcare systems, enabling them to adapt and respond effectively to evolving health challenges. To achieve this vision, a coordinated approach is necessary—aligning policies, financial investments, technical capacity building, and infrastructure development. By fostering a commitment to sustainability, African countries can establish healthcare systems that not only address immediate needs but also pave the way for a healthier, more sustainable future.

References

Andrade, L.A., & Silva, M.A. (2020). Green logistics in the healthcare sector: Insights from Brazil. *Journal of Supply Chain Management, 56*(4), 67–85.

Bill & Melinda Gates Foundation. (2021). *Expanding solar-powered vaccine storage in Nigeria*. www.gatesfoundation.org.

eWaste Africa. (2020). *Managing healthcare e-waste in Ghana*. www.ewasteafrica.com.

eWaste Africa. (2022). *E-waste management in Africa: Challenges and opportunities*. www.ewasteafrica.org.

Freeman, R.E. (1984). *Strategic management: A stakeholder approach*. Boston, MA: Pitman.

Gibson, R.B., & Black, L. (2020). Telemedicine in the US: The impact on healthcare sustainability and patient care. *Journal of Telemedicine and Telecare, 26*(3), 168–176.

Green Healthcare Partnership. (2019). *Sustainable healthcare practices in the U.S.* www.greenhealthcare.org.

GreenCape. (2022a). *Green transport strategy and health sector logistics in South Africa*. www.greencape.co.za.

GreenCape (2022b). *Electric vehicles in South Africa: Potential for the healthcare sector*. GreenCape. www.greencape.co.za

Gupta, P., & Sharma, R. (2019). Solar-powered healthcare solutions: The Indian experience. *Renewable and Sustainable Energy Reviews, 101*, 528–536.

Health-E-Waste Nigeria. (2021). *Promoting safe e-waste disposal in Nigerian healthcare facilities*. www.healthewaste.ng.

Hempel, M., & Henze, J. (2020). Green health care and energy efficiency in the healthcare sector: An analysis of the German healthcare system. *Journal of Environmental Economics and Management, 60*(1), 45–67.

KEMSA. (2021). *Promoting sustainable healthcare packaging in Kenya*. www.kemsa.co.ke.

Kenya Red Cross. (2020). *Solar-powered mobile clinics in Kenya*. www.redcross.or.ke.

Kenya Solar Energy Association (KSEA). (2020). *Promoting solar energy in healthcare*. www.ksea.or.ke.

Mara Group. (2021). *Solar-powered health centers in Rwanda*. www.mara.com.

Miller, M., & Thompson, J. (2021). Sustainable hospital design and construction in the United States: Trends and innovations. *Sustainability, 13*(10), 1830–1850.

NHS Sustainability (2021). *NHS green plan: Sustainable healthcare solutions*. National Health Service. www.nhs.uk

North, D.C. (1990). *Institutions, institutional change and economic performance*. Cambridge University Press.

Rogers, E.M. (2003). *Diffusion of innovations* (5th ed.). Free Press.

South African Department of Health (2021). *Solar-powered vaccination storage solutions*. South African Department of Health. www.health.gov.za

Takahashi, T., & Yamaguchi, Y. (2021). Implementing green technologies in healthcare supply chains: Lessons from Japan. *International Journal of Sustainable Healthcare, 17*(4), 234–245.

UNICEF. (2020a). *Solar-powered cold chain for vaccines in Africa*. United Nations Children's Fund. www.unicef.org.

UNICEF. (2020b). *Solar-powered cold chain for vaccines in India*. United Nations Children's Fund. www.unicef.org.

United Nations Environment Programme (UNEP) (2020). *E-waste management and recycling in Africa*. UNEP. www.unep.org

UPS Healthcare. (2021). *UPS green delivery solutions in Africa*. www.ups.com.

WHO (2021). *Sustainable procurement guidelines for health products*. World Health Organization. www.who.int

WHO & UNEP (2020). *Training programs for e-waste management in Africa*. World Health Organization. www.who.int

World Bank. (2020). *Addressing infrastructure gaps in healthcare: A focus on rural Africa*. www.worldbank.org.

World Health Organization. (2020). *Health systems and COVID-19: Lessons learned*. www.who.int.

Zeman, D. (2021). E-waste management in Germany: A national case study. *International Journal of Environmental Sciences, 72*(5), 345–360.

Index

Green Innovations in Supply Chain Management: Case Studies and Applications, First Edition. Edited by Marcia Mkansi, Godfrey Mugurusi, Aaron L. Nsakanda, Temidayo Akenroye, and Frank Tietze.

Published 2026 by John Wiley & Sons, Inc.

d